First Edition, Feb. 1892. Reprinted, May, 1892.
Second Edition, 1896. Reprinted, 1905.

MATHEMATICAL RECREATIONS AND ESSAYS

BY

W.W. ROUSE BALL

Fellow and Tutor of Trinity College, Cambridge.

FOURTH EDITION

London:
MACMILLAN AND CO., Limited
NEW YORK: THE MACMILLAN COMPANY
1905

[*All rights reserved.*]

Produced by Joshua Hutchinson, David Starner, David Wilson and the Online Distributed Proofreading Team at http://www.pgdp.net

Transcriber's notes

Most of the open questions discussed by the author were settled during the twentieth century.

The author's footnotes are labelled using printer's marks[*]; footnotes showing where corrections to the text have been made are labelled numerically[1].

Minor typographical corrections are documented in the LaTeX source.

This document is designed for two-sided printing. Consequently, the many hyperlinked cross-references are not visually distinguished. The document can be recompiled for more comfortable on-screen viewing: see comments in source LaTeX code.

PREFACE TO THE FIRST EDITION.

THE following pages contain an account of certain mathematical recreations, problems, and speculations of past and present times. I hasten to add that the conclusions are of no practical use, and most of the results are not new. If therefore the reader proceeds further he is at least forewarned.

At the same time I think I may assert that many of the diversions—particularly those in the latter half of the book—are interesting, not a few are associated with the names of distinguished mathematicians, while hitherto several of the memoirs quoted have not been easily accessible to English readers.

The book is divided into two parts, but in both parts I have included questions which involve advanced mathematics.

The *first part* consists of seven chapters, in which are included various problems and amusements of the kind usually called *mathematical recreations*. The questions discussed in the first of these chapters are connected with arithmetic; those in the second with geometry; and those in the third relate to mechanics. The fourth chapter contains an account of some miscellaneous problems which involve both number and situation; the fifth chapter contains a concise account of magic squares; and the sixth and seventh chapters deal with some unicursal problems. Several of the questions mentioned in the first three chapters are of a somewhat trivial character, and had they been treated in any standard English work to which I could have referred the reader, I should have pointed them out. In the absence of such a work, I thought it best to insert them and trust to the judicious reader to omit them altogether or to skim them as he feels inclined.

The *second part* consists of five chapters, which are mostly *historical*. They deal respectively with three classical problems in geometry—namely, the duplication of the cube, the trisection of an angle, and the quadrature of the circle—astrology, the hypotheses as to the nature of space and mass, and a means of measuring time.

I have inserted detailed references, as far as I know, as to the sources of the various questions and solutions given; also, wherever I have given only the result of a theorem, I have tried to indicate authorities where a proof may be found. In general, unless it is stated otherwise, I have taken the references direct from the original works; but, in spite of considerable time spent in verifying them, I dare not suppose that they are free from all errors or misprints.

I shall be grateful for notices of additions or corrections which may occur to any of my readers.

W.W. ROUSE BALL

TRINITY COLLEGE, CAMBRIDGE.
    February, 1892.

NOTE TO THE FOURTH EDITION.

IN this edition I have inserted in the earlier chapters descriptions of several additional Recreations involving elementary mathematics, and I have added in the second part chapters on the *History of the Mathematical Tripos at Cambridge, Mersenne's Numbers,* and *Cryptography and Ciphers.*

It is with some hesitation that I include in the book the chapters on *Astrology* and *Ciphers,* for these subjects are only remotely connected with Mathematics, but to afford myself some latitude I have altered the title of the second part to *Miscellaneous Essays and Problems.*

W.W.R.B.

TRINITY COLLEGE, CAMBRIDGE.
13 *May,* 1905.

TABLE OF CONTENTS.

PART I.

Mathematical Recreations.

    CHAPTER I.   SOME ARITHMETICAL QUESTIONS.

	PAGE
Elementary Questions on Numbers (Miscellaneous)	4
Arithmetical Fallacies	20
Bachet's Weights Problem	27
Problems in Higher Arithmetic	29
Fermat's *Theorem on Binary Powers*	31
Fermat's *Last Theorem*	32

    CHAPTER II.   SOME GEOMETRICAL QUESTIONS.

Geometrical Fallacies	35
Geometrical Paradoxes	42
Colouring Maps	44
Physical Geography	46
Statical Games of Position	48
Three-in-a-row. Extension to p-in-a-row	48
Tesselation. Cross-Fours	50
Colour-Cube Problem	51

	PAGE
Dynamical Games of Position	52
Shunting Problems	53
Ferry-Boat Problems	55
Geodesic Problems	57
Problems with Counters placed in a row	58
Problems on a Chess-board with Counters or Pawns	60
Guarini's Problem	63
Geometrical Puzzles (rods, strings, &c.)	64
Paradromic Rings	64

CHAPTER III. SOME MECHANICAL QUESTIONS.

Paradoxes on Motion	67
Force, Inertia, Centrifugal Force	70
Work, Stability of Equilibrium, &c.	72
Perpetual Motion	75
Models	78
Sailing quicker than the Wind	79
Boat moved by a rope inside the boat	81
Results dependent on Hauksbee's Law	82
Cut on a tennis-ball. Spin on a cricket-ball	83
Flight of Birds	85
Curiosa Physica	86

CHAPTER IV. SOME MISCELLANEOUS QUESTIONS.

The Fifteen Puzzle	88
The Tower of Hanoï	91
Chinese Rings	93
The Eight Queens Problem	97
Other Problems with Queens and Chess-pieces	102
The Fifteen School-Girls Problem	103

	PAGE
Problems connected with a pack of cards	109
Monge on shuffling a pack of cards	109
Arrangement by rows and columns	111
Determination of one out of $\frac{1}{2}n(n+1)$ given couples	113
Gergonne's Pile Problem	115
The Mouse Trap. Treize	119

CHAPTER V.   MAGIC SQUARES.

Notes on the History of Magic Squares	122
Construction of Odd Magic Squares	123
Method of De la Loubère	124
Method of Bachet	125
Method of De la Hire	126
Construction of Even Magic Squares	128
First Method	129
Method of De la Hire and Labosne	132
Composite Magic Squares	134
Bordered Magic Squares	135
Hyper-Magic Squares	136
Pan-diagonal or Nasik Squares	136
Doubly Magic Squares	137
Magic Pencils	137
Magic Puzzles	140
Card Square	140
Euler's Officers Problem	140
Domino Squares	141
Coin Squares	141

CHAPTER VI.   UNICURSAL PROBLEMS.

Euler's Problem	143
Definitions	145
Euler's Theorems	145
Examples	148

	PAGE
Mazes	149
Rules for completely traversing a Maze	150
Notes on the History of Mazes	150
Geometrical Trees	154
The Hamiltonian Game	155
Knight's Path on a Chess-Board	158
Method of De Montmort and De Moivre	159
Method of Euler	159
Method of Vandermonde	163
Method of Warnsdorff	164
Method of Roget	164
Method of Moon	167
Method of Jaenisch	168
Number of possible routes	168
Paths of other Chess-Pieces	168

PART II.

Miscellaneous Essays and Problems.

CHAPTER VII.   THE MATHEMATICAL TRIPOS.

	PAGE
Medieval Course of Studies: Acts	171
The Renaissance at Cambridge	172
Rise of a Mathematical School	172
Subject-Matter of Acts at different periods	172
Degree Lists	174
Oral Examinations always possible	174
Public Oral Examinations become customary, 1710–30	175
Additional work thrown on Moderators. Stipends raised	175
Facilitates order of merit	176
Scheme of Examination in 1750	176
Right of M.A.s to take part in it	176
Scheme of Examination in 1763	177
Foundations of Smith's Prizes, 1768	178
Introduction of a Written Examination, circ. 1770	179
Description of the Examination in 1772	179
Scheme of Examination in 1779	182
System of Brackets	182
Problem Papers in 1785 and 1786	183
Description of the Examination in 1791	184
The Poll Part of the Examination	185
A Pass Standard introduced	186
Problem Papers from 1802 onwards	186
Description of the Examination in 1802	187
Scheme of Reading in 1806	189
Introduction of modern analytical notation	192
Alterations in Schemes of Study, 1824	195
Scheme of Examination in 1827	195
Scheme of Examination in 1833	197
All the papers marked	197
Scheme of Examination in 1839	197

	PAGE
Scheme of Examination in 1848	198
Creation of a Board of Mathematical Studies	198
Scheme of Examination in 1873	199
Scheme of Examination in 1882	200
Fall in number of students reading mathematics	201
Origin of term Tripos	201
Tripos Verses	202

CHAPTER VIII. THREE GEOMETRICAL PROBLEMS.

The Three Problems	204
The Duplication of the Cube	205
Legendary origin of the problem	205
Lemma of Hippocrates	206
Solutions of Archytas, Plato, Menaechmus, Apollonius, and Sporus	207
Solutions of Vieta, Descartes, Gregory of St Vincent, and Newton	209
The Trisection of an Angle	210
Solutions quoted by Pappus (three)	210
Solutions of Descartes, Newton, Clairaut, and Chasles	211
The Quadrature of the Circle	212
Incommensurability of π	212
Definitions of π	213
Origin of symbol π	214
Methods of approximating to the numerical value of π	214
Geometrical methods of approximation	214
Results of Egyptians, Babylonians, Jews	215
Results of Archimedes and other Greek writers	215
Results of Roman surveyors and Gerbert	216
Results of Indian and Eastern writers	216
Results of European writers, 1200–1630	217
Theorems of Wallis and Brouncker	220
Analytical methods of approximation. Gregory's series	220
Results of European writers, 1699–1873	220
Geometrical approximations	222
Approximations by the theory of probability	222

CHAPTER IX. MERSENNE'S NUMBERS.

	PAGE
Mersenne's Enunciation of the Theorem	224
List of known results	225
Cases awaiting verification	225
History of Investigations	226
Methods used in attacking the problem	230
By trial of divisors of known forms	231
By indeterminate equations	233
By properties of quadratic forms	234
By the use of a *Canon Arithmeticus*	234
By properties of binary powers	235
By the use of the binary scale	235
By the use of Fermat's Theorem	236
Mechanical methods of Factorizing Numbers	236

CHAPTER X. ASTROLOGY.

Astrology. Two branches: natal and horary astrology	238
Rules for casting and reading a horoscope	238
Houses and their significations	238
Planets and their significations	240
Zodiacal signs and their significations	242
Knowledge that rules were worthless	243
Notable instances of horoscopy	246
Lilly's prediction of the Great Fire and Plague	246
Flamsteed's guess	246
Cardan's horoscope of Edward VI	247

CHAPTER XI. CRYPTOGRAPHS AND CIPHERS.

A Cryptograph. Definition. Illustration	251
A Cipher. Definition. Illustration	252
Essential Features of Cryptographs and Ciphers	252
Cryptographs of Three Types. Illustrations	253
Order of letters re-arranged	253
Use of non-significant symbols. The Grille	256
Use of broken symbols. The Scytale	258
Ciphers. Use of arbitrary symbols unnecessary	259

	PAGE
Ciphers of Four Types	259
Ciphers of the First Type. Illustrations	260
Ciphers of the Second Type. Illustrations	263
Ciphers of the Third Type. Illustrations	265
Ciphers of the Fourth Type. Illustrations	267
Requisites in a good Cipher	268
Cipher Machines	269
Historical Ciphers	269
Julius Caesar, Augustus	269
Bacon	269
Charles I	269
Pepys	271
De Rohan	272
Marie Antoinette	272
The Code Dictionary	274
Poe's Writings	275

CHAPTER XII.   HYPER-SPACE.

Two subjects of speculation on Hyper-space	278
Space of two dimensions and of one dimension	278
Space of four dimensions	279
Existence in such a world	279
Arguments in favour of the existence of such a world	280
Non-Euclidean Geometries	284
Euclid's axioms and postulates. The parallel postulate	284
Hyperbolic Geometry of two dimensions	285
Elliptic Geometry of two dimensions	285
Elliptic, Parabolic and Hyperbolic Geometries compared	285
Non-Euclidean Geometries of three or more dimensions	287

CHAPTER XIII.   TIME AND ITS MEASUREMENT.

Units for measuring durations (days, weeks, months, years)	289
The Civil Calendar (Julian, Gregorian, &c.)	292
The Ecclesiastical Calendar (date of Easter)	294
Day of the week corresponding to a given date	297

	PAGE
Means of measuring Time	297
Styles, Sun-dials, Sun-rings	297
Water-clocks, Sand-clocks, Graduated Candles	301
Clocks and Watches	301
Watches as Compasses	303

CHAPTER XIV. MATTER AND ETHER THEORIES.

Hypothesis of Continuous Matter	306
Atomic Theories	306
Popular Atomic Hypothesis	306
Boscovich's Hypothesis	307
Hypothesis of an Elastic Solid Ether. Labile Ether	307
Dynamical Theories	308
The Vortex Ring Hypothesis	308
The Vortex Sponge Hypothesis	309
The Ether-Squirts Hypothesis	310
The Electron Hypothesis	311
Speculations due to investigations on Radio-activity	311
The Bubble Hypothesis	313
Conjectures as to the cause of Gravity	314
Conjectures to explain the finite number of species of Atoms	318
Size of the molecules of bodies	320
INDEX	323
NOTICES OF SOME WORKS—CHIEFLY HISTORICO-MATHEMATICAL	335
PROJECT GUTENBERG LICENSING INFORMATION	355

PART I.

𝔐athematical 𝔑ecreations.

"Les hommes ne sont jamais plus ingénieux que dans l'invention des jeux; l'esprit s'y trouve à son aise.... Après les jeux qui dépendent uniquement des nombres viennent les jeux où entre la situation.... Après les jeux où n'entrent que le nombre et la situation viendraient les jeux où entre le mouvement.... Enfin il serait a souhaiter qu'on eût un cours entier des jeux, traités mathématiquement."
(Leibnitz: letter to De Montmort, July 29, 1715.)

CHAPTER I.

SOME ARITHMETICAL QUESTIONS.

THE interest excited by statements of the relations between numbers of certain forms has been often remarked. The majority of works on mathematical recreations include several such problems, which are obvious to any one acquainted with the elements of algebra, but which to many who are ignorant of that subject possess the same kind of charm that some mathematicians find in the more recondite propositions of higher arithmetic. I shall devote the bulk of this chapter to these elementary problems, but I append a few remarks on one or two questions in the theory of numbers.

Before entering on the subject of the chapter, I may add that a large proportion of the elementary questions mentioned here and in the following two chapters are taken from one of two sources. The first of these is the classical *Problèmes plaisans et délectables*, by C.G. Bachet, sieur de Méziriac, of which the first edition was published in 1612 and the second in 1624: it is to the edition of 1624 that the references hereafter given apply. Several of Bachet's problems are taken from the writings of Alcuin, Pacioli di Burgo, Tartaglia, or Cardan, and possibly some of them are of oriental origin, but I have made no attempt to add such references. The other source to which I alluded above is Ozanam's *Récréations mathématiques et physiques*. The greater portion of the original edition, published in two volumes at Paris in 1694, was a compilation from the works of Bachet, Leurechon, Mydorge, van Etten, and Oughtred: this part is excellent, but the same cannot be said of the additions due to Ozanam. In the *Biographie Universelle* allusion is made to subsequent editions issued in 1720, 1735, 1741, 1778, and 1790; doubtless these references are correct, but the following editions, all of which I have seen, are the only ones of which I have any knowledge. In 1696 an edition was issued at Amsterdam. In 1723—six years after the death of Ozanam—one was issued in three volumes, with a supplementary fourth volume, containing (among other things) an appendix on puzzles: I believe that it would be difficult to find in any of the books current in England on mathematical amusements as many as a dozen puzzles which are not contained in one of these four volumes. Fresh editions were issued in 1741, 1750 (the second volume of which bears the date 1749), 1770, and 1790. The edition of 1750 is said to have been corrected by Montucla on condition that his name should not be associated with it; but the edition of 1790 is the earliest one in which reference is made to these corrections, though the editor is referred to only as Monsieur M***. Montucla expunged most of what was actually incorrect in the older editions, and added several historical notes, but unfortunately his scruples prevented him from striking out the accounts of numerous trivial experiments and truisms which overload the work. An English translation of the original edition appeared in 1708, and I believe ran through four editions, the last of them being published in Dublin in 1790. Montucla's revision of 1790 was translated by C. Hutton, and editions of this were issued in 1803, in 1814, and (in one volume) in 1840: my references are to the editions of 1803 and 1840.

I proceed now to enumerate some of the elementary questions connected with numbers which for nearly three centuries have formed a large part of most compilations of mathematical amusements. They are given here mainly for their historical—not for their arithmetical—interest; and perhaps a mathematician may well omit them, and pass at once to the latter part of this chapter.

These questions are of the nature of tricks or puzzles and I follow the usual course and present them in that form. I may note however that most of them are not worth proposing, even as tricks, unless either the *modus operandi* is disguised or the result arrived at is different from that expected; but, as I am not writing on conjuring, I refrain from alluding to the means of disguising the operations indicated, and give merely a bare enumeration of the steps essential to the success of the method used, though I may recall the fundamental rule that no trick, however good, will bear immediate repetition, and that, if it is necessary to appear to repeat it, a different method of obtaining the result should be used.

TO FIND A NUMBER SELECTED BY SOME ONE. There are innumerable ways of finding a number chosen by some one, provided the result of certain operations on it is known. I confine myself to methods typical of those commonly used. Any one acquainted with algebra will find no difficulty in modifying the rules here given or framing new ones of an analogous nature.

First Method[*]. (i) Ask the person who has chosen the number to treble it. (ii) Enquire if the product is even or odd: if it is even, request him to take half of it; if it is odd, request him to add unity to it and then to take half of it. (iii) Tell him to multiply the result of the second step by 3. (iv) Ask how many integral times 9 divides into the latter product: suppose the answer to be n. (v) Then the number thought of was $2n$ or $2n+1$, according as the result of step (i) was even or odd.

The demonstration is obvious. Every even number is of the form $2n$, and the successive operations applied to this give (i) $6n$, which is even; (ii) $\frac{1}{2}6n = 3n$; (iii) $3 \times 3n = 9n$; (iv) $\frac{1}{9}9n = n$; (v) $2n$. Every odd number is of the form $2n+1$, and the successive operations applied to this give (i) $6n+3$, which is odd; (ii) $\frac{1}{2}(6n+3+1) = 3n+2$; (iii) $3(3n+2) = 9n+6$; (iv) $\frac{1}{9}(9n+6) = n +$ a remainder; (v) $2n+1$. These results lead to the rule given above.

Second Method[†]. Ask the person who has chosen the number to perform in succession the following operations. (i) To multiply the number by 5. (ii) To add 6 to the product. (iii) To multiply the sum by 4. (iv) To add 9 to the product. (v) To multiply the sum by 5. Ask to be told the result of the last operation: if from this product 165 is subtracted, and then the remainder is divided by 100, the quotient will be the number thought of originally.

For let n be the number selected. Then the successive operations applied to it give (i) $5n$; (ii) $5n + 6$; (iii) $20n + 24$; (iv) $20n + 33$; (v) $100n + 165$. Hence the rule.

Third Method[‡]. Request the person who has thought of the number to perform the following operations. (i) To multiply it by any number you like, say, a. (ii) To divide the product by any number,

---

[*] Bachet, *Problèmes plaisans*, Lyons, 1624, problem I, p. 53.
[†] A similar rule was given by Bachet, problem IV, p. 74.
[‡] Bachet, problem V, p. 80.

say, b. (iii) To multiply the quotient by c. (iv) To divide this result by d. (v) To divide the final result by the number selected originally. (vi) To add to the result of operation (v) the number thought of at first. Ask for the sum so found: then, if ac/bd is subtracted from this sum, the remainder will be the number chosen originally.

For, if n was the number selected, the result of the first four operations is to form nac/bd; operation (v) gives ac/bd; and (vi) gives $n + (ac/bd)$, which number is mentioned. But ac/bd is known; hence, subtracting it from the number mentioned, n is found. Of course a, b, c, d may have any numerical values it is liked to assign to them. For example, if $a = 12$, $b = 4$, $c = 7$, $d = 3$ it is sufficient to subtract 7 from the final result in order to obtain the number originally selected.

Fourth Method[*]. Ask some one to select a number less than 90. (i) Request him to multiply it by 10, and to add any number he pleases, a, which is less than 10. (ii) Request him to divide the result of step (i) by 3, and to mention the remainder, say, b. (iii) Request him to multiply the quotient obtained in step (ii) by 10, and to add any number he pleases, c, which is less than 10. (iv) Request him to divide the result of step (iii) by 3, and to mention the remainder, say d, and the third digit (from the right) of the quotient; suppose this digit is e. Then, if the numbers a, b, c, d, e are known, the original number can be at once determined. In fact, if the number is $9x + y$, where $x \not> 9$ and $y \not> 8$, and if r is the remainder when $a - b + 3(c - d)$ is divided by 9, we have $x = e$, $y = 9 - r$.

The demonstration is not difficult. For if the selected number is $9x+y$, step (i) gives $90x+10y+a$; (ii) let $y+a = 3n+b$, then the quotient obtained in step (ii) is $30x+3y+n$; step (iii) gives $300x+30y+10n+c$; (iv) let $n + c = 3m + d$, then the quotient obtained in step (iv) is $100x + 10y + 3n + m$, which I will denote by Q. Now the third digit in Q must be x, because, since $y \not> 8$ and $a \not> 9$, we have $n \not> 5$; and since $n \not> 5$ and $c \not> 9$, we have $m \not> 4$; therefore $10y + 3n + m \not> 99$. Hence the third or hundreds digit in Q is x.

Again, from the relations $y + a = 3n + b$ and $n + c = 3m + d$, we have $9m - y = a - b + 3(c - d)$: hence, if r is the remainder when $a-b+3(c-d)$ is divided by 9, we have $y = 9-r$. [This is always true, if we make r positive; but if $a-b+3(c-d)$ is negative, it is simpler to take y as equal to its numerical value; or we may prevent the occurrence of

---

[*] *Educational Times*, London, May 1, 1895, vol. XLVIII, p. 234.

this case by assigning proper values to a and c.] Thus x and y are both known, and therefore the number selected, namely $9x + y$, is known.

Fifth Method[*]. Ask any one to select a number less than 60. (i) Request him to divide it by 3 and mention the remainder; suppose it to be a. (ii) Request him to divide it by 4, and mention the remainder; suppose it to be b. (iii) Request him to divide it by 5, and mention the remainder; suppose it to be c. Then the number selected is the remainder obtained by dividing $40a + 45b + 36c$ by 60.

This method can be generalized and then will apply to any number chosen. Let $a', b', c', \ldots$ be a series of numbers prime to one another, and let p be their product. Let n be any number less than p, and let $a, b, c, \ldots$ be the remainders when n is divided by $a', b', c', \ldots$ respectively. Find a number A which is a multiple of the product $b'c'd' \cdots$ and which exceeds by unity a multiple of a'. Find a number B which is a multiple of $a'c'd' \cdots$ and which exceeds by unity a multiple of b'; and similarly find analogous numbers $C, D, \ldots$. Rules for the calculation of $A, B, C, \ldots$ are given in the theory of numbers, but in general, if the numbers $a', b', c', \ldots$ are small, the corresponding numbers, $A, B, C, \ldots$ can be found by inspection. I proceed to show that n is equal to the remainder when $Aa + Bb + Cc + \cdots$ is divided by p.

Let $N = Aa + Bb + Cc + \cdots$, and let $M(x)$ stand for a multiple of x.

Now $A = M(a') + 1$, therefore $Aa = M(a') + a$. Hence, if the first term in N, that is Aa, is divided by a', the remainder is a. Again, B is a multiple of $a'c'd' \cdots$. Therefore Bb is exactly divisible by a'. Similarly $Cc, Dd, \ldots$ are each exactly divisible by a'. Thus every term in N, except the first, is exactly divisible by a'. Hence, if N is divided by a', the remainder is a. But if n is divided by a', the remainder is a.

Therefore $\qquad N - n = M(a')$.
Similarly $\qquad N - n = M(b')$,
$\qquad\qquad N - n = M(c')$,
$\qquad\qquad \ldots\ldots\ldots\ldots\ldots\ldots$

But $a', b', c', \ldots$ are prime to one another.
$\qquad\therefore\ N - n = M(a'b'c' \cdots) = M(p)$,
that is, $\qquad N = M(p) + n$.

---

[*] Bachet, problem VI, p. 84: Bachet added, on p. 87, a note on the previous history of the problem.

Now n is less than p, hence if N is divided by p, the remainder is n.

The rule given by Bachet corresponds to the case of $a' = 3$, $b' = 4$, $c' = 5$, $p = 60$, $A = 40$, $B = 45$, $C = 36$. If the number chosen is less than 420, we may take $a' = 3$, $b' = 4$, $c' = 5$, $d' = 7$, $p = 420$, $A = 280$, $B = 105$, $C = 336$, $D = 120$.

TO FIND THE RESULT OF A SERIES OF OPERATIONS PERFORMED ON ANY NUMBER (*unknown to the questioner*) WITHOUT ASKING ANY QUESTIONS. All rules for solving such problems ultimately depend on so arranging the operations that the number disappears from the final result. Four examples will suffice.

First Example[*]. Request some one to think of a number. Suppose it to be n. Ask him (i) to multiply it by any number you please (say) a; (ii) then to add (say) b; (iii) then to divide the sum by (say) c. (iv) Next, tell him to take a/c of the number originally chosen; and (v) to subtract this from the result of the third operation. The result of the first three operations is $(na + b)/c$, and the result of operation (iv) is na/c: the difference between these is b/c, and therefore is known to you. For example, if $a = 6$, $b = 12$, $c = 4$, and $a/c = 1\frac{1}{2}$, then the final result is 3.

Second Example[†]. Ask A to take any number of counters that he pleases: suppose that he takes n counters. (i) Ask some one else, say B, to take p times as many, where p is any number you like to choose. (ii) Request A to give q of his counters to B, where q is any number you like to select. (iii) Next, ask B to transfer to A a number of counters equal to p times as many counters as A has in his possession. Then there will remain in B's hands $q(p+1)$ counters: this number is known to you; and the trick can be finished either by mentioning it or in any other way you like.

The reason is as follows. The result of operation (ii) is that B has $pn + q$ counters, and A has $n - q$ counters. The result of (iii) is that B transfers $p(n - q)$ counters to A: hence he has left in his possession $(pn + q) - p(n - q)$ counters, that is, he has $q(p + 1)$.

For example, if originally A took any number of counters, then (if you chose p equal to 2), first you would ask B to take twice as many counters as A had done; next (if you chose q equal to 3) you would ask

---

[*] Bachet, problem VIII, p. 102.
[†] Bachet, problem XIII, p. 123: Bachet presented the above trick in a somewhat more general form, but one which is less effective in practice.

A to give 3 counters to B; and then you would ask B to give to A a number of counters equal to twice the number then in A's possession; after this was done you would know that B had $3(2+1)$, that is, 9 left.

This trick (as also some of the following problems) may be performed equally well with one person, in which case A may stand for his right hand and B for his left hand.

Third Example. Ask some one to perform in succession the following operations. (i) Take any number of three digits. (ii) Form a new number by reversing the order of the digits. (iii) Find the difference of these two numbers. (iv) Form another number by reversing the order of the digits in this difference. (v) Add together the results of (iii) and (iv). Then the sum obtained as the result of this last operation will be 1089.

An illustration and the explanation of the rule are given below.

$$
\begin{array}{rll}
\text{(i)} & 237 & 100a + 10b + c \\
\text{(ii)} & 732 & 100c + 10b + a \\
\text{(iii)} & 495 & 100(a - c - 1) + 90 + (10 + c - a) \\
\text{(iv)} & 594 & 100(10 + c - a) + 90 + (a - c - 1) \\
\text{(v)} & 1089 & 900 \quad\quad + 180 + 9
\end{array}
$$

Fourth Example*. The following trick depends on the same principle. Ask some one to perform in succession the following operations. (i) To write down any sum of money less than £12; the number of pounds not being the same as the number of pence. (ii) To *reverse* this sum, that is, to write down a sum of money obtained from it by interchanging the numbers of pounds and pence. (iii) To find the difference between the results of (i) and (ii). (iv) To reverse this difference. (v) To add together the results of (iii) and (iv). Then this sum will be £12. 18*s*. 11*d*.

---

* *Educational Times Reprints*, 1890, vol. LIII, p. 78.

For instance, take the sum £10. 17s. 5d.; we have

	£.	s.	d.
(i)	10	17	5
(ii)	5	17	10
(iii)	4	19	7
(iv)	7	19	4
(v)	12	18	11

The following work explains the rule, and shows that the final result is independent of the sum written down initially.

	£.	s.	d.
(i)	a	b	c
(ii)	c	b	a
(iii)	$a - c - 1$	19	$c - a + 12$
(iv)	$c - a + 12$	19	$a - c - 1$
(v)	11	38	11

The rule can be generalized to cover any system of monetary units.

PROBLEMS INVOLVING TWO NUMBERS.   I proceed next to give a couple of examples of a class of problems which involve two numbers.

First Example[*]. Suppose that there are two numbers, one even and the other odd, and that a person A is asked to select one of them, and that another person B takes the other. It is desired to know whether A selected the even or the odd number. Ask A to multiply his number by 2 (or any even number) and B to multiply his by 3 (or any odd number). Request them to add the two products together and tell you the sum. If it is even, then originally A selected the odd number, but if it is odd, then originally A selected the even number. The reason is obvious.

Second Example[†]. The above rule was extended by Bachet to any two numbers, provided they were prime to one another and one of them

---

[*] Bachet, problem IX, p. 107.
[†] Bachet, problem XI, p. 113.

was not itself a prime. Let the numbers be m and n, and suppose that n is exactly divisible by p. Ask A to select one of these numbers, and B to take the other. Choose a number prime to p, say q. Ask A to multiply his number by q, and B to multiply his number by p. Request them to add the products together and state the sum. Then A originally selected m or n, according as this result is not or is divisible by p. For example, $m = 7$, $n = 15$, $p = 3$, $q = 2$.

PROBLEMS DEPENDING ON THE SCALE OF NOTATION. Many of the rules for finding two or more numbers depend on the fact that in arithmetic an integral number is denoted by a succession of digits, where each digit represents the product of that digit and a power of ten, and the number is equal to the sum of these products. For example, 2017 signifies $(2 \times 10^3) + (0 \times 10^2) + (1 \times 10) + 7$; that is, the 2 represents 2 thousands, *i.e.* the product of 2 and 10^3, the 0 represents 0 hundreds, *i.e.* the product of 0 and 10^2; the 1 represents 1 ten, *i.e.* the product of 1 and 10, and the 7 represents 7 units. Thus every digit has a local value.

The application to tricks connected with numbers will be understood readily from three illustrative examples.

First Example[*]. A common conjuring trick is to ask a boy among the audience to throw two dice, or to select at random from a box a domino on each half of which is a number. The boy is then told to recollect the two numbers thus obtained, to choose either of them, to multiply it by 5, to add 7 to the result, to double this result, and lastly to add to this the other number. From the number thus obtained, the conjurer subtracts 14, and obtains a number of two digits which are the two numbers chosen originally.

For suppose that the boy selected the numbers a and b. Each of these is less than ten—dice or dominoes ensuring this. The successive operations give (i) $5a$; (ii) $5a + 7$; (iii) $10a + 14$; (iv) $10a + 14 + b$. Hence, if 14 is subtracted from the final result, there will be left a number of two digits, and these digits are the numbers selected originally. An analogous trick might be performed in other scales of notation if it was thought necessary to disguise the process further.

---

[*] Some similar questions were given by Bachet in problem XII, p. 117; by Oughtred in his *Mathematicall Recreations* (translated from or founded on van Etten's work of 1633), London, 1653, problem XXXIV; and by Ozanam, part I, chapter X.

Second Example*. Similarly, if three numbers, say, a, b, c, are chosen, then, if each of them is less than ten, they can be found by the following rule. (i) Take one of the numbers, say, a, and multiply it by 2. (ii) Add 3 to the product; the result is $2a + 3$. (iii) Multiply this by 5, and add 7 to the product; the result is $10a + 22$. (iv) To this sum add the second number. (v) Multiply the result by 2. (vi) Add 3 to the product. (vii) Multiply by 5, and add the third number to the product. The result is $100a + 10b + c + 235$. Hence, if the final result is known, it is sufficient to subtract 235 from it, and the remainder will be a number of three digits. These digits are the numbers chosen originally.

I have seen a similar rule applied to determine the birthday and age of some one in the audience. The result is a number of six digits, of which the first two digits give the day of the month, the middle two digits the number of the month, and the last two digits the present age.

Third Example†. The following rule for finding a man's age is of the same kind. Take the tens digit of the year of birth; (i) multiply it by 5; (ii) to the product add 2; (iii) multiply the result by 2; (iv) to this product add the units digit of the birth-year; (v) subtract the sum from 110. The result is the man's age in 1906.

The algebraic proof of the rule is obvious. Let a and b be the tens and units digits of the birth-year. The successive operations give (i) $5a$; (ii) $5a + 2$; (iii) $10a + 4$ (iv) $10a + 4 + b$; (v) $106 - (10a + b)$, which is his age in 1906. The rule can be easily adapted to give the age in any specified year.

OTHER PROBLEMS WITH NUMBERS IN THE DENARY SCALE. I may mention here two or three other slight problems dependent on the common scale of notation, which, as far as I am aware, are unknown to most compilers of books of puzzles.

First Problem. The first of them is as follows. Take any number of three digits: reverse the order of the digits: subtract the number so formed from the original number: then, if the last digit of the difference is mentioned, all the digits in the difference are known.

For let a be the hundreds digit of the number chosen, b be the tens digit, and c be the units digit. Therefore the number is $100a + 10b + c$. The number obtained by reversing the digits is $100c + 10b + a$. The

---

* Bachet gave some similar questions in problem XII, p. 117.
† A similar question was given by Laisant and Perrin in their *Algèbre*, Paris, 1892; and in *L'Illustration* for July 13, 1895.

difference of these numbers is equal to $(100a+c)-(100c+a)$, that is, to $99(a-c)$. But $a-c$ is not greater than 9, and therefore the remainder can only be 99, 198, 297, 396, 495, 594, 693, 792, or 891—in each case the middle digit being 9 and the digit before it (if any) being equal to the difference between 9 and the last digit. Hence, if the last digit is known, so is the whole of the remainder.

Second Problem. The second problem is somewhat similar and is as follows. (i) Take any number; (ii) reverse the digits; (iii) find the difference between the number formed in (ii) and the given number; (iv) multiply this difference by any number you like to name; (v) cross out any digit except a nought; (vi) read the remainder. Then the sum of the digits in the remainder subtracted from the next highest multiple of nine will give the figure struck out.

This follows at once from the fact that the result of operation (iii)—and therefore also of operation (iv)—is necessarily a multiple of nine, and it is known that the sum of the digits of every multiple of nine is itself a multiple of nine.

Miscellaneous Questions. Besides these problems, properly so called, there are numerous questions on numbers which can be solved empirically, but which are of no special mathematical interest.

As an instance I may quote a question which attracted some attention in London in 1893, and may be enunciated as follows. With the seven digits 9, 8, 7, 6, 5, 4, 0 express three numbers whose sum is 82: each digit, being used only once, and the use of the usual notations for fractions being allowed. One solution is $80.6\dot{9} + .7\dot{4} + .\dot{5}$. Similar questions are with the ten digits, 9, 8, 7, 6, 5, 4, 3, 2, 1, 0, to express numbers whose sum is unity; a solution is $35/70$ and $148/296$. If the sum were 100, a solution would be 50, 49, 1/2, and 38/76. A less straightforward question would be, with the nine digits, 9, 8, 7, 6, 5, 4, 3, 2, 1, to express four numbers whose sum is 100; a solution is 78, 15, $\sqrt[2]{9}$, and $\sqrt[3]{64}$.

PROBLEMS WITH A SERIES OF THINGS WHICH ARE NUMBERED. Any collection of things which can be distinguished one from the other—especially if numbered consecutively—afford easy concrete illustrations of questions depending on these elementary properties of numbers. As examples I proceed to enumerate a few familiar tricks. The first two of these are commonly shown by the use of a *watch*, the last three are best exemplified by the use of a *pack of playing cards*, which readily lend themselves to such illustrations, and I present them in these forms.

First Example[*]. The first of these examples is connected with the hours marked on the face of a watch. In this puzzle some one is asked to think of some hour, say, m, and then to touch a number that marks another hour, say, n. Then if, beginning with the number touched, he taps each successive hour marked on the face of the watch, going in the opposite direction to that in which the hands of the watch move, and reckoning to himself the taps as m, $(m + 1)$, &c., the $(n + 12)$th tap will be on the hour he thought of. For example, if he thinks of V and touches IX, then, if he taps successively IX, VIII, VII, VI, ..., going backwards and reckoning them respectively as 5, 6, 7, 8, ..., the tap which he reckons as 21 will be on the V.

The reason of the rule is obvious, for he arrives finally at the $(n + 12 - m)$th hour from which he started. Now, since he goes in the opposite direction to that in which the hands of the watch move, he has to go over $(n - m)$ hours to reach the hour m: also it will make no difference if in addition he goes over 12 hours, since the only effect of this is to take him once completely round the circle. Now $(n + 12 - m)$ is always positive, since $m < 12$, and therefore if we make him pass over $(n + 12 - m)$ hours we can give the rule in a form which is equally valid whether m is greater or less than n.

Second Example. The following is another well-known way of indicating on a watch-dial an hour selected by some one. I do not know who first invented it. If the hour is tapped by a pencil beginning at VII and proceeding backwards round the dial to VI, V, &c., and if the person who selected the number counts the taps, reckoning from the hour selected (thus, if he selected X, he would reckon the first tap as the 11th), then the 20th tap as reckoned by him will be on the hour chosen.

For suppose he selected the nth hour. Then the 8th tap is on XII and is reckoned by him as the $(n + 8)$th. The tap which he reckons as $(n + 9)$th is on XI, and generally the tap which he reckons as $(n + p)$th is on the hour $(20 - p)$. Hence, putting $p - 20 - n$, the tap which he reckons as 20th is on the hour n. Of course the hours indicated by the first seven taps are immaterial.

---

[*] Bachet, problem XX, p. 155; Oughtred, *Mathematicall Recreations*, London, 1653, p. 28.

Extension. It is obvious that the same trick can be performed with any collection of m things, such as cards or dominoes, which are distinguishable one from the other, provided $m < 20$. For suppose the m things are arranged on a table in some numerical order, and the nth thing is selected by a spectator. Then the first $(19-m)$ taps are immaterial, the $(20-m)$th tap must be on the mth thing and be reckoned by the spectator as the $(n+20-m)$th, the $(20-m+1)$th tap must be on the $(m-1)$th thing and be reckoned as the $(n+20-m+1)$th, and finally the $(20-n)$th tap will be on the nth thing and is reckoned as the 20th tap.

Third Example. The following example rests on an extension of the method used in the last question; it is very simple, but I have never seen it previously described in print. Suppose that a pack of n cards is given to some one who is asked to select one out of the first m cards and to remember (but not to mention) what is its number from the top of the pack (say it is actually the xth card in the pack). Then take the pack, reverse the order of the top m cards (which can be easily effected by shuffling), and transfer y cards (where $y < n - m$) from the bottom to the top of the pack. The effect of this is that the card originally chosen is now the $(y + m - x + 1)$th from the top. Return to the spectator the pack so rearranged, and ask that the top card be counted as the $(x+1)$th, the next as the $(x+2)$th, and so on, in which case the card originally chosen will be the $(y+m+1)$th. Now y and m can be chosen as we please, and may be varied every time the trick is performed; thus any one unskilled in arithmetic will not readily detect the *modus operandi*.

Fourth Example*. Place a card on the table, and on it place as many other cards from the pack as with the number of pips on the card will make a total of twelve. For example, if the card placed first on the table is the five of clubs, then seven additional cards must be placed on it. The court cards may have any values assigned to them, but usually they are reckoned as tens. This is done again with another card, and thus another pile is formed. The operation may be repeated either only three or four times or as often as the pack will permit of such piles being formed. If finally there are p such piles, and if the number of cards left over is r, then the sum of the number of pips on the bottom cards of all the piles will be $13(p-4)+r$.

---

* A particular case of this problem was given by Bachet, problem XVII, p. 138.

For, if x is the number of pips on the bottom card of a pile, the number of cards in that pile will be $13 - x$. A similar argument holds for each pile. Also there are 52 cards in the pack; and this must be equal to the sum of the cards in the p piles and the r cards left over.

$$\therefore (13 - x_1) + (13 - x_2) + \cdots + (13 - x_p) + r = 52,$$
$$\therefore 13p - (x_1 + x_2 + \cdots + x_p) + r = 52,$$
$$\therefore x_1 + x_2 + \cdots + x_p = 13p - 52 + r$$
$$= 13(p - 4) + r.$$

More generally, if a pack of n cards is taken, and if in each pile the sum of the pips on the bottom card and the number of cards put on it is equal to m, then the sum of the pips on the bottom cards of the piles will be $(m + 1)p + r - n$. In an écarté pack $n = 32$, and it is convenient to take $m = 15$.

Fifth Example. It may be noticed that cutting a pack of cards never alters the relative position of the cards provided that, if necessary, we regard the top card as following immediately after the bottom card in the pack. This is used in the following trick[*]. Take a pack, and deal the cards face upwards on the table, calling them one, two, three, &c. as you put them down, and noting in your own mind the card first dealt. Ask some one to select a card and recollect its number. Turn the pack over, and let it be cut (not shuffled) as often as you like. Enquire what was the number of the card chosen. Then, if you deal, and as soon as you come to the original first card begin (silently) to count, reckoning this as one, the selected card will appear at the number mentioned. Of course, if all the cards are dealt before reaching this number, you must turn the cards over and go on counting continuously.

Another similar trick is performed by handing the pack face upwards to some one, and asking him to select a card and state its number, reckoning from the top; suppose it to be the nth. Next, ask him to choose a number at which it shall appear in the pack; suppose he selects the mth. Take the pack and secretly move $m - n$ cards from the bottom to the top (or if n is greater than m, then $n - m$ from the top to the bottom) and of course the card will be in the required position.

---

[*] Bachet, problem XIX, p. 152.

MEDIEVAL PROBLEMS IN ARITHMETIC.   Before leaving the subject of these elementary questions, I may mention a few problems which for centuries have appeared in nearly every collection of mathematical recreations, and therefore may claim what is almost a prescriptive right to a place here.

First Example[*]. The following is a sample of one class of these puzzles. Three men robbed a gentleman of a vase, containing 24 ounces of balsam. Whilst running away they met in a wood with a glass-seller, of whom in a great hurry they purchased three vessels. On reaching a place of safety they wished to divide the booty, but they found that their vessels contained 5, 11, and 13 ounces respectively. How could they divide the balsam into equal portions?

Problems like this can be worked out only by trial: there are several solutions, of which one is as follows.

	24 oz.	13 oz.	11 oz.	5 oz.
The vessels can contain				
Their contents originally are	24	0	0	0
First, make their contents	0	8	11	5
Second, " "	16	8	0	0
Third, " "	16	0	8	0
Fourth, " "	3	13	8	0
Fifth, " "	3	8	8	5
Sixth, " "	8	8	8	0

Second Example[†]. The next of these is a not uncommon game, played by two people, say A and B. A begins by mentioning some number not greater than (say) six, B may add to that any number not greater than six, A may add to that again any number not greater than six, and so on. He wins who is the first to reach (say) 50. Obviously, if A calls 43, then whatever B adds to that, A can win next time. Similarly, if A calls 36, B cannot prevent A's calling 43 the next time. In this way it is clear that the key numbers are those forming the arithmetical progression 43, 36, 29, 22, 15, 8, 1; and whoever plays first ought to win.

Similarly, if no number greater than m may be added at any one time, and n is the number to be called by the victor, then the key num-

---

[*] Some similar problems were given by Bachet, appendix, problem III, p. 206; problem IX, p. 233; by Oughtred in his *Recreations*, p. 22: and by Ozanam, 1803 edition, vol. I, p. 174; 1840 edition, p. 79. Earlier instances occur in Tartaglia's writings.

[†] Bachet, problem XXII, p. 170.

bers will be those forming the arithmetical progression whose common difference is $m+1$ and whose smallest term is the remainder obtained by dividing n by $m+1$.

The same game may be played in another form by placing n coins, matches, or other objects on a table, and directing each player in turn to take away not more than m of them. Whoever takes away the last coin wins. Obviously the key numbers are multiples of $m+1$, and the first player who is able to leave an exact multiple of $(m+1)$ coins can win. Perhaps a better form of the game is to make that player lose who takes away the last coin, in which case each of the key numbers exceeds by unity a multiple of $m+1$.

Mr Loyd has also suggested* a modification which is equivalent to placing n counters in the form of a circle, and allowing each player in succession to take away not more than m of them which are in unbroken sequence: m being less than n and greater than unity. In this case the second of the two players can always win.

Recent Extension of this Problem. The games last described are very simple, but if we impose on the original problem the additional restriction that each player may not add the same number more than three times, the analysis becomes by no means easy. It is difficult in this case to say whether it is an advantage to begin or not. I have never seen this extension described in print, and I will therefore enunciate it at length.

Suppose that each player is given eighteen cards, three of them marked 6, three marked 5, three marked 4, three marked 3, three marked 2, and three marked 1. They play alternately; A begins by playing one of his cards; then B plays one of his, and so on. He wins who first plays a card which makes the sum of the points or numbers on all the cards played exactly equal to 50, but he loses if he plays a card which makes this sum exceed 50. The game can be played mentally or by noting the numbers on a piece of paper, and in practice it is unnecessary to use cards.

Thus, if they play as follows A, 4; B, 3; A, 1; B, 6; A, 3; B, 4; A, 4; B, 5; A, 4; B, 4; A, 5; the game stands at 43. B can now win, for he may safely play 3, since A has not another 4 wherewith to follow it; and if A plays less than 4, B will win the next time. Again, if they play

---

* *Tit-Bits*, London, July 17, Aug. 7, 1897.

thus, A, 6; B, 3; A, 1; B, 6; A, 3; B, 4; A, 2; B, 5; A, 1; B, 5; A, 2; B, 5; A, 2; B, 3; A is now forced to play 1, and B wins by playing 1.

The game can be also played if each player is given only two cards of each kind.

Third Example. The following medieval problem is somewhat more elaborate. Suppose that three people, P, Q, R, select three things, which we may denote by a, e, i, respectively, and that it is desired to find by whom each object was selected*.

Place 24 counters on a table. Ask P to take one counter, Q to take two counters, and R to take three counters. Next, ask the person who selected a to take as many counters as he has already, whoever selected e to take twice as many counters as he has already, and whoever selected i to take four times as many counters as he has already. Note how many counters remain on the table. There are only six ways of distributing the three things among P, Q, and R; and the number of counters remaining on the table is different for each way. The remainders may be 1, 2, 3, 5, 6, or 7.

Bachet summed up the results in the mnemonic line *Par fer* (1) *César* (2) *jadis* (3) *devint* (5) *si grand* (6) *prince* (7). Corresponding to any remainder is a word or words containing two syllables: for instance, to the remainder 5 corresponds the word *devint*. The vowel in the first syllable indicates the thing selected by P, the vowel in the second syllable indicates the thing selected by Q, and of course R selected the remaining thing. *Salve certa animae semita vita quies* was suggested by Oughtred[†] as an alternative mnemonic line.

Extension. M. Bourlet, in the course of a very kindly notice[‡] of the second edition of this work, has given a much neater solution of the above question, and has extended the problem to the case of n people, $P_0, P_1, P_2, \ldots, P_{n-1}$, each of whom selects one object, out of a collection of n objects, such as dominoes or cards. It is required to know which domino or card was selected by each person.

Let us suppose the dominoes to be denoted or marked by the numbers $0, 1, \ldots, n-1$, instead of by vowels. Give one counter to P_1, two counters to P_2, and generally k counters to P_k. Note the number of counters left on the table. Next ask the person who had chosen the

---

[*] Bachet, problem XXV, p. 187.
[†] *Mathematicall Recreations*, London, 1653, p. 20.
[‡] *Bulletin des sciences mathématiques*, Paris, 1893, vol. XVII, pp. 105–107.

domino 0 to take as many counters as he had already, and generally whoever had chosen the domino h to take n^h times as many dominoes as he had already: thus if P_k had chosen the domino numbered h, he would take $n^h k$ counters. Note the total number of counters taken, *i.e.* $\sum n^h k$. Divide it by n, then the remainder will be the number on the domino selected by P_0; divide the quotient by n, and the remainder will be the number on the domino selected by P_1; divide this quotient by n, and the remainder will be the number on the domino selected by P_2; and so on. In other words, if the number of counters taken is expressed in the scale of notation whose radix is n, then the $(h+1)$th digit from the right will give the number on the domino selected by P_h.

Thus in Bachet's problem with 3 people and 3 dominoes, we should first give one counter to Q, and two counters to R, while P would have no counters; then we should ask the person who selected the domino marked 0 or a to take as many counters as he had already, whoever selected the domino marked 1 or e to take three times as many counters as he had already, and whoever selected the domino marked 2 or i to take nine times as many counters as he had already. By noticing the original number of counters, and observing that 3 of these had been given to Q and R, we should know the total number taken by P, Q, and R. If this number were divided by 3, the remainder would be the number of the domino chosen by P; if the quotient were divided by 3 the remainder would be the number of the domino chosen by Q; and the final quotient would be the number of the domino chosen by R.

I may add that Bachet also discussed the case when $n=4$, which had been previously considered by Diego Palomino in 1599, but as M. Bourlet's method is general, it is unnecessary to discuss further particular cases.

Decimation. The last of these antique problems to which I referred consists in placing men round a circle so that if every nth man is killed the remainder shall be certain specified individuals. When decimation was a not uncommon punishment a knowledge of this kind may have had practical interest.

Hegesippus[*] says that Josephus saved his life by such a device. According to his account, after the Romans had captured Jotopat, Josephus and forty other Jews took refuge in a cave. Josephus, much to his disgust, found that all except himself and one other man were

---

[*] *De Bello Judaico*, bk. III, chaps. 16–18.

resolved to kill themselves, so as not to fall into the hands of their conquerors. Fearing to show his opposition too openly he consented, but declared that the operation must be carried out in an orderly way, and suggested that they should arrange themselves round a circle and that every third person should be killed until but one man was left, who must then commit suicide. It is alleged that he placed himself and the other man in the 31st and 16th place respectively, with a result which will be easily foreseen.

The question is usually presented in the following form. A ship, carrying as passengers fifteen Turks and fifteen Christians, encountered a storm, and the pilot declared that, in order to save the ship and crew, one-half of the passengers must be thrown into the sea. To choose the victims the passengers were placed round a circle, and it was agreed that every ninth man should be cast overboard, reckoning from a certain point. It is desired to find an arrangement by which all the Christians should be saved.*

Problems like this can be easily solved by counting, but it is impossible to give a general rule. In this case, the Christians, reckoning from the man first counted, must occupy the places 1, 2, 3, 4, 10, 11, 13, 14, 15, 17, 20, 21, 25, 28, 29. This arrangement can be recollected by the positions of the vowels in the following doggerel rhyme,

From numbers' aid and art, never will fame depart,

where a stands for 1, e for 2, i for 3, o for 4, and u for 5. Hence (looking only at the vowels in the verse) the order is 4 Christians, 5 Turks, 2 Christians, 1 Turk, 3 Christians, 1 Turk, 1 Christian, 2 Turks, 2 Christians, 3 Turks, 1 Christian, 2 Turks, 2 Christians, 1 Turk. Other similar mnemonic lines in French and in Latin were given by Bachet and by Ozanam respectively.

ARITHMETICAL FALLACIES. I insert next some instances of demonstrations† leading to arithmetical results which are obviously

---

* Bachet, problem XXIII, p. 174. The same problem had been previously enunciated by Tartaglia.

† Of the fallacies given in the text, the first, second, and third, are well known; the fourth is not new, but the earliest work in which I recollect seeing it is my *Algebra*, Cambridge, 1890, p. 430; the fifth is given in G.C. Chrystal's *Algebra*,

impossible. I include algebraical proofs as well as arithmetical ones. The fallacies are so patent that in preparing the first and second editions I did not think such questions worth printing, but, as some correspondents have expressed a contrary opinion, I give them for what they are worth.

First fallacy. One of the oldest of these—and not a very interesting specimen—is as follows. Suppose that $a = b$, then

$$ab = a^2.$$
$$\therefore ab - b^2 = a^2 - b^2.$$
$$\therefore b(a - b) = (a + b)(a - b).$$
$$\therefore b = a + b.$$
$$\therefore b = 2b.$$
$$\therefore 1 = 2.$$

Second Fallacy. Another instance, almost as puerile, is as follows. Let a and b be two unequal numbers, and let c be their arithmetic mean, hence

$$a + b = 2c.$$
$$\therefore (a + b)(a - b) = 2c(a - b).$$
$$\therefore a^2 - 2ac = b^2 - 2bc.$$
$$\therefore a^2 - 2ac + c^2 = b^2 - 2bc + c^2.$$
$$\therefore (a - c)^2 = (b - c)^2.$$
$$\therefore a = b.$$

Edinburgh, 1889, vol. II, p. 159; the eighth is due to G.T. Walker, and, as far as I know, has not appeared in any other book; the ninth is due to D'Alembert; and the tenth to F. Galton. A mechanical demonstration that $1 = 2$ was given by R. Chartres in *Knowledge*, July, 1891. J.L.F. Bertrand pointed out that a demonstration that $1 = -1$ can be also obtained from the proposition in the Integral Calculus that, if the limits are constant, the order of integration is indifferent; hence the integral to x (from $x = 0$ to $x = 1$) of the integral to y (from $y = 0$ to $y = 1$) of a function φ should be equal to the integral to y (from $y = 0$ to $y = 1$) of the integral to x (from $x = 0$ to $x = 1$) of φ, but if $\varphi = (x^2 - y^2)/(x^2 + y^2)^2$, this gives $\frac{1}{4}\pi = -\frac{1}{4}\pi$.

Third Fallacy. Another example, the idea of which is due to John Bernoulli, may be stated as follows.

We have $$(-1)^2 = 1.$$
Take logarithms, $$\therefore 2\log(-1) = \log 1 = 0.$$
$$\therefore \log(-1) = 0.$$
$$\therefore -1 = e^0.$$
$$\therefore -1 = 1.$$

The same argument may be expressed thus. Let x be a quantity which satisfies the equation
$$e^x = -1.$$
Square both sides, $$\therefore e^{2x} = 1.$$
$$\therefore 2x = 0.$$
$$\therefore x = 0.$$
$$\therefore e^x = e^0.$$
But $e^x = -1$ and $e^0 = 1$, $\quad \therefore -1 = 1$.

Fourth Fallacy. As yet another instance, we know that
$$\log(1+x) = x - \tfrac{1}{2}x^2 + \tfrac{1}{3}x^3 - \cdots.$$

If $x = 1$, the resulting series is convergent; hence we have
$$\log 2 = 1 - \tfrac{1}{2} + \tfrac{1}{3} - \tfrac{1}{4} + \tfrac{1}{5} - \tfrac{1}{6} + \tfrac{1}{7} - \tfrac{1}{8} + \tfrac{1}{9} - \cdots.$$
$$\therefore 2\log 2 = 2 - 1 + \tfrac{2}{3} - \tfrac{1}{2} + \tfrac{2}{5} - \tfrac{1}{3} + \tfrac{2}{7} - \tfrac{1}{4} + \tfrac{2}{9} - \cdots.$$

Taking those terms together which have a common denominator, we obtain
$$2\log 2 = 1 + \tfrac{1}{3} - \tfrac{1}{2} + \tfrac{1}{5} + \tfrac{1}{7} - \tfrac{1}{4} + \tfrac{1}{9} - \cdots$$
$$= 1 - \tfrac{1}{2} + \tfrac{1}{3} - \tfrac{1}{4} + \tfrac{1}{5} - \cdots$$
$$= \log 2.$$
Hence $$2 = 1.$$

Fifth Fallacy. This fallacy is very similar to that last given. We have

$$\log 2 = 1 - \tfrac{1}{2} + \tfrac{1}{3} - \tfrac{1}{4} + \tfrac{1}{5} - \tfrac{1}{6} + \cdots$$
$$= \left(1 + \tfrac{1}{3} + \tfrac{1}{5} + \cdots\right) - \left(\tfrac{1}{2} + \tfrac{1}{4} + \tfrac{1}{6} + \cdots\right)$$
$$= \left\{\left(1 + \tfrac{1}{3} + \tfrac{1}{5} + \cdots\right) + \left(\tfrac{1}{2} + \tfrac{1}{4} + \tfrac{1}{6} + \cdots\right)\right\} - 2\left(\tfrac{1}{2} + \tfrac{1}{4} + \tfrac{1}{6} + \cdots\right)$$
$$= \left\{1 + \tfrac{1}{2} + \tfrac{1}{3} + \cdots\right\} - \left(1 + \tfrac{1}{2} + \tfrac{1}{3} + \cdots\right)$$
$$= 0.$$

The error in each of the foregoing examples is obvious, but the fallacies in the next examples are concealed somewhat better.

Sixth Fallacy. We can write the identity $\sqrt{-1} = \sqrt{-1}$ in the form

$$\sqrt{\tfrac{-1}{1}} = \sqrt{\tfrac{1}{-1}},$$

hence
$$\frac{\sqrt{-1}}{\sqrt{1}} = \frac{\sqrt{1}}{\sqrt{-1}},$$

therefore
$$(\sqrt{-1})^2 = (\sqrt{1})^2,$$

that is,
$$-1 = 1.$$

Seventh Fallacy. Again, we have

$$\sqrt{a} \times \sqrt{b} = \sqrt{ab}.$$

Hence
$$\sqrt{-1} \times \sqrt{-1} = \sqrt{(-1)(-1)},$$

therefore
$$(\sqrt{-1})^2 = \sqrt{1},$$

that is,
$$-1 = 1.$$

Eighth Fallacy. The following demonstration depends on the fact that an algebraical identity is true whatever be the symbols used in it, and it will appeal only to those who are familiar with this fact.

We have, as an identity,

$$\sqrt{x - y} = i\sqrt{y - x} \qquad \ldots\ldots(i),$$

where i stands either for $+\sqrt{-1}$ or for $-\sqrt{-1}$. Now an *identity* in x and y is necessarily true whatever numbers x and y may represent. First put $x = a$ and $y = b$,

$$\therefore \sqrt{a - b} = i\sqrt{b - a} \qquad \ldots\ldots(ii).$$

Next put $x = b$ and $y = a$,

$$\therefore \sqrt{b-a} = i\sqrt{a-b} \qquad \ldots\ldots \text{(iii)}.$$

Also since (i) is an identity, it follows that in (ii) and (iii) the symbol i must be the same, that is, it represents $+\sqrt{-1}$ or $-\sqrt{-1}$ in both cases. Hence, from (ii) and (iii), we have

$$\sqrt{a-b}\,\sqrt{b-a} = i^2\sqrt{b-a}\,\sqrt{a-b},$$
$$\therefore 1 = i^2,$$
that is $\qquad\qquad\qquad 1 = -1.$

Ninth Fallacy. The following fallacy is due to D'Alembert*. We know that if the product of two numbers is equal to the product of two other numbers, the numbers will be in proportion, and from the definition of a proportion it follows that if the first term is greater than the second, then the third term will be greater than the fourth: thus, if $ad = bc$, then $a : b = c : d$, and if in this proportion $a > b$, then $c > d$. Now if we put $a = d = 1$ and $b = c = -1$ we have four numbers which satisfy the relation $ad = bc$ and such that $a > b$; hence, by the proposition, $c > d$, that is, $-1 > 1$, which is absurd.

Tenth Fallacy. The mathematical theory of probability leads to various paradoxes: of these one specimen† will suffice. Suppose three coins to be thrown up and the fact whether each comes down head or tail to be noticed. The probability that all three coins come down head is clearly $(\frac{1}{2})^3$, that is, is $\frac{1}{8}$; similarly the probability that all three come down tail is $\frac{1}{8}$: hence the probability that all the coins come down alike (*i.e.* either all of them heads or all of them tails) is $\frac{1}{4}$. But, of three coins thus thrown up, at least two must come down alike; now the probability that the third coin comes down head is $\frac{1}{2}$ and the probability that it comes down tail is $\frac{1}{2}$, thus the probability that it comes down the same as the other two coins is $\frac{1}{2}$: hence the probability that all the coins come down alike is $\frac{1}{2}$. I leave to my readers to say whether either of these conflicting conclusions is right and if so, which.

Arithmetical Problems. To the above examples I may add the following questions, which I have often propounded in past years: though not fallacies, they may serve to illustrate the fact that the answer to

---

* *Opuscules mathématiques*, Paris, 1761, vol. I, p. 201.
† See *Nature*, Feb. 15, March 1, 1894, vol. XLIX, pp. 365–366, 413.

an arithmetical question is frequently different to what a hasty reader might suppose.

The first of these questions is as follows. Two clerks are engaged, one at a salary commencing at the rate of (say) £100 a year with a rise of £20 every year, the other at a salary commencing at the same rate (£100 a year) with a rise of £5 every half-year, in each case payments being made half-yearly: which has the larger income? The answer is the latter; for in the first year the first clerk receives £100, but the second clerk receives £50 and £55 as his two half-yearly payments and thus receives in all £105. In the second year the first clerk receives £120, but the second clerk receives £60 and £65 as his two half-yearly payments and thus receives in all £125. In fact the second clerk will always receive £5 a year more than the first clerk.

As another question take the following. A man bets $1/n$th of his money on an even chance (say tossing heads or tails with a penny): he repeats this again and again, each time betting $1/n$th of all the money then in his possession. If, finally, the number of times he has won is equal to the number of times he has lost, has he gained or lost by the transaction? He has, in fact, lost.

Here is another simple question to which not unfrequently I have received incorrect answers. One tumbler is half-full of wine, another is half-full of water: from the first tumbler a teaspoonful of wine is taken out and poured into the tumbler containing the water: a teaspoonful of the mixture in the second tumbler is then transferred to the first tumbler. As the result of this double transaction, is the quantity of wine removed from the first tumbler greater or less than the quantity of water removed from the second tumbler? Nineteen people out of twenty will say it is greater, but this is not the case.

Routes on a Chess-Board. A not uncommon problem can be generalised as follows[*]. Construct a rectangular board of mn cells (or small squares) by ruling $m+1$ vertical lines and $n+1$ horizontal lines. It is required to know how many routes can be taken from the top left-hand corner to the bottom right-hand corner, the motion being along the ruled lines and its direction being always either vertically downwards or horizontally from left to right. The answer is $(m+n)!/m!n!$: thus on a square board containing 16 cells (*i.e.* one-quarter of a chess-board),

---

[*] The substance of the problem was given in a scholarship paper set at Cambridge about 30 years ago, and possibly was not new then.

where $m = n = 4$, there are 70 such routes; while on a common chessboard, where $m = n = 8$, there are no less than 12870 such routes. A similar theorem can be enunciated for a parallelopiped.

Another problem of a somewhat similar type is the determination of the number of closed routes through mn points arranged in m rows and n columns, following the lines of the quadrilateral net-work, and passing once and only once through each point[*].

Permutation Problems. As other simple illustrations of the very large number of ways in which combinations of even a few things can be arranged, I may note that as many as $19,958400$ distinct skeleton cubes can be formed with twelve differently coloured rods of equal length[†]; again there are $3,979614,965760$ ways of arranging a set of twenty-eight dominoes (*i.e.* a set from double zero to double six) in a line, with like numbers in contact[‡]; while there are no less than $53644, 737765, 488792, 839237, 440000$ possible different distributions of hands at whist with a pack of fifty-two cards[§].

Voting Problems. Here is a simple example on combinations dealing with the cumulative vote as affecting the representation of a minority. If there are p electors each having r votes of which not more than s may be given to one candidate, and n men are to be elected, then the least number of supporters who can secure the election of a candidate must exceed $pr/(ns + r)$.

Exploration Problems. Another common question is concerned with the maximum distance into a desert which could be reached from a frontier settlement by the aid of a party of n explorers, each capable of carrying provisions that would last one man for a days. The answer is that the man who reaches the greatest distance will occupy $na/(n+1)$ days before he returns to his starting point. If in the course of their journey they may make depôts, the longest possible journey will occupy $\frac{1}{2}a(1 + \frac{1}{2} + \frac{1}{3} + \cdots + 1/n)$ days. Further extensions by the use of horses and cycles will suggest themselves.

Here I conclude my account of such of these easy problems on

---

[*] See C.F. Sainte-Marie in *L'Intermédiaire des mathématiciens*, Paris, vol. XI, March, 1904, pp. 86–88.
[†] *Mathematical Tripos*, Cambridge, Part I, 1894.
[‡] Reiss in *Annali di matematica*, Milan, November, 1871, vol. V, pp. 63–120.
[§] That is $(52!)/(13!)^4$.

numbers or elementary algebra as seemed worth reproducing. It will be noticed that the majority of them either are due to Bachet or were collected by him in his classical *Problèmes*; but it should be added that besides the questions I have mentioned he enunciated, even if he did not always solve, some other problems of greater interest. One instance will suffice.

BACHET'S WEIGHTS PROBLEM[*]. Among the more difficult problems proposed by Bachet was the determination of the least number of weights which would serve to weigh any integral number of pounds from 1 lb. to 40 lbs. inclusive. Bachet gave two solutions: namely, (i) the series of weights of 1, 2, 4, 8, 16, and 32 lbs.; (ii) the series of weights of 1, 3, 9, and 27 lbs.

If the weights may be placed in only one of the scale-pans, the first series gives a solution, as had been pointed out in 1556 by Tartaglia[†].

Bachet, however, assumed that any weight might be placed in either of the scale-pans. In this case the second series gives the least possible number of weights required. His reasoning is as follows. To weigh 1 lb. we must have a 1 lb. weight. To weigh 2 lbs. we must have in addition either a 2 lb. weight or a 3 lb. weight; but, if we are confined to only one new weight (in addition to the 1 lb. we have got already), then with no weight greater than 3 lbs. could we weigh 2 lbs.: if we use a 2 lb. weight we then can weigh 1 lb., 2 lbs., and 3 lbs., but if we use a 3 lb. weight we then can weigh 1 lb., $(3-1)$ lbs., 3 lbs., and $(3+1)$ lbs.; hence a 3 lb. weight is preferable. Similarly, to enable us to weigh 5 lbs. we must have another weight not greater than 9 lbs., and a weight of 9 lbs. enables us to weigh every weight from 1 lb. to 13 lbs.; hence it is the best to choose. The next weight required will be $2(1+3+9)+1$ lb., that is, will be 27 lbs.; and this enables us to weigh from 1 lb. to 40 lbs. Thus only four weights are required, namely, 1 lb., 3 lbs., 3^2 lbs., and 3^3 lbs.

We can show similarly that the series of weights of $1, 3, 3^2, \ldots, 3^{n-1}$ lbs. will enable us to weigh any integral number of pounds from 1 lb. to $(1+3+3^2+\cdots 3^{n-1})$ lbs., that is, to $\frac{1}{2}(3^n-1)$ lbs. This is the least number with which the problem can be effected.

To determine the arrangement of the weights to weigh any given mass we have only to express the number of pounds in it as a number in the ternary scale of notation, except that in finding the successive

---

[*] Bachet, Appendix, problem V, p. 215.
[†] *Trattato de' numeri e misure*, Venice, 1556, vol. II, bk. I, chap. XVI, art. 32.

digits we must make every remainder either 0, 1, or −1: to effect this a remainder 2 must be written as 3 − 1, that is, the quotient must be increased by unity, in which case the remainder is −1. This is explained in most text-books on algebra.

Bachet's argument does not prove that his result is unique or that it gives the least possible number of weights required. These omissions have been supplied by Major MacMahon, who has discussed the far more difficult problem (of which Bachet's is a particular case) of the determination of all possible sets of weights, not necessarily unequal, which enable us to weigh any integral number of pounds from 1 to n inclusive, (i) when the weights may be placed in only one scale-pan, and (ii) when any weight may be placed in either scale-pan. He has investigated also the modifications of the results which are necessary when we impose either or both of the further conditions (a) that no other weighings are to be possible, and (b) that each weighing is to be possible in only one way, that is, is to be unique*.

The method for case (i) consists in resolving $1 + x + x^2 + \cdots + x^n$ into factors, each factor being of the form $1 + x^a + x^{2a} + \cdots + x^{ma}$; the number of solutions depends on the composite character of $n+1$. The method for case (ii) consists in resolving the expression $x^{-n} + x^{-n+1} + \cdots + x^{-1} + 1 + x + \cdots + x^{n-1} + x^n$ into factors, each factor being of the form $x^{-ma} + \cdots + x^{-a} + 1 + x^a + \cdots + x^{ma}$; the number of solutions depends on the composite character of $2n+1$.

Bachet's problem falls under case (ii), $n = 40$. MacMahon's analysis shows that there are eight such ways of factorizing $x^{-40} + x^{-39} + \cdots + 1 + x^{39} + x^{40}$. First, there is the expression itself in which $a = 1$, $m = 40$. Second, the expression is equal to $(1 − x^{81})/x^{40}(1 − x)$, which can be resolved into the product of $(1 − x^3)/x(1 − x)$ and $(1 − x^{81})/x^{39}(1 − x^3)$; hence it can be resolved into two factors of the form given above, in one of which $a = 1$, $m = 1$, and in the other, $a = 3$, $m = 13$. Third, similarly, it can be resolved into two such factors, in one of which $a = 1$, $m = 4$, and in the other $a = 9$, $m = 4$. Fourth, it can be resolved into three such factors, in one of which $a = 1$, $m = 1$, in another $a = 3$, $m = 1$, and in the other, $a = 9$, $m = 4$. Fifth, it can be resolved into two such factors, in one of which $a = 1$, $m = 13$,

---

* See his article in the *Quarterly Journal of Mathematics*, 1886, vol. XXI, pp. 367–373. An account of the method is given in *Nature*, Dec. 4, 1890, vol. XLII, pp. 113–114.

and in the other $a = 27$, $m = 1$. Sixth, it can be resolved into three such factors, in one of which $a = 1$, $m = 1$, in another $a = 3$, $m = 4$, and in the other $a = 27$, $m = 1$. Seventh, it can be resolved into three such factors, in one of which $a = 1$, $m = 4$, in another $a = 9$, $m = 1$, and in the other $a = 27$, $m = 1$. Eighth, it can be resolved into four such factors, in one of which $a = 1$, $m = 1$, in another $a = 3$, $m = 1$, in another $a = 9$, $m = 1$, and in the other $a = 27$, $m = 1$.

These results show that there are eight possible sets of weights with which any integral number of pounds from 1 to 40 can be weighed subject to the conditions (ii), (a), and (b). If we denote p weights each equal to w by w^p, these eight solutions are 1^{40}; 1, 3^{13}; 1^4, 9^4; 1, 3, 9^4; 1^{13}, 27; 1, 3^4, 27; 1^4, 9, 27; 1, 3, 9, 27. The last of these is Bachet's solution: not only is it that in which the least number of weights are employed, but it is also the only unique one in which all the weights are unequal.

PROBLEMS IN HIGHER ARITHMETIC. At the commencement of this chapter I alluded to the special interest which many mathematicians find in the theorems of higher arithmetic: such, for example, as that every prime of the form $4n + 1$ and every power of it is expressible as the sum of two squares[*], and the first and second powers can be expressed thus in only one way. For instance, $13 = 3^2 + 2^2$, $13^2 = 12^2 + 5^2$, $13^3 = 46^2 + 9^2$, and so on. Similarly $41 = 5^2 + 4^2$, $41^2 = 40^2 + 9^2$, $41^3 = 236^2 + 115^2$, and so on.

Propositions such as the one just quoted may be found in textbooks on the theory of numbers and therefore lie outside the limits of this work, but there are one or two questions in higher arithmetic involving points not yet quite cleared up which may find a place here.

PRIMES. The first of these is concerned with the possibility of determining readily whether a given number is prime or not. Euler and Gauss attached great importance to this problem, but failed to establish any conclusive test. It would seem, however, that Fermat possessed some means of finding from its form whether a given number (at any rate if one of certain known forms) was prime or not. Thus, in answer to Mersenne who asked if he could tell without much trouble whether the number $100895,598169$ was a prime, Fermat wrote on April 7, 1643, that it was the product of 898423 and 112303, both of which were

---

[*] Fermat's *Diophantus*, Toulouse, 1670, bk. III, prop. 22, p. 127; or Brassinne's *Précis*, Paris, 1853, p. 65.

primes. I have indicated elsewhere one way by which this result can be found, and Mr F.W. Laurence has indicated another which may have been that used by Fermat in this particular case.

MERSENNE'S NUMBERS*. Another illustration, confirmatory of the opinion that Fermat or some of his contemporaries had a test by which it was possible to find out whether certain numbers were prime, may be drawn from Mersenne's *Cogitata Physico-Mathematica* which was published in 1644. In the preface to that work it is asserted that in order that $2^p - 1$ may be prime, the only values of p, not greater than 257, which are possible are 1, 2, 3, 5, 7, 13, 17, 19, 31, 67, 127, and 257: I conjecture that the number 67 is a misprint for 61. With this correction the statement appears to be true, and it has been verified for all except nineteen values of p: namely, 71, 101, 103, 107, 109, 137, 139, 149, 157, 163, 167, 173, 181, 193, 199, 227, 229, 241, and 257. Of these values, Mersenne asserted that $p = 257$ makes $2^p - 1$ a prime, and that the other values make $2^p - 1$ a composite number. The demonstrations for the cases when $p = 89, 127$ have not been published; nor have the actual factors of $2^p - 1$ when $p = 89$ been as yet determined: the discovery of these factors may be commended to those interested in the theory of numbers.

Mersenne's result could not be obtained empirically, and it is impossible to suppose that it was worked out for every case; hence it would seem that whoever first enunciated it was acquainted with certain theorems in higher arithmetic which have not been re-discovered.

PERFECT NUMBERS†. The theory of *perfect numbers* depends directly on that of Mersenne's Numbers. A number is said to be perfect if it is equal to the sum of all its integral subdivisors. Thus the subdivisors of 6 are 1, 2, and 3; the sum of these is equal to 6; hence 6 is a perfect number.

It is probable that all perfect numbers are included in the formula $2^{p-1}(2^p - 1)$, where $2^p - 1$ is a prime. Euclid proved that any number of this form is perfect; Euler showed that the formula includes all even perfect numbers; and there is reason to believe—though a rigid demonstration is wanting—that an odd number cannot be perfect. If

---

* For references, see chapter ix below.
† On the theory of perfect numbers, see bibliographical references by H. Brocard, *L'Intermédiaire des mathématiciens*, Paris, 1895, vol. II, pp. 52–54; and 1905, vol. XII, p. 19.

we assume that the last of these statements is true, then every perfect number is of the above form. It is easy to establish that every number included in this formula (except when $p = 2$) is congruent to unity to the modulus 9, that is, when divided by 9 leaves a remainder 1; also that either the last digit is a 6 or the last two digits are 28.

Thus, if $p = 2, 3, 5, 7, 13, 17, 19, 31, 61$, then by Mersenne's rule the corresponding values of $2^p - 1$ are prime; they are 3, 7, 31, 127, 8191, 131071, 524287, 2147483647, 2305843009213693951; and the corresponding perfect numbers are 6, 28, 496, 8128, 33550336, 8589869056, 137438691328, 2305843008139952128, and 2658455991569831744654692615953842176.

GOLDBACH'S THEOREM. Another interesting problem in higher arithmetic is the question whether there are any even integers which cannot be expressed as a sum of two primes. Probably there are none. The expression of all even[1] integers not greater than 5000 in the form of a sum of two primes has been effected*, but a general demonstration that all even integers can be so expressed is wanting.

LAGRANGE'S THEOREM[†]. Another theorem in higher arithmetic which, as far as I know, is still unsolved, is to the effect that every prime of the form $4n - 1$ is the sum of a prime of the form $4n + 1$ and of double a prime of the form $4n + 1$; for example, $23 = 13 + 2 \times 5$. Lagrange, however, added that it was only by induction that he arrived at the result.

FERMAT'S THEOREM ON BINARY POWERS. Fermat enriched mathematics with a multitude of new propositions. With two exceptions all these have been proved subsequently to be true. The first of these exceptions is his *theorem on binary powers*, in which he asserted that all numbers of the form $2^m + 1$, where $m = 2^n$, are primes[‡], but he added that, though he was convinced of the truth of this proposition, he could not obtain a valid demonstration.

---

* *Transactions of the Halle Academy (Naturforschung)*, vol. LXXII, Halle, 1897, pp. 5–214: see also *L'Intermédiaire des mathématiciens*, 1903, vol. X, and 1904, vol. XI.

† *Nouveaux Mémoires de l'Académie Royale des Sciences*, Berlin, 1775, p. 356.

‡ Letter of Oct. 18, 1640, *Opera*, Toulouse, 1679, p. 162: or Brassinne's *Précis*, p. 143.

---

1. 'even' inserted as per errata sheet

It may be shown that $2^m + 1$ is composite if m is not a power of 2, but of course it does not follow that $2^m + 1$ is a prime if m is a power of 2. As a matter of fact the theorem is not true. In 1732 Euler[*] showed that if $n = 5$ the formula gives $4294,967297$, which is equal to $641 \times 6,700417$: curiously enough, these factors can be deduced at once from Fermat's remark on the possible factors of numbers of the form $2^m \pm 1$, from which it may be shown that the prime factors (if any) of $2^{32} + 1$ must be primes of the form $64n + 1$.

During the last thirty years it has been shown[†] that the resulting numbers are composite when $n = 6, 9, 11, 12, 18, 23, 36$, and 38: the two last numbers contain many thousands of millions of digits. I believe that Eisenstein asserted that the number of primes of the form $2^m + 1$, where $m = 2^n$, is infinite: the proof has not been published, but perhaps it might throw some light on the general theory.

FERMAT'S LAST THEOREM. I pass now to the only other assertion made by Fermat which has not been proved hitherto. This, which is sometimes known as *Fermat's Last Theorem*, is to the effect[‡] that no integral values of x, y, z can be found to satisfy the equation $x^n + y^n = z^n$, if n is an integer greater than 2. This proposition has acquired extraordinary celebrity from the fact that no general demonstration of it has been given, but there is no reason to doubt that it is true.

Fermat seems to have discovered its truth first[§] for the case $n = 3$, and then for the case $n = 4$. His proof for the former of these cases is lost, but that for the latter is extant[‖], and a similar proof for the case of $n = 3$ was given by Euler[¶]. These proofs depend upon showing that, if three integral values of x, y, z can be found which satisfy the equation,

---

[*] *Commentarii Academiae Scientiarum Petropolitanae*, St Petersburg, 1738, vol. VI, p. 104; see also *Novi Comm. Acad. Sci. Petrop.*, St Petersburg, 1764, vol. IX, p. 101: or *Commentationes Arithmeticae Collectae*, St Petersburg, 1849, vol. I, pp. 2, 357.

[†] For the factors and bibliographical references, see the memoir by A.J.C. Cunningham and A.E. Western, *Transactions of the London Mathematical Society*, May 14, 1903, series 2, vol. I, p. 175.

[‡] Fermat's enunciation will be found in his edition of *Diophantus*, Toulouse, 1670, bk. II, qu. 8, p. 61; or Brassinne's *Précis*, Paris, 1853, p. 53. For bibliographical references, see *L'Intermédiaire des mathématiciens*, 1905, vol. XII, pp. 11, 12.

[§] See a letter from Fermat quoted in my *History of Mathematics*, London, chapter XV.

[‖] Fermat's *Diophantus*, note on p. 339; or Brassinne's *Précis*, p. 127.

[¶] Euler's Algebra (English trans. 1797), vol. II, chap. XV, p. 247.

then it will be possible to find three other and smaller integers which also satisfy it: in this way finally we show that the equation must be satisfied by three values which obviously do not satisfy it. Thus no integral solution is possible. It would seem that this method is inapplicable except when $n = 3$ and $n = 4$.

Fermat's discovery of the general theorem was made later. An easy demonstration can be given on the assumption that every number can be resolved into prime (complex) factors in one and only one way. That assumption has been made by some writers, but it is not universally true. It is possible that Fermat made some such supposition, though it is perhaps more probable that he discovered a rigorous demonstration. At any rate he asserts definitely that he had a valid proof—demonstratio mirabilis sane—and the fact that every other theorem on the subject which he stated he had proved has been subsequently verified must weigh strongly in his favour; especially as in making the one statement in his writings which is not correct he was scrupulously careful to add that he could not obtain a satisfactory demonstration of it.

It must be remembered that Fermat was a mathematician of quite the first rank who had made a special study of the theory of numbers. That subject is in itself one of peculiar interest and elegance, but its conclusions have little practical importance, and since his time it has been discussed by only a few mathematicians, while even of them not many have made it their chief study. This is the explanation of the fact that it took more than a century before some of the simpler results which Fermat had enunciated were proved, and thus it is not surprising that a proof of the theorem which he succeeded in establishing only towards the close of his life should involve great difficulties.

In 1823 Legendre[*] obtained a proof for the case of $n = 5$; in 1832 Lejeune Dirichlet[†] gave one for $n = 14$, and in 1840 Lamé and Lebesgue[‡] gave proofs for $n = 7$.

The proposition appears to be true universally, and in 1849 Kummer[§], by means of ideal primes, proved it to be so for all numbers except those (if any) which satisfy three conditions. It is not known

---

[*] Reprinted in his *Théorie des Nombres*, Paris, 1830, vol. II, pp. 361–368: see also pp. 5, 6.
[†] *Crelle's Journal*, 1832, vol. IX, pp. 390–393.
[‡] *Liouville's Journal*, 1841, vol. V, pp. 195–215, 276–9, 348–9.
[§] References to Kummer's Memoirs are given in Smith's Report to the British Association on the Theory of Numbers, London, 1860.

whether any number can be found to satisfy these conditions, but it seems unlikely, and it has been shown that there is no number less than 100 which does so. The proof is complicated and difficult, and there can be little doubt is based on considerations unknown to Fermat. I may add that to prove the truth of the proposition when n is greater than 4, it obviously is sufficient to confine ourselves to cases where n is a prime, and the first step in Kummer's demonstration is to show that in such cases one of the numbers x, y, z must be divisible by n.

Naturally there has been much speculation as to how Fermat arrived at the result. The modern treatment of higher arithmetic is founded on the special notation and processes introduced by Gauss, who pointed out that the theory of discrete magnitude is essentially different from that of continuous magnitude, but until the end of the last century the theory of numbers was treated as a branch of algebra, and such proofs by Fermat as are extant involve nothing more than elementary geometry and algebra, and indeed some of his arguments do not involve any symbols. This has led some writers to think that Fermat used none but elementary algebraic methods. This may be so, but the following remark, which I believe is not generally known, rather points to the opposite conclusion. He had proposed, as a problem to the English mathematicians, to show that there was only one integral solution of the equation $x^2 + 2 = y^3$: the solution evidently being $x = 5, y = 3$. On this he has a note[*] to the effect that there was no difficulty in finding a solution in rational fractions, but that he had discovered an entirely new method—sane pulcherrima et subtilissima—which enabled him to solve such questions in integers. It was his intention to write a work[†] on his researches in the theory of numbers, but it was never completed, and we know but little of his methods of analysis. I venture however to add my private suspicion that continued fractions played a not unimportant part in his researches, and as strengthening this conjecture I may note that some of his more recondite results—such as the theorem that a prime of the form $4n + 1$ is expressible as the sum of two squares—may be established with comparative ease by properties of such fractions.

---

[*] Fermat's *Diophantus*, bk. VI, prop. 19, p. 320; or Brassinne's *Précis*, p. 122.
[†] Fermat's *Diophantus*, bk. IV, prop. 31, p. 181; or Brassinne's *Précis*, p. 82.

CHAPTER II.

SOME GEOMETRICAL QUESTIONS.

In this chapter I propose to enumerate certain geometrical questions the discussion of which will not involve necessarily any considerable use of algebra or arithmetic. Unluckily no writer like Bachet has collected and classified problems of this kind, and I take the following instances from my note-books with the feeling that they represent the subject but imperfectly.

The first part of the chapter is devoted to questions which are of the nature of formal propositions: the last part contains a description of various trivial puzzles and games, which the older writers would have termed geometrical, but which the reader of to-day may omit without loss.

In accordance with the rule I laid down for myself in the preface, I exclude the detailed discussion of theorems which involve advanced mathematics. Moreover (with one possible exception) I exclude also any mention of the numerous geometrical paradoxes which depend merely on the inability of the eye to compare correctly the dimensions of figures when their relative position is changed. This apparent deception does not involve the conscious reasoning powers, but rests on the inaccurate interpretation by the mind of the sensations derived through the eyes, and I do not consider such paradoxes as coming within the domain of mathematics.

GEOMETRICAL FALLACIES. Most educated Englishmen are acquainted with the series of logical propositions in geometry associated with the name of Euclid, but it is not known so generally that these propositions were supplemented originally by certain exercises. Of such exercises Euclid issued three series: two containing easy theorems or problems, and the third consisting of geometrical fallacies, the errors in which the student was required to find.

The collection of fallacies prepared by Euclid is lost, and tradition has not preserved any record as to the nature of the erroneous reasoning or conclusions; but, as an illustration of such questions, I append two or three demonstrations, leading to obviously impossible results, which perhaps may amuse any one to whom they are new. I leave the discovery of the errors to the ingenuity of my readers.

First Fallacy. *To prove that a right angle is equal to an angle which is greater than a right angle.* Let $ABCD$ be a rectangle. From A draw a line AE outside the rectangle, equal to AB or DC and making an acute angle with AB, as indicated in the diagram. Bisect CB in

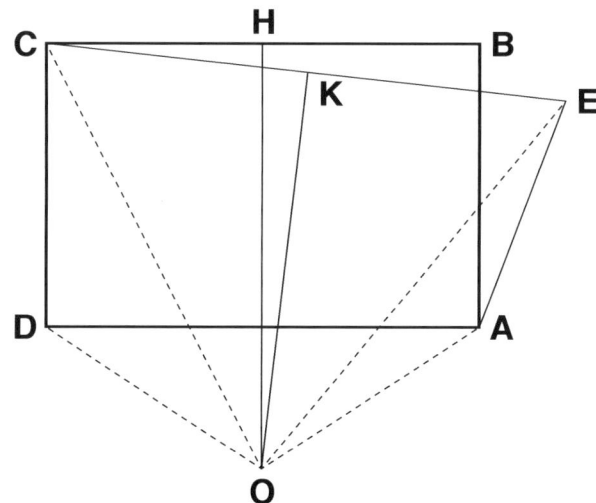

H, and through H draw HO at right angles to CB. Bisect CE in K, and through K draw KO at right angles to CE. Since CB and CE are not parallel the lines HO and KO will meet (say) at O. Join OA, OE, OC, and OD.

The triangles ODC and OAE are equal in all respects. For, since KO bisects CE and is perpendicular to it, we have $OC = OE$. Similarly, since HO bisects CB and DA and is perpendicular to them, we have $OD = OA$. Also, by construction, $DC = AE$. Therefore the three sides of the triangle ODC are equal respectively to the three sides of the triangle OAE. Hence, by Euc. I. 8, the triangles are equal; and therefore the angle ODC is equal to the angle OAE.

Again, since HO bisects DA and is perpendicular to it, we have the angle ODA equal to the angle OAD.

Hence the angle ADC (which is the difference of ODC and ODA) is equal to the angle DAE (which is the difference of OAE and OAD). But ADC is a right angle, and DAE is necessarily greater than a right angle. Thus the result is impossible.

Second Fallacy[*]. *To prove that a part of a line is equal to the whole line.* Let ABC be a triangle; and, to fix our ideas, let us suppose that the triangle is scalene, that the angle B is acute, and that the

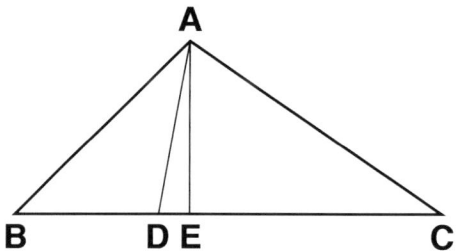

angle A is greater than the angle C. From A draw AD making the angle BAD equal to the angle C, and cutting BC in D. From A draw AE perpendicular to BC.

The triangles ABC, ABD are equiangular; hence, by Euc. VI. 19,

$$\triangle ABC : \triangle ABD = AC^2 : AD^2.$$

Also the triangles ABC, ABD are of equal altitude: hence, by Euc. VI. 1,

$$\triangle ABC : \triangle ABD = BC : BD,$$
$$\therefore AC^2 : AD^2 = BC : BD.$$
$$\therefore \frac{AC^2}{BC} = \frac{AD^2}{BD}.$$

---

[*] See a note by M. Coccoz in *L'Illustration*, Paris, Jan. 12, 1895.

Hence, by Euc. II. 13,

$$\frac{AB^2 + BC^2 - 2BC \cdot BE}{BC} = \frac{AB^2 + BD^2 - 2BD \cdot BE}{BD}.$$

$$\therefore \frac{AB^2}{BC} + BC - 2BE = \frac{AB^2}{BD} + BD - 2BE.$$

$$\therefore \frac{AB^2}{BC} - BD = \frac{AB^2}{BD} - BC.$$

$$\therefore \frac{AB^2 - BC \cdot BD}{BC} = \frac{AB^2 - BC \cdot BD}{BD}.$$

$$\therefore BC = BD,$$

a result which is impossible.

Third Fallacy. *To prove that every triangle is isosceles.* Let ABC be any triangle. Bisect BC in D, and through D draw DO perpendicular to BC. Bisect the angle BAC by AO.

First. If DO and AO do not meet, then they are parallel. Therefore AO is at right angles to BC. Therefore $AB = AC$.

Second. If DO and AO meet, let them meet in O. Draw OE perpendicular to AC. Draw OF perpendicular to AB. Join OB, OC.

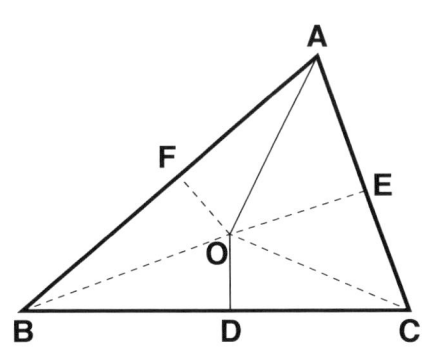

Let us begin by taking the case where O is inside the triangle, in which case E falls on AC and F on BC.

The triangles AOF and AOE are equal, since the side AO is common, angle OAF = angle OAE, and angle OFA = angle OEA. Hence $AF = AE$. Also, the triangles BOF and COE are equal. For since OD bisects BC at right angles, we have $OB = OC$; also, since the triangles AOF and AOE are equal, we have $OF = OE$; lastly, the angles at F and E are right angles. Therefore, by Euc. I. 47 and I. 8, the triangles BOF and COE are equal. Hence $FB = EC$.

Therefore $AF + FB = AE + EC$, that is, $AB = AC$.

The same demonstration will cover the case where DO and AO meet at D, as also the case where they meet outside BC but so near it that E and F fall on AC and AB and not on AC and AB produced.

Next take the case where DO and AO meet outside the triangle, and E and F fall on AC and AB produced. Draw OE perpendicular to AC produced. Draw OF perpendicular to AB produced. Join OB, OC.

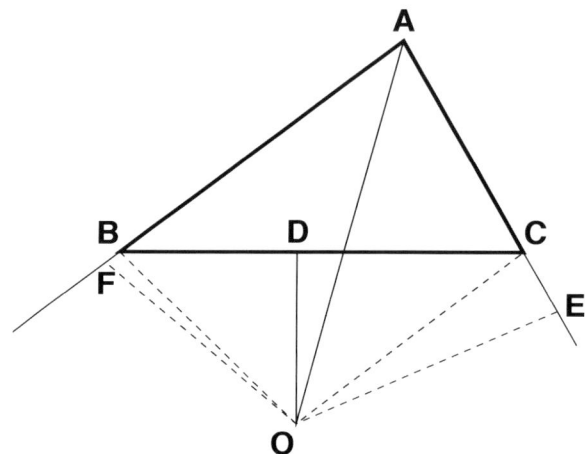

Following the same argument as before, from the equality of the triangles AOF and AOE, we obtain $AF = AE$; and, from the equality of the triangles BOF and COE, we obtain $FB = EC$. Therefore $AF - FB = AE - EC$, that is, $AB = AC$.

Thus in all cases, whether or not DO and AO meet, and whether they meet inside or outside the triangle, we have $AB = AC$: and therefore every triangle is isosceles, a result which is impossible.

Fourth Fallacy. I am indebted to Captain Turton for the following ingenious fallacy; it appeared for the first time in the third edition of this work.

On the hypothenuse, BC, of an isosceles right-angled triangle, DBC, describe an equilateral triangle ABC, the vertex A being on the same side of the base as D is. On CA take a point H so that $CH = CD$. Bisect BD in K. Join HK and let it cut CB (produced) in L. Join DL. Bisect DL at M, and through M draw MO perpendicular to DL. Bisect HL at N, and through N draw NO perpendicular to HL. Since DL and HL intersect, therefore MO and NO will also intersect; moreover, since BDC is a right angle, MO and NO both slope away from DC and therefore they will meet on the side of DL remote from A. Join OC, OD, OH, OL.

The triangles OMD and OML are equal, hence $OD = OL$. Similarly the triangles ONL and ONH are equal, hence $OL = OH$. Therefore $OD = OH$. Now in the triangles OCD and OCH, we have $OD = OH$, $CD = CH$ (by construction), and OC common, hence (by Euc. I. 8) the angle OCD is equal to the angle OCH, which is absurd.

Fifth Fallacy[*]. To prove that, if two opposite sides of a quadrilateral are equal, the other two sides must be parallel. Let $ABCD$ be a quadrilateral such that AB is equal to DC. Bisect AD in M, and through M draw MO at right angles to AD. Bisect BC in N, and draw NO at right angles to BC.

If MO and NO are parallel, then AD and BC (which are at right angles to them) are also parallel.

If MO and NO are not parallel, let them meet in O; then O must be either inside the quadrilateral as in the left-hand diagram or outside

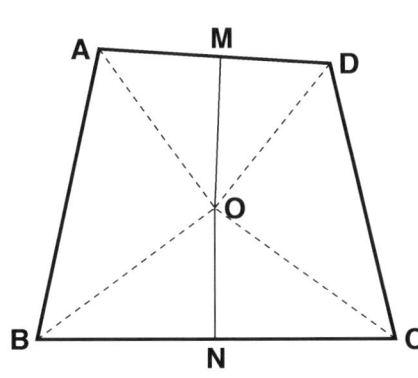
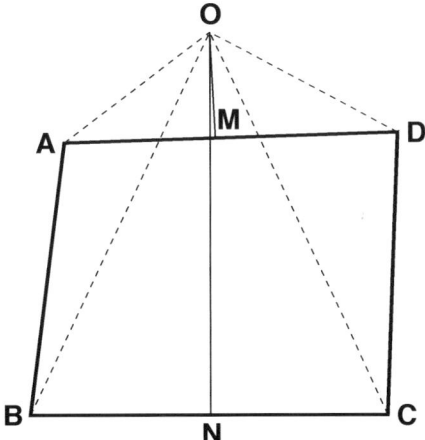

the quadrilateral as in the right-hand diagram. Join OA, OB, OC, OD.

Since OM bisects AD and is perpendicular to it, we have $OA = OD$, and the angle OAM equal to the angle ODM. Similarly $OB = OC$, and the angle OBN equal to the angle OCN. Also by hypothesis $AB = DC$, hence, by Euc. I. 8, the triangles OAB and ODC are equal in all respects, and therefore the angle AOB is equal to the angle DOC.

Hence in the left-hand diagram the sum of the angles AOM, AOB is equal to the sum of the angles DOM, DOC; and in the right-hand

---

[*] *Mathesis*, October, 1893, series 2, vol. III, p. 224.

diagram the difference of the angles AOM, AOB is equal to the difference of the angles DOM, DOC; and therefore in both cases the angle MOB is equal to the angle MOC, i.e. OM (or OM produced) bisects the angle BOC. But the angle NOB is equal to the angle NOC, i.e. ON bisects the angle BOC; hence OM and ON coincide in direction. Therefore AD and BC, which are perpendicular to this direction, must be parallel. This result is not universally true, and the above demonstration contains a flaw.

Sixth Fallacy. The following argument is taken from a text-book on electricity, published in 1889 by two distinguished mathematicians, in which it was presented as valid. A given vector OP of length l can be resolved in an infinite number of ways into two vectors OM, MP, of lengths l', l'', and we can make l'/l'' have any value we please from nothing to infinity. Suppose that the system is referred to rectangular axes Ox, Oy; and that OP, OM, MP make respectively angles θ, θ', θ'' with Ox. Hence, by projection on Oy and on Ox, we have

$$l \sin \theta = l' \sin \theta' + l'' \sin \theta'',$$
$$l \cos \theta = l' \cos \theta' + l'' \cos \theta''.$$

Therefore
$$\tan \theta = \frac{n \sin \theta' + \sin \theta''}{n \cos \theta' + \cos \theta''},$$

where $n = l'/l''$. This result is true whatever be the value of n. But n may have any value (*ex. gr.* $n = \infty$, or $n = 0$), hence $\tan \theta = \tan \theta' = \tan \theta''$, which obviously is impossible.

Seventh Fallacy. Here is a fallacious investigation, to which Mr Chartres first called my attention, of the value of π: it is founded on well-known quadratures. The area of the semi-ellipse bounded by the minor axis is (in the usual notation) equal to $\frac{1}{2}\pi ab$. If the centre is moved off to an indefinitely great distance along the major axis, the ellipse degenerates into a parabola, and therefore in this particular limiting position the area is equal to two-thirds of the circumscribing rectangle. But the first result is true whatever be the dimensions of the curve.

$$\therefore \tfrac{1}{2}\pi ab = \tfrac{2}{3}a \times 2b,$$
$$\therefore \pi = 8/3,$$

a result which is obviously untrue.

GEOMETRICAL PARADOXES. To the above examples I may add the following questions, which, though not exactly fallacious, lead to results which at a hasty glance appear impossible.

First Paradox. The first is a problem, sent to me by Mr Renton, to rotate a plane lamina (say, for instance, a sheet of paper) through four right angles so that the effect is equivalent to turning it through only one right angle.

If it is desired that the effect shall be equivalent to turning it through a right angle about a point O, the solution is as follows. Describe on the lamina a square $OABC$. Rotate the lamina successively through two right angles about the diagonal OB as axis and through two right angles about the side OA as axis, and the required result will be attained.

Second Paradox. As in arithmetic, so in geometry, the theory of probability lends itself to numerous paradoxes. Here is a very simple illustration. A stick is broken at random into three pieces. It is possible to put them together into the shape of a triangle provided the length of the longest piece is less than the sum of the other two pieces (*cf.* Euc. I. 20), that is, provided the length of the longest piece is less than half the length of the stick. But the probability that a fragment of a stick shall be half the original length of the stick is $\frac{1}{2}$. Hence the probability that a triangle can be constructed out of the three pieces into which the stick is broken would appear to be $\frac{1}{2}$. This is not true, for actually the probability is $\frac{1}{4}$.

Third Paradox. The following example illustrates how easily the eye may be deceived in demonstrations obtained by actually dissecting the figures and re-arranging the parts. In fact proofs by superposition should be regarded with considerable distrust unless they are supplemented by mathematical reasoning. The well-known proofs of the propositions Euclid I. 32 and Euclid I. 47 can be so supplemented and are valid. On the other hand, as an illustration of how deceptive a non-mathematical proof may be, I here mention the familiar paradox that a square of paper, subdivided like a chessboard into 64 small squares, can be cut into four pieces which being put together form a figure containing 65 such small squares[*]. This is effected by cutting the original

---

[*] I do not know who discovered this paradox. It is given in various modern books, but I cannot find an earlier reference to it than one by Prof. G.H. Darwin, *Messenger of Mathematics*, 1877, vol. VI, p. 87.

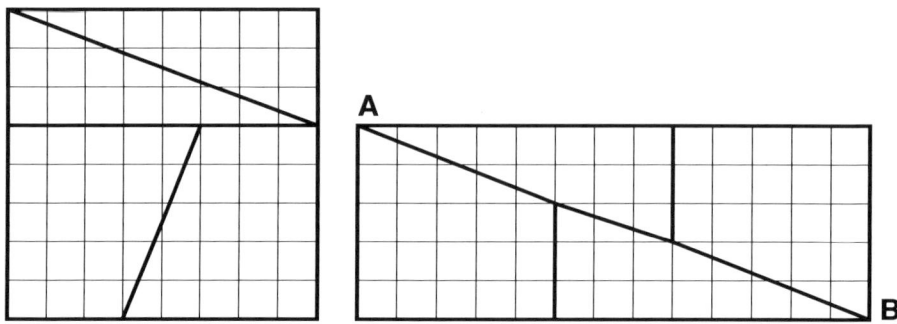

square into four pieces in the manner indicated by the thick lines in the first figure. If these four pieces are put together in the shape of a rectangle in the way shown in the second figure it will appear as if this rectangle contains 65 of the small squares.

This phenomenon, which in my experience non-mathematicians find perplexing, is due to the fact that the edges of the four pieces of paper, which in the second figure lie along the diagonal AB, do not coincide exactly in direction. In reality they include a small lozenge or diamond-shaped figure, whose area is equal to that of one of the 64 small squares in the original square, but whose length AB is much greater than its breadth. The diagrams show that the angle between the two sides of this lozenge which meet at A is $\tan^{-1}\frac{2}{5} - \tan^{-1}\frac{3}{8}$, that is, is $\tan^{-1}\frac{1}{46}$, which is less than $1\frac{1}{4}°$. To enable the eye to distinguish so small an angle as this the dividing lines in the first figure would have to be cut with extreme accuracy and the pieces placed together with great care.

The paradox depends upon the relation $5 \times 13 - 8^2 = 1$. Similar results can be obtained from the formulae $13 \times 34 - 21^2 = 1$, $34 \times 89 - 55^2 = 1, \ldots$; or from the formulae $5^2 - 3 \times 8 = 1$, $13^2 - 8 \times 21 = 1$, $34^2 - 21 \times 55 = 1, \ldots$. These numbers are obtained by finding convergents to the continued fraction

$$1 + \frac{1}{1+}\frac{1}{1+}\frac{1}{1+} \cdots .$$

A similar paradox for a square of 17 cells, by which it was shown that 289 was equal to 288, was alluded to by Ozanam[*] who gave also the diagram for dividing a rectangle of 11 by 3 into two rectangles whose dimensions appear to be 5 by 4 and 7 by 2.

---

[*] Ozanam, 1803 edition, vol. I, p. 299.

Turton's Seventy-Seven Puzzle. A far better dissection puzzle was invented by Captain Turton. In this a piece of cardboard, 11 inches by 7 inches, subdivided into 77 small equal squares, each 1 inch by 1 inch, can be cut up and re-arranged so as to give 78 such equal squares, each 1 inch by 1 inch, of which 77 are arranged in a rectangle of the same dimensions as the original rectangle from one side of which projects a small additional square. The construction is ingenious, but cannot be described without the use of a model. The trick consists in utilizing the fact that cardboard has a sensible thickness. Hence the edges of the cuts can be bevelled, but in the model the bevelling is so slight as to be imperceptible save on a very close scrutiny. The play thus given in fitting the pieces together permits the apparent production of an additional square.

COLOURING MAPS. I proceed next to mention the geometrical proposition that *not more than four colours are necessary in order to colour a map of a country (divided into districts) in such a way that no two contiguous districts shall be of the same colour.* By contiguous districts are meant districts having a common *line* as part of their boundaries: districts which touch only at points are not contiguous in this sense.

The problem was mentioned by A.F. Möbius[*] in his Lectures in 1840, but it was not until Francis Guthrie[†] communicated it to De Morgan about 1850 that attention was generally called to it: it is said that the fact had been familiar to practical map-makers for a long time previously. Through De Morgan the proposition became generally known; and in 1878 Cayley[‡] recalled attention to it by stating that he did not know of any rigorous proof of it.

Probably the following argument, though not a formal demonstration, will satisfy the reader that the result is true.

Let A, B, C be three contiguous districts, and let X be any other district contiguous with all of them. Then X must lie either wholly outside the external boundary of the area ABC or wholly inside the internal boundary, that is, it must occupy a position either like X or

---

[*] *Leipzig Transactions* (*Math.-phys. Classe*), 1885, vol. XXXVII, pp. 1–6.
[†] *Proceedings of the Royal Society of Edinburgh*, July 19, 1880, vol. X, p. 728.
[‡] *Proceedings of the London Mathematical Society*, 1878, vol. IX, p. 148, and *Proceedings of the Royal Geographical Society*, 1879, N.S., vol. I, p. 259.

like X'. In either case every remaining occupied area in the figure is enclosed by the boundaries of not more than three districts: hence there is no possible way of drawing another area Y which shall be contiguous with A, B, C, and X. In other words, it is possible to draw on a plane four areas which are contiguous, but it is not possible to draw five such areas.

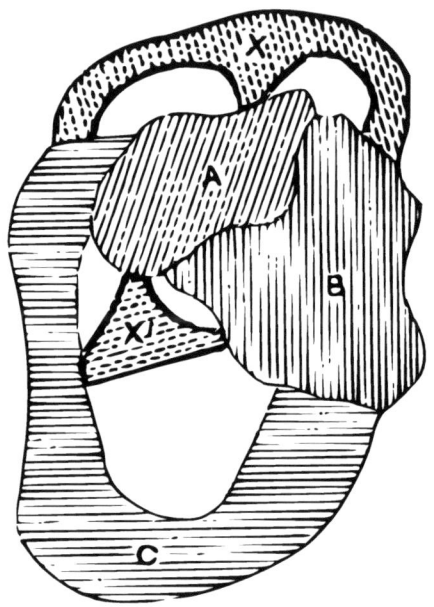

If A, B, C are not contiguous, each with the other, or if X is not contiguous with A, B, and C, it is not necessary to colour them all differently, and thus the most unfavourable case is that already treated. Moreover any of the above areas may diminish to a point and finally disappear without affecting the argument.

That we may require at least four colours is obvious from the diagram on this page, since in that case the areas A, B, C, and X would have to be coloured differently.

A proof of the proposition involves difficulties of a high order, which as yet have baffled all attempts to surmount them.

The argument by which the truth of the proposition was formerly supposed to be demonstrated was given by A.B. Kempe[*] in 1879, but

---

[*] He sent his first demonstration across the Atlantic to the *American Journal of Mathematics*, 1879, vol. II, pp. 193–200; but subsequently he communicated it

there is a flaw[*] in it.

In 1880, Tait published a solution[†] depending on the theorem that if a closed network of lines joining an even number of points is such that three and only three lines meet at each point then three colours are sufficient to colour the lines in such a way that no two lines meeting at a point are of the same colour; a closed network being supposed to exclude the case where the lines can be divided into two groups between which there is but one connecting line. His deduction therefrom that four colours will suffice for a map was given in the last edition of this work. The demonstration appeared so straightforward that at first it was generally accepted, but it would seem that it too involves a fallacy[‡]. The proof however leads to the interesting corollary that four colours may not suffice for a map drawn on a multiply-connected surface such as an anchor ring.

Although a proof of the theorem is still wanting, no one has succeeded in constructing a plane map which requires more than four tints to colour it, and there is no reason to doubt the correctness of the statement that it is not necessary to have more than four colours for any plane map. The number of ways which such a map can be coloured with four tints has been also considered[§], but the results are not sufficiently interesting to require mention here.

PHYSICAL CONFIGURATION OF A COUNTRY. As I have been alluding to maps, I may here mention that the theory of the representation of the physical configuration of a country by means of lines drawn on a map was discussed, by Cayley and Clerk Maxwell[‖]. They

---

in simplified forms to the London Mathematical Society, *Transactions*, 1879, vol. X, pp. 229–231, and to *Nature*, Feb. 26, 1880, vol. XXI, pp. 399–400.

[*] See articles by P.J. Heawood in the *Quarterly Journal of Mathematics*, London, 1890, vol. XXIV, pp. 332–338; and 1897, vol. XXXI, pp. 270–285.

[†] *Proceedings of the Royal Society of Edinburgh*, July 19, 1880, vol. X, p. 729; and PHILOSOPHICAL MAGAZINE, January, 1884, series 5, vol. XVII, p. 41.

[‡] See J. Peterson of Copenhagen, *L'Intermédiaire des mathématiciens*, vol. V, 1898, pp. 225–227; and vol. VI, 1899, pp. 36–38.

[§] See A.C. Dixon, *Messenger of Mathematics*, Cambridge, 1902–3, vol. XXXII, pp. 81–83.

[‖] Cayley on 'Contour and Slope Lines,' *Philosophical Magazine*, London, October, 1859, series 4, vol. XVIII, pp. 264–268; *Collected Works*, vol. IV, pp. 108–111. J. Clerk Maxwell on 'Hills and Dales,' *Philosophical Magazine*, December, 1870, series 4, vol. XL, pp. 421–427; *Collected Works*, vol. II, pp. 233–240.

showed that a certain relation exists between the number of hills, dales, passes, &c. which can co-exist on the earth or on an island. I proceed to give a summary of their nomenclature and conclusions.

All places whose heights above the mean sea level are equal are on the same level. The locus of such points on a map is indicated by a *contour-line*. Roughly speaking, an island is bounded by a contour-line. It is usual to draw the successive contour-lines on a map so that the difference between the heights of any two successive lines is the same, and thus the closer the contour-lines the steeper is the slope, but the heights are measured dynamically by the amount of work to be done to go from one level to the other and not by linear distances.

A contour-line in general will be a closed curve. This curve may enclose a region of elevation: if two such regions meet at a point, that point will be a crunode (*i.e.* a real double point) on the contour-line through it, and such a point is called a *pass*. The contour-line may enclose a region of depression: if two such regions meet at a point, that point will be a crunode on the contour-line through it, and such a point is called a *fork* or bar. As the heights of the corresponding level surfaces become greater, the areas of the regions of elevation become smaller, and at last become reduced to points: these points are the *summits* of the corresponding mountains. Similarly as the level surface sinks the regions of depression contract, and at last are reduced to points: these points are the *bottoms* (or immits) of the corresponding valleys.

Lines drawn so as to be everywhere at right angles to the contour-lines are called *lines of slope*. If we go up a line of slope generally we shall reach a summit, and if we go down such a line generally we shall reach a bottom: we may come however in particular cases either to a pass or to a fork. Districts whose lines of slope run to the same summit are *hills*. Those whose lines of slope run to the same bottom are *dales*. A *watershed* is the line of slope from a summit to a pass or a fork, and it separates two dales. A *watercourse* is the line of slope from a pass or a fork to a bottom, and it separates two hills.

If $n + 1$ regions of elevation or of depression meet at a point, the point is a multiple point on the contour-line drawn through it; such a point is called a pass or a fork of the nth order, and must be counted as n separate passes (or forks). If one region of depression meets another in several places at once, one of these must be taken as a fork and the rest as passes.

Having now a definite geographical terminology we can apply geometrical propositions to the subject. Let h be the number of hills on the earth (or an island), then there will be also h summits; let d be the number of dales, then there will be also d bottoms; let p be the whole number of passes, p_1 that of single passes, p_2 of double passes, and so on; let f be the whole number of forks, f_1 that of single forks, f_2 of double forks, and so on; let w be the number of watercourses, then there will be also w watersheds. Hence, by the theorems of Cauchy and Euler,

$$h = 1 + p_1 + 2p_2 + \cdots,$$
$$d = 1 + f_1 + 2f_2 + \cdots,$$
and
$$w = 2(p_1 + f_1) + 3(p_2 + f_2)) + \cdots.$$

The above results can be extended to the case of a multiply-connected closed surface.

GAMES. Leaving now the question of formal geometrical propositions, I proceed to enumerate a few games or puzzles which depend mainly on the relative position of things, but I postpone to chapter IV the discussion of such amusements of this kind as necessitate any considerable use of arithmetic or algebra. Some writers regard draughts, solitaire, chess and such like games as subjects for geometrical treatment in the same way as they treat dominoes, backgammon, and games with dice in connection with arithmetic: but these discussions require too many artificial assumptions to correspond with the games as actually played or to be interesting.

The amusements to which I refer are of a more trivial description, and it is possible that a mathematician may like to omit the remainder of the chapter. In some cases it is difficult to say whether they should be classified as mainly arithmetical or geometrical, but the point is of no importance.

STATICAL GAMES OF POSITION. Of the innumerable statical games involving geometry of position I shall mention only three or four.

Three-in-a-row. First, I may mention the game of three-in-a-row, of which noughts and crosses, one form of merrilees, and go-bang are well-known examples. These games are played on a board—generally in the form of a square containing n^2 small squares or cells. The common practice is for one player to place a white counter or piece or to make a cross on each small square or cell which he occupies: his opponent similarly uses black counters or pieces or makes a nought on each square which he occupies. Whoever first gets three (or any other assigned number) of his pieces in three adjacent cells and in a straight line wins. The mathematical theory for a board of 9 cells has been worked out completely, and there is no difficulty in extending it to one of 16 cells: but the analysis is lengthy and not particularly interesting. Most of these games were known to the ancients*, and it is for that reason I mention them here.

Three-in-a-row. Extension. I may, however, add an elegant but difficult extension which has not previously found its way, so far as I am aware, into any book of mathematical recreations. The problem is to place n counters on a plane so as to form as many rows as possible, each of which shall contain three and only three counters†.

It is easy to arrange the counters in a number of rows equal to the integral part of $\frac{1}{8}(n-1)^2$. This can be effected by the following construction. Let P be any point on a cubic. Let the tangent at P cut the curve again in Q. Let the tangent at Q cut the curve in A. Let PA cut the curve in B, QB cut it in C, PC cut it in D, QD cut it in E, and so on. Then the counters must be placed at the points $P, Q, A, B, \ldots$. Thus 9 counters can be placed in 8 such rows; 10 counters in 10 rows; 15 counters in 24 rows; 81 counters in 800 rows; and so on.

If however the point P is a pluperfect point of the nth order on the cubic, then Sylvester proved that the above construction gives a number of rows equal to the integral part of $\frac{1}{6}(n-1)(n-2)$. Thus 9 counters can be arranged in 9 rows, which is a well-known and easy puzzle; 10 counters in 12 rows; 15 counters in 30 rows; and so on.

Even this however is an inferior limit and may be exceeded—for instance, Sylvester stated that 9 counters can be placed in 10 rows, each containing three counters; I do not know how he placed them, but one way of so arranging them is by putting them at points whose coordinates are $(2,0)$, $(2,2)$, $(2,4)$, $(4,0)$, $(4,2)$, $(4,4)$, $(0,0)$, $(3,2)$, $(6,4)$; another way is by putting them at the points $(0,0)$, $(0,2)$, $(0,4)$, $(2,1)$, $(2,2)$, $(2,3)$, $(4,0)$, $(4,2)$, $(4,4)$; more generally, the angular points of

---

* Becq de Fouquières, *Les jeux des anciens*, second edition, Paris, 1873, chap. XVIII.

† *Educational Times Reprints*, 1868, vol. VIII, p. 106; *Ibid.* 1886, vol. XLV, pp. 127–128.

a regular hexagon and the three points of intersection of opposite sides form such a group, and therefore any projection of that figure will give a solution.

Thus at present it is not possible to say what is the maximum number of rows of three which can be formed from n counters placed on a plane.

Extension to p-in-a-row. The problem mentioned above at once suggests the extension of placing n counters so as to form as many rows as possible, each of which shall contain p and only p counters. Such problems can be often solved immediately by placing at infinity the points of intersection of some of the lines, and (if it is so desired) subsequently projecting the diagram thus formed so as to bring these points to a finite distance. One instance of such a solution is given above.

As easy examples I may give the arrangement of 16 counters in 15 rows[1], each containing 4 counters; and the arrangement of 19 counters in 10 rows, each containing 5 counters. A solution of the second of these problems can be obtained by placing counters at the 19 points of intersection of the 10 lines $x = \pm a$, $x = \pm b$, $y = \pm a$, $y = \pm b$, $y = \pm x$: of these points two are at infinity. The first problem I leave to the ingenuity of my readers.

Tesselation. Another of these statical recreations is known as tesselation and consists in the formation of geometrical designs or mosaics by means of tesselated tiles.

To those who have never looked into the matter it may be surprising that patterns formed by the use of square tiles (of which one-half bounded by a diagonal is white and the other half black) should be subject to mathematical analysis. In view of the discussion of this subject by Montucla[*], Lucas[†], and other writers it would be hard to refuse to call the formation of such patterns a mathematical amusement, but the treatment is (perhaps necessarily) somewhat empirical, and though there are some interesting puzzles of this kind, I do not propose to describe them here.

---

[*] See Ozanam, 1803 edition, vol. I, p. 100; 1840 edition, p. 46.
[†] Lucas, *Récréations Mathématiques*, Paris, 1882–3, vol. II, part 4: hereafter I shall refer to this work by the name of the author.

---

1. '13' corrected to '15' as per errata sheet

Sylvester* proposed a modified tesselation problem which consists in forming anallagmatic squares, that is, squares such that in every row and every column the number of changes of colour or the number of permanences is constant, the tiles used being square white tiles and square black tiles.

If more than two colours are used, the problems become increasingly difficult. As a simple instance of this class of problems I may mention one, sent to me by a correspondent who termed it *Cross-Fours*, wherein sixteen square counters are used, the upper half of each being yellow, red, pink, or blue, and the lower half being gold, green, black, or white, no two counters being coloured alike. Such counters can be arranged in the form of a square so that in each vertical, horizontal, and diagonal line there shall be 8 colours and no more: they can be also arranged so that in each of these ten lines there shall be 6 colours and no more, or 5 colours and no more, or 4 colours and no more. Puzzles of this kind are but little known; they are however not uninstructive.

Colour-Cube Problem. As an example of a recreation analogous to tesselation I will mention the colour-cube problem; I select this partly because it is one of the most difficult of such puzzles, but chiefly because it has been subjected† to mathematical analysis.

Stripped of mathematical technicalities the problem may be enunciated as follows. A cube has six faces, and if six colours are chosen we can paint each face with a different colour. By permuting the order of the colours we can obtain thirty such cubes, no two of which are coloured alike. Take any one of these cubes, K, then it is desired to select eight out of the remaining twenty-nine cubes, such that they can be arranged in the form of a cube (whose linear dimensions are double those of any of the separate cubes) coloured like the cube K, and placed so that where any two cubes touch each other the faces in contact are coloured alike.

Only one collection of eight cubes can be found to satisfy these conditions. To pick out these eight cubes empirically would be out of the question, but the mathematical analysis enables us to select them by the following rule. Take any face of the cube K: it has four

---

* *Ex. gr.* see the *Educational Times Reprints*, London, 1868, vol. x, pp. 74–76, 112: see also vol. XLV, p. 127; vol. LVI, pp. 97–99.

† By Major MacMahon; an abstract of his paper, read before the London Mathematical Society on Feb. 9, 1893, was given in *Nature*, Feb. 23, 1893, vol. XLVII, p. 406.

angles, and at each angle three colours meet. By permuting the colours cyclically we can obtain from each angle two other cubes, and the eight cubes so obtained are those required.

For instance suppose that the six colours are indicated by the letters a, b, c, d, e, f. Let the cube K be put on a table, and to fix our ideas suppose that the face coloured f is at the bottom, the face coloured a is at the top, and the faces coloured b, c, d, and e front respectively the east, north, west, and south points of the compass. I may denote such an arrangement by $(f; a; b, c, d, e)$. One cyclical permutation of the colours which meet at the north-east corner of the top face gives the cube $(f; c; a, b, d, e)$, and a second cyclical permutation gives the cube $(f; b; c, a, d, e)$. Similarly cyclical permutations of the colours which meet at the north-west corner of the top face of K give the cubes $(f; d; b, a, c, e)$ and $(f; c; b, d, a, e)$. Similarly from the top south-west corner of K we get the cubes $(f; e; b, c, a, d)$ and $(f; d; b, c, e, a)$: and from the top south-east corner we get the cubes $(f; e; a, c, d, b)$ and $(f; b; e, c, d, a)$.

The eight cubes being thus determined it is not difficult to arrange them in the form of a cube coloured similarly to K, and subject to the condition that faces in contact are coloured alike; in fact they can be arranged in two ways to satisfy these conditions. One such way, taking the cubes in the numerical order given above, is to put the cubes 3, 6, 8, and 2 at the SE, NE, NW, and SW corners of the bottom face; of course each placed with the colour f at the bottom, while 3 and 6 have the colour b to the east, and 2 and 8 have the colour d to the west: the cubes 7, 1, 4, and 5 will then form the SE, NE, NW, and SW corners of the top face; of course each placed with the colour a at the top, while 7 and 1 have the colour b to the east, and 5 and 4 have the colour d to the west. If however K is not given, then, without the aid of mathematical analysis, it is a difficult puzzle to arrange the eight cubes in the form of a cube coloured similarly to one of the other twenty-two cubes and subject to the condition that faces in contact are coloured alike.

It is easy to make similar puzzles in two dimensions which are fairly difficult; it is somewhat surprising that none are to be bought, but I have never seen any except those that I have made myself.

DYNAMICAL GAMES OF POSITION. Games which are played by moving pieces on boards of various shapes—such as merrilees, fox and geese, solitaire, backgammon, draughts, and chess—present more interest. In general, however, they permit of so many movements of the pieces that any mathematical analysis of them becomes too intricate to follow out completely. Probably this is obvious, but it may emphasize the impossibility of discussing such games effectively if I add that it has been shown that in a game of chess there may be as many as 197299 ways of playing the first four moves, and nearly 72000 different positions at the end of the first four moves (two on each side), of which 16556 arise when the players move pawns only*.

Games in which the possible movements are very limited may be susceptible of mathematical treatment. One or two of these are given in the next chapter: here I shall confine myself mainly to puzzles and simple amusements.

Shunting Problems. The first I will mention is a little puzzle which I bought some years ago and which was described as the "Great Northern puzzle." It is typical of a good many problems connected with the shunting of trains, and though it rests on a most improbable hypothesis, I give it as a specimen of its kind.

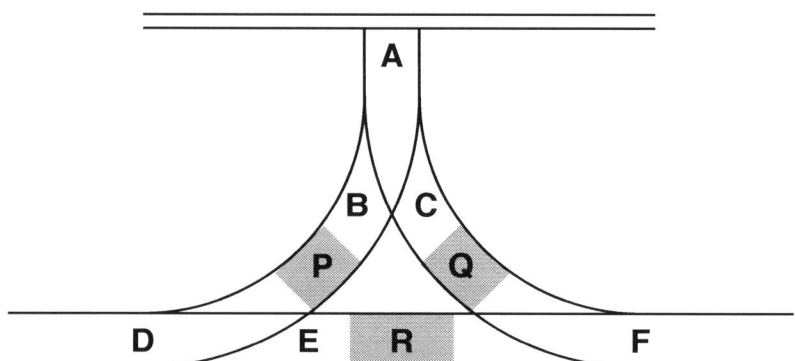

The puzzle shows a railway, DEF, with two sidings, DBA and FCA, connected at A. The portion of the rails at A which is common to the two sidings is long enough to permit of a single wagon, like P or Q, running in or out of it; but is too short to contain the whole of an engine, like R. Hence, if an engine runs up one siding, such as DBA, it must come back the same way.

---

* *L'Intermédiaire des mathématiciens*, Paris, December, 1903, vol. x, pp. 305–308: also *Royal Engineers Journal*, London, August–November, 1889; or *British Association Transactions*, 1890, p. 745.

Initially a small block of wood, P, coloured to represent a wagon, is placed at B; a similar block, Q, is placed at C; and a longer block of wood, R, representing an engine, is placed at E. The problem is to use the engine R to interchange the wagons P and Q, without allowing any flying shunts.

This is effected thus. (i) R pushes P into A. (ii) R returns, pushes Q up to P in A, couples Q to P, draws them both out to F, and then pushes them to E. (iii) P is now uncoupled, R takes Q back to A, and leaves it there. (iv) R returns to P, pulls P back to C, and leaves it there. (v) R running successively through F, D, B comes to A, draws Q out, and leaves it at B.

A somewhat similar puzzle, on sale in the streets in 1905, is made as follows. A loop-line BGE connects two points B and E on a railway track AF, which is supposed blocked at both ends, as shown in the diagram. In the model, the track AF is 9 inches long, $AB = EF = 1\frac{5}{6}$ inches, and $AH = FK = BC = DE = \frac{1}{4}$ inch. On the track and loop are eight wagons, numbered successively 1 to 8, each one inch long and

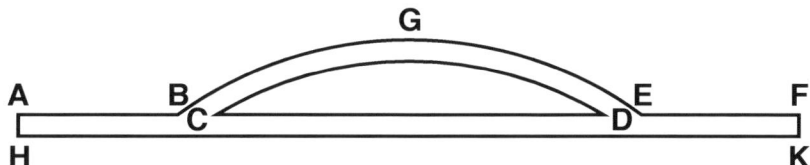

one-quarter of an inch broad, and an engine of the same dimensions. Originally the wagons are on the track from A to F and in the order 1, 2, 3, 4, 5, 6, 7, 8, and the engine is on the loop. The construction and the initial arrangement ensure that at any one time there cannot be more than eight vehicles on the track. Also if eight vehicles are on it only the penultimate vehicle at either end can be moved on to the loop, but if less than eight are on the track then the last two vehicles at either end can be moved on to the loop. If the points at each end of the loop-line are clear, it will hold four, but not more than four, vehicles. The object is to reverse the order of the wagons on the track, so that from A to F they will be numbered successively 8 to 1; and to do this by means which will involve as few transferences of the engine or a wagon to or from the loop as is possible.

Other shunting problems are not uncommon, but these two examples will suffice.

Ferry-Boat Problems. Everybody is familiar with the story of the showman who was travelling with a wolf, a goat, and a basket of cabbages; and for obvious reasons was unable to leave the wolf alone with the goat, or the goat alone with the cabbages. The only means of transporting them across a river was a boat so small that he could take in it only one of them at a time. The problem is to show how the passage could be effected[*].

A similar problem, given by Alcuin, Tartaglia, and others, is as follows[†]. Three beautiful ladies have for husbands three men, who are as jealous as they are young, handsome, and gallant. The party are travelling, and find on the bank of a river, over which they have to pass, a small boat which can hold no more than two persons. How can they pass, it being agreed that, in order to avoid scandal, no woman shall be left in the society of a man unless her husband is present?

The method of transportation to be used in the above cases is obvious, and can be illustrated practically by using six court cards out of a pack. Another problem similar to the one last mentioned is the case of n married couples who have to cross a river by means of a boat which can be rowed by one person and will carry $n-1$ people, but not more, with the condition that no woman is to be in the society of a man unless her husband is present. Alcuin's problem is the case of $n=3$. Let y denote the number of passages from one bank to the other which will be necessary. Then it has been shown that if $n=3$, $y=11$; if $n=4$, $y=9$; and if $n>4$, $y=7$; the demonstration presents no difficulty.

The following analogous problem is due to the late Prof. Lucas[‡]. To find the smallest number x of persons that a boat must be able to carry in order that n married couples may by its aid cross a river in such a manner that no woman shall remain in the company of any man unless her husband is present; it being assumed that the boat can be rowed by one person only. Also to find the least number of passages, say y, from one bank to the other which will be required. M. Delannoy has shown that if $n=2$, then $x=2$, and $y=5$. If $n=3$, then $x=2$, and $y=11$. If $n=4$, then $x=3$, and $y=9$. If $n=5$, then $x=3$, and $y=11$. And finally if $n>5$, then $x=4$, and $y=2n-1$.

M. De Fonteney has remarked that, if there was an island in the

---

[*] Ozanam, 1803 edition, vol. I, p. 171; 1840 edition, p. 77.
[†] Bachet, Appendix, problem IV, p. 212.
[‡] Lucas, vol. I, pp. 15–18, 237–238.

middle of the river, the passage might be always effected by the aid of a boat which could carry only two persons. If there are only two or only three couples the island is unnecessary, and the case is covered by the preceding method. If $n > 3$ then the least number of passages from land to land which will be required is $8(n-1)$.

His solution is as follows. The first nine passages will be the same, no matter how many couples there may be: the result is to transfer one couple to the island and one couple to the second bank. The result of the next eight passages is to transfer one couple from the first bank to the second bank: this series of eight operations must be repeated as often as necessary until there is left only one couple on the first bank, only one couple on the island, and all the rest on the second bank. The result of the last seven passages is to transfer all the couples to the second bank.

The solution for the case when there are four couples may be represented as follows. Let A and a, B and b, C and c, D and d, be the four couples. The letters in the successive lines indicate the positions of the men and their respective wives after different passages of the boat.

		First Bank		Island		Second Bank	
Initially		$ABCD$	$abcd$	....	....	....	....
After 1st passage		$ABCD$	$..cd$	....	$ab..$	....	....
" 2nd "		$ABCD$	$.bcd$	....	$a...$	....	....
" 3rd "		$ABCD$	$...d$	....	$abc.$	....	....
" 4th "		$ABCD$	$..cd$	....	$ab..$	....	....
" 5th "		$..CD$	$..cd$	$AB..$	$ab..$	....	....
" 6th "		$..CD$	$..cd$	$AB..$	....	....	$ab..$
" 7th "		$..CD$	$..cd$	$AB..$	$.b..$	....	$a...$
" 8th "		$..CD$	$..cd$	....	$.b..$	$AB..$	$a...$
" 9th "		$..CD$	$..cd$	$.B..$	$.b..$	$A...$	$a...$
" 10th "		$.BCD$	$..cd$	....	$.b..$	$A...$	$a...$
" 11th "		$.BCD$	....	....	$.bcd$	$A...$	$a...$
" 12th "		$.BCD$	$...d$	....	$.bc.$	$A...$	$a...$
" 13th "		$...D$	$...d$	$.BC.$	$.bc.$	$A...$	$a...$
" 14th "		$...D$	$...d$	....	$.bc.$	$ABC.$	$a...$
" 15th "		$...D$	$...d$	....	$abc.$	$ABC.$	....
" 16th "		$...D$	$...d$	....	$.b..$	$ABC.$	$a.c.$
" 17th "		$...D$	$...d$	$.B..$	$.b..$	$A.C.$	$a.c.$
" 18th "		$.B.D$	$...d$	....	$.b..$	$A.C.$	$a.c.$

		First Bank	Island	Second Bank	
After 19th	passage	. . . . . . .d	.B.D .b..	A.C.	a.c.
" 20th	"	. . . . . . .d	. . . . .b..	ABCD	a.c.
" 21st	"	. . . . . . .d	. . . . .bc.	ABCD	a...
" 22nd	"	. . . . . . .d	. . . .	ABCD	abc.
" 23rd	"	. . . . ..cd	. . . .	ABCD	ab..
" 24th	"	. . . .	. . . .	ABCD	abcd

Prof. G. Tarry has suggested an extension of the problem, which still further complicates its solution. He supposes that each husband travels with a harem of m wives or concubines; moreover, as Mohammedan women are brought up in seclusion, it is reasonable to suppose that they would be unable to row a boat by themselves without the aid of a man. But perhaps the difficulties attendant on the travels of one wife may be deemed sufficient for Christians, and I content myself with merely mentioning the increased anxieties experienced by Mohammedans in similar circumstances.

Geodesics. Geometrical problems connected with finding the shortest routes from one point to another on a curved surface are often difficult, but geodesics on a flat surface or flat surfaces are in general readily determinable.

I append an instance[*], but I should have hesitated to do so had not experience shown that some readers do not readily see the solution. It is as follows: A room is 30 feet long, 12 feet wide, and 12 feet high. On the middle line of one of the smaller side walls and one foot from the ceiling is a wasp. On the middle line of the opposite wall and 11 feet from the ceiling is a fly. The wasp catches the fly by crawling all the way to it: the fly, paralysed by fear, remaining still. The problem is to find the shortest route that the wasp can follow.

To obtain a solution we observe that we can cut a sheet of paper so that, when folded properly, it will make a model to scale of the room. This can be done in several ways. If, when the paper is again spread out flat, we can join the points representing the wasp and the fly by a straight line lying wholly on the paper we shall obtain a geodesic route between them. Thus the problem is reduced to finding the way of cutting out the paper which gives the shortest route of the kind.

---

[*] I heard a similar question propounded at Cambridge in 1903, but the only place where I have seen it in print is the *Daily Mail*, London, February 1, 1905.

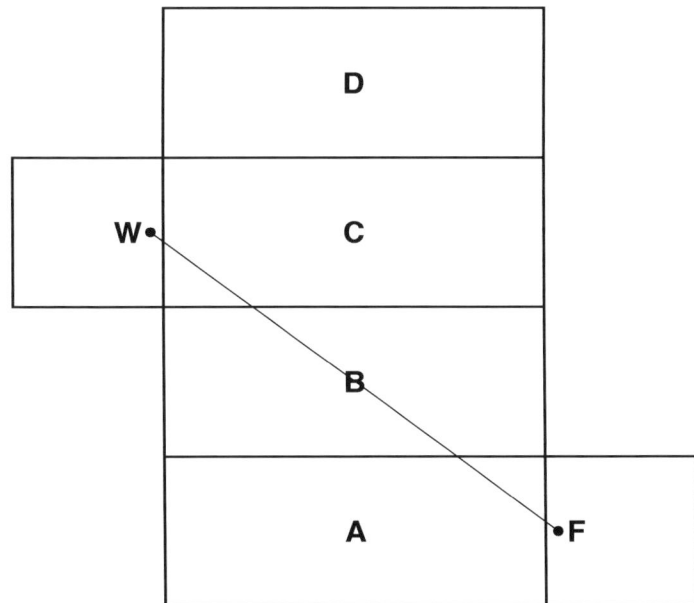

Here is the diagram corresponding to a solution of the above question, where A represents the floor, B and D the longer side-walls, C the ceiling, and W and F the positions on the two smaller side-walls occupied initially by the wasp and fly. In the diagram the square of the distance between W and F is $(32)^2+(24)^2$; hence the distance is 40 feet.

Problems with Counters placed in a row. Numerous dynamical problems and puzzles may be illustrated with a box of counters, especially if there are counters of two colours. Of course coins or pawns or cards will serve equally well. I proceed to enumerate a few of these played with counters placed in a row.

First Problem with Counters. The following problem must be familiar to many of my readers. Ten counters (or coins) are placed in a row. Any counter may be moved over two of those adjacent to it on the counter next beyond them. It is required to move the counters according to the above rule so that they shall be arranged in five equidistant couples.

If we denote the counters in their initial positions by the numbers 1, 2, 3, 4, 5, 6, 7, 8, 9, 10, we proceed as follows. Put 7 on 10, then 5 on 2, then 3 on 8, then 1 on 4, and lastly 9 on 6. Thus they are arranged in pairs on the places originally occupied by the counters 2, 4, 6, 8, 10.

Similarly by putting 4 on 1, then 6 on 9, then 8 on 3, then 10 on

7, and lastly 2 on 5, they are arranged in pairs on the places originally occupied by the counters 1, 3, 5, 7, 9.

If two superposed counters are reckoned as only one, solutions analogous to those given above will be obtained by putting 7 on 10, then 5 on 2, then 3 on 8, then 1 on 6, and lastly 9 on 4; or by putting 4 on 1, then 6 on 9, then 8 on 3, then 10 on 5, and lastly 2 on 7[*].

There is a somewhat similar game played with eight counters, but in this case the four couples finally formed are not equidistant. Here the transformation will be effected if we move 5 on 2, then 3 on 7, then 4 on 1, and lastly 6 on 8. This form of the game is applicable equally to $(8 + 2n)$ counters, for if we move 4 on 1 we have left on one side of this couple a row of $(8 + 2n - 2)$ counters. This again can be reduced to one of $(8 + 2n - 4)$ counters, and in this way finally we have left 8 counters which can be moved in the way explained above.

A more complete generalization would be the case of n counters, where each counter might be moved over the m counters adjacent to it on to the one beyond them.

Second Problem with Counters. Another problem of a somewhat similar kind is due to Tait[†]. Place four florins (or white counters) and four halfpence (or black counters) alternately in a line in contact with one another. It is required in four moves, each of a pair of two contiguous pieces, without altering the relative position of the pair, to form a continuous line of four halfpence followed by four florins.

His solution is as follows. Let a florin be denoted by a and a halfpenny by b, and let ×× denote two contiguous blank spaces. Then the successive positions of the pieces may be represented thus:

```
Initially ................  ×  ×  a  b  a  b  a  b  a  b.
After the first move .....  b  a  a  b  a  b  a  ×  ×  b.
After the second move ..    b  a  a  b  ×  ×  a  a  b  b.
After the third move ....   b  ×  ×  b  a  a  a  a  b  b.
After the fourth move ...   b  b  b  b  a  a  a  a  ×  ×.
```

The operation is conducted according to the following rule. Suppose the pieces to be arranged originally in circular order, with two contiguous blank spaces, then we always move to the blank space for

---

[*] Note by J. Fitzpatrick to a French translation of the third edition of this work, Paris, 1898.
[†] *Philosophical Magazine*, London, January, 1884, series 5, vol. XVII, p. 39.

the time being that pair of coins which occupies the places next but one and next but two to the blank space on one assigned side of it.

A similar problem with $2n$ counters—n of them being white and n black—will at once suggest itself, and, if n is greater than 4, it can be solved in n moves. I have however failed to find a simple rule which covers all cases alike, but solutions, due to M. Delannoy, have been given[*] for the four cases where n is of the form $4m$, $4m+2$, $4m+1$, or $4m+3$; in the first two cases the first $\frac{1}{2}n$ moves are of pairs of dissimilar counters and the last $\frac{1}{2}n$ moves are of pairs of similar counters; in the last two cases, the first move is similar to that given above, namely, of the penultimate and antepenultimate counters to the beginning of the row, the next $\frac{1}{2}(n-1)$ moves are of pairs of dissimilar counters, and the final $\frac{1}{2}(n-1)$ moves are of similar counters.

The problem is also capable of solution if we substitute the restriction that at each move the pair of counters taken up must be moved to one of the two ends of the row instead of the condition that the final arrangement is to be continuous.

Tait suggested a variation of the problem by making it a condition that the two coins to be moved shall also be made to interchange places; in this form it would seem that 5 moves are required; or, in the general case, $n+1$ moves are required.

Problems on a Chess-board with Counters or Pawns. The following three problems require the use of a chess-board as well as of counters or pieces of two colours. It is more convenient to move a pawn than a counter, and if therefore I describe them as played with pawns it is only as a matter of convenience and not that they have any connection with chess. The first is characterized by the fact that in every position not more than two moves are possible; in the second and third problems not more than four moves are possible in any position. With these limitations, analysis is possible. I shall not discuss the similar problems in which more moves are possible.

First Problem with Pawns[†]. On a row of seven squares on a chess-board 3 white pawns (or counters), denoted in the diagram by "a"s, are placed on the 3 squares at one end, and 3 black pawns (or counters), denoted by "b"s, are placed on the 3 squares at the other end—the middle square being left vacant. Each piece can move only in one

---

[*] *La Nature*, June, 1887, p. 10.
[†] Lucas, vol. II, part 5, pp. 141-143.

direction; the "a" pieces can move from left to right, and the "b" pieces from right to left. If the square next to a piece is unoccupied, it can

move on to that; or if the square next to it is occupied by a piece of the opposite colour and the square beyond that is unoccupied, then it can, like a queen in draughts, leap over that piece on to the unoccupied square beyond it. The object is to get all the white pawns in the places occupied initially by the black pawns and vice versa.

The solution requires 15 moves. It may be effected by moving first a white pawn, then successively two black pawns then three white pawns, then three black pawns, then three white pawns, then two black pawns, and then one white pawn. We can express this solution by saying that if we number the cells (a term used to describe each of the small squares on a chess-board) consecutively, then initially the vacant space occupies the cell 4 and in the successive moves it will occupy the cells 3, 5, 6, 4, 2, 1, 3, 5, 7, 6, 4, 2, 3, 5, 4. Of these moves, six are simple and nine are leaps.

Similarly if we have n white pawns at one end of a row of $2n+1$ cells, and n black pawns at the other end, they can be interchanged in $n(n+2)$ moves, by moving in succession 1 pawn, 2 pawns, 3 pawns, ..., $n-1$ pawns, n pawns, n pawns, n pawns, $n-1$ pawns, ..., 2 pawns, and 1 pawn—all the pawns in each group being of the same colour and different from that of the pawns in the group preceding it. Of these moves $2n$ are simple and n^2 are leaps.

*Second Problem with Pawns**. A similar game may be played on a rectangular or square board. The case of a square board containing 49 cells, or small squares, will illustrate this sufficiently: in this case the initial position is shown in the annexed diagram where the "a"s denote the pawns or pieces of one colour, and the "b"s those of the other colour. The "a" pieces can move horizontally from left to right or vertically down, and the "b" pieces can move horizontally from right to left or vertically up, according to the same rules as before.

The solution reduces to the preceding case. The pieces in the middle column can be interchanged in 15 moves. In the course of these moves every one of the seven cells in that column is at some time or

---

* Lucas, vol. II, part 5, p. 144.

a	a	a	a	b	b	b
a	a	a	a	b	b	b
a	a	a	a	b	b	b
a	a	a		b	b	b
a	a	a	b	b	b	b
a	a	a	b	b	b	b
a	a	a	b	b	b	b

other vacant, and whenever that is the case the pieces in the row containing the vacant cell can be interchanged. To interchange the pieces in each of the seven rows will require 15 moves. Hence to interchange all the pieces will require $15 + (7 \times 15)$ moves, that is, 120 moves.

If we place $2n(n + 1)$ white pawns and $2n(n + 1)$ black pawns in a similar way on a square board of $(2n + 1)^2$ cells, we can transpose them in $2n(n + 1)(n + 2)$ moves: of these $4n(n + 1)$ are simple and $2n^2(n + 1)$ are leaps.

Third Problem with Pawns. The following analogous, though somewhat more complicated, game was I believe originally published in the first edition of this work: but I find that it has been since widely

a	b	c		
d	e	f		
g	h	*	H	G
		F	E	D
		C	B	A

distributed in connexion with an advertisement and probably now is well-known. On a square board of 25 cells, place eight white pawns or counters on the cells denoted by small letters in the annexed diagram, and eight black pawns or counters on the cells denoted by capital letters: the cell marked with an asterisk (*) being left blank. Each pawn can move according to the laws already explained—the white pawns being able to move only horizontally from left to right or vertically downwards, and the black pawns being able to move only horizontally from right to left or vertically upwards. The object is to get all the white pawns in the places initially occupied by the black pawns and vice versa. No moves outside the dark line are permitted.

Since there is only one cell on the board which is unoccupied, and since no diagonal moves and no backward moves are permitted, it follows that at each move not more than two pieces of either colour are capable of moving. There are however a very large number of solutions. The following empirical solution in forty-eight moves is one way of effecting the transfer—the letters indicating the cells *from* which the pieces are successively moved:

$$h\,H\,*\,f\,F\,E\,H\,G\,*\,c\,b\,h\,g\,d\,f\,F\,C\,*\,h\,H\,B\,A\,C\,*$$
$$c\,a\,b\,h\,H\,*\,c\,f\,F\,D\,G\,H\,B\,C\,*\,g\,h\,e\,f\,F\,*\,h\,H\,*.$$

It will be noticed that the first twenty-four moves lead to a symmetrical position, and that the next twenty-three moves can be at once obtained by writing the first twenty-three moves in reverse order and interchanging small and capital letters.

Probably, were it worth the trouble, the mathematical theory of games such as that just described might be worked out by the use of Vandermonde's notation, described later in chapter VI, or by the analogous method employed in the theory of the game of solitaire[*]. I believe that this has not been done, and I do not think it would repay the labour involved.

Problems on a Chess-board with Chess-pieces. There are several mathematical recreations with chess-pieces, other than pawns, somewhat similar to those given above. One of these, on the determination of the ways in which eight queens can be placed on a board so that no queen can take any other, is given later in chapter IV. Another, on the path to be followed by a knight which is moved on a chess-board so that it shall occupy every cell once and only once, is given in chapter VI. Here I will mention one of the simplest of such problems, which is interesting from the fact that it is given in Guarini's manuscript written in 1512; it was quoted by Lucas, but so far as I know has not been otherwise published.

Guarini's Problem. On a board of nine cells, such as that drawn below, the two white knights are placed on the two top corner cells

---

[*] On the theory of the solitaire, see Reiss, '*Beiträge zur Theorie des Solitär-Spiels*,' *Crelle's Journal*, Berlin, 1858, vol. LIV, pp. 344–379; and Lucas, vol. I, part V, pp. 89–141.

a	C	d
D		B
b	A	c

(a, d), and the two black knights on the two bottom corner cells (b, c): the other cells are left vacant. It is required to move the knights so that the white knights shall occupy the cells b and c, while the black shall occupy the cells a and d.

The solution is tolerably obvious. First, move the pieces from a to A, from b to B, from c to C, and from d to D. Next, move the pieces from A to d, from B to a, from C to b, and from D to c. The effect of these eight moves is the same as if the original square had been rotated through one right angle. Repeat the above process, that is, move the pieces successively from a to A, from b to B, from c to C, from d to D; from A to d, from B to a, from C to b, and from D to c. The required result is then attained.

GEOMETRICAL PUZZLES WITH RODS, ETC. Another species of geometrical puzzles, to which here I will do no more than allude, are made of steel rods, or of wire, or of wire and string. Numbers of these are often sold in the streets of London for a penny each, and some of them afford ingenious problems in the geometry of position. Most of them could hardly be discussed without the aid of diagrams, but they are inexpensive to construct, and in fact innumerable puzzles on geometry of position can be made with a couple of stout sticks and a ball of string, or even with only a box of matches: several examples are given in the appendix to the fourth volume of the 1723 edition of Ozanam's work. I will mention, as an easy example, analogous to one group of the string puzzles, that any one can take off his waistcoat (which may be unbuttoned) without taking off his coat, and without pulling the waistcoat over the head like a jersey.

This last feat may serve to show the difficulty of mentally realizing the effect of geometrical alterations in a figure unless they are of the simplest character.

PARADROMIC RINGS. The fact just stated is illustrated by the familiar experiment of making *paradromic rings* by cutting a paper ring prepared in the following manner.

Take a strip of paper or piece of tape, say, for convenience, an inch or two wide and at least nine or ten inches long, rule a line in the middle down the length AB of the strip, gum one end over the other end B, and we get a ring like a section of a cylinder. If this ring is cut by a pair of scissors along the ruled line we obtain two rings exactly like the first, except that they are only half the width. Next suppose that the end A is twisted through two right angles before it is gummed to B (the result of which is that the back of the strip at A is gummed over the front of the strip at B), then a cut along the line will produce only one ring. Next suppose that the end A is twisted once completely round (*i.e.* through four right angles) before it is gummed to B, then a similar cut produces two interlaced rings. If any of my readers think that these results could be predicted off-hand, it may be interesting to them to see if they can predict correctly the effect of again cutting the rings formed in the second and third experiments down their middle lines in a manner similar to that above described.

The theory is due to J.B. Listing[*] who discussed the case when the end A receives m half-twists, that is, is twisted through $m\pi$, before it is gummed to B.

If m is even we obtain a surface which has two sides and two edges, which are termed paradromic. If the ring is cut along a line midway between the edges, we obtain two rings, each of which has m half-twists, and which are linked together $\frac{1}{2}m$ times.

If m is odd we obtain a surface having only one side and one edge. If this ring is cut along its mid-line, we obtain only one ring, but it has $2m$ half-twists, and if m is greater than unity it is knotted.

---

[*] *Vorstudien zur Topologie, Die Studien*, Göttingen, 1847, part x.

CHAPTER III.

SOME MECHANICAL QUESTIONS.

I PROCEED now to enumerate a few questions connected with mechanics which lead to results that seem to me interesting from a historical point of view or paradoxical. Problems in mechanics generally involve more difficulties than problems in arithmetic, algebra, or geometry, and the explanations of some phenomena—such as those connected with the flight of birds—are still incomplete, while the explanations of many others of an interesting character are too difficult to find a place in a non-technical work. Here, however, I shall confine myself to questions which, like those treated in the two preceding chapters, are of an elementary, not to say trivial, character; and the conclusions are well-known to mathematicians.

I assume that the reader is acquainted with the fundamental ideas of kinematics and dynamics, and is familiar with the three Newtonian laws; namely, first that a body will continue in its state of rest or of uniform motion in a straight line unless compelled to change that state by some external force: second, that the change of momentum per unit of time is proportional to the external force and takes place in the direction of it: and third, that the action of one body on another is equal in magnitude but opposite in direction to the reaction of the second body on the first. The first and second laws state the principles required for solving any question on the motion of a particle under the action of given forces. The third law supplies the additional principle required for the solution of problems in which two or more particles influence one another.

MOTION. The difficulties connected with the idea of *motion* have been for a long time a favourite subject for paradoxes, some of which bring us into the realm of the philosophy of mathematics.

Zeno's Paradoxes on Motion. One of the earliest of these is the remark of Zeno to the effect that since an arrow cannot move where it is not, and since also it cannot move where it is (*i.e.* in the space it exactly fills), it follows that it cannot move at all. The answer that the very idea of the motion of the arrow implies the passage from where it is to where it is not was rejected by Zeno, who seems to have thought that the appearance of motion of a body was a phenomenon caused by the successive appearances of the body at rest but in different positions.

Zeno also asserted that the idea of motion was itself inconceivable, for what moves must reach the middle of its course before it reaches the end. Hence the assumption of motion presupposes another motion, and that in turn another, and so ad infinitum. His objection was in fact analogous to the biological difficulty expressed by Swift:—

"So naturalists observe, a flea hath smaller fleas that on him prey.
    And these have smaller fleas to bite 'em. And so proceed ad
        infinitum."

Or as De Morgan preferred to put it

"Great fleas have little fleas upon their backs to bite 'em,
And little fleas have lesser fleas, and so ad infinitum.
And the great fleas themselves, in turn, have greater fleas to go
    on;
While these have greater still, and greater still, and so on."

Achilles and the Tortoise. Zeno's paradox about Achilles and the tortoise is known even more widely. The assertion was that if Achilles ran ten times as fast as a tortoise, yet if the tortoise had (say) 1000 yards start it could never be overtaken. To establish this, Zeno argued that when Achilles had gone the 1000 yards, the tortoise would still be 100 yards in front of him; by the time he had covered these 100 yards, it would still be 10 yards in front of him; and so on for ever. Thus Achilles would get nearer and nearer to the tortoise but would never overtake it. Zeno regarded this as confirming his view that the popular idea of motion is self-contradictory.

Zeno's Paradox on Time. The fallacy of Achilles and the Tortoise is usually explained by saying that though the time required to overtake the tortoise can be divided into an infinite number of intervals, as stated in the argument, yet these intervals get smaller and smaller in geometrical progression, and the sum of them all is a finite time: after the lapse of that time Achilles would be in front of the tortoise. Probably Zeno would have replied that this explanation rests on the assumption that space and time are infinitely divisible, propositions which he would not admit. He seems further to have contended that while, to an accurate thinker, the notion of the infinite divisibility of time was impossible, it was equally impossible to think of a minimum measure of time. For suppose, he argued, that τ is the smallest conceivable interval, and suppose that three horizontal lines composed of three consecutive spans abc, $a'b'c'$, $a''b''c''$ are placed so that $aa'a''$, $bb'b''$, $cc'c''$ are vertically over one another. Imagine the second line moved as a whole one span to the right in the time τ, and simultaneously the third line moved as a whole one span to the left. Then b, a', c'' will be vertically over one another. And in this duration τ (which by hypothesis is indivisible) c' must have passed vertically over a''. Hence the duration is divisible, contrary to the hypothesis.

The Paradox of Tristram Shandy. Mr Russell has enunciated[*] a paradox somewhat similar to that of Achilles and the Tortoise, save that the intervals of time considered get longer and longer during the course of events. Tristram Shandy, as we know, took two years writing the history of the first two days of his life, and lamented that, at this rate, material would accumulate faster than he could deal with it, so that he could never come to an end, however long he lived. But had he lived long enough, and not wearied of his task, then, even if his life had continued as eventfully as it began, no part of his biography would have remained unwritten. For if he wrote the events of the first day in the first year, he would write the events of the nth day in the nth year, hence in time the events of any assigned day would be written, and therefore no part of his biography would remain unwritten. This argument might be put in the form of a demonstration that the part of a magnitude may be equal to the whole of it.

Questions, such as those given above, which are concerned with the continuity and extent of space and time involve difficulties of a

---

[*] B.A.W. Russell, *Principles of Mathematics*, Cambridge, 1903, vol. I, p. 358.

high order.

Angular Motion. A non-mathematician finds additional difficulties in the idea of angular motion. For instance, here is a well-known proposition on motion in an equiangular spiral (of which the result is true on the ordinary conventions of mathematics) which shows that a body, moving with uniform velocity and as slowly as we please, may in a finite time whirl round a fixed point an infinite number of times.

The equiangular spiral is the trace of a point P, which moves along a line OP, the line OP turning round a fixed point O with uniform angular velocity while the distance of P from O decreases with the time in geometrical progression. If the radius vector rotates through four right angles we have one convolution of the curve. All convolutions are similar, and the length of each convolution is a constant fraction, say $1/n$th, that of the convolution immediately outside it. Inside any given convolution, there are an infinite number of convolutions which get smaller and smaller as we get nearer the pole. Now suppose a point Q to move uniformly along the spiral from any point towards the pole. If it covers the first convolution in a seconds, it will cover the next in a/n seconds, the next in a/n^2 seconds, and so on, and will finally reach the pole in $(a + a/n + a/n^2 + a/n^3 + \cdots)$ seconds, that is, in $an/(n-1)$ seconds. The velocity is uniform, and yet in a finite time, Q will have traversed an infinite number of convolutions and therefore have circled round the pole an infinite number of times[*].

Simple Relative Motion. Even if the philosophical difficulties suggested by Zeno are settled or evaded, the mere idea of relative motion has been often found to present difficulties, and Zeno himself failed to explain a simple phenomenon involving the principle. As one of the easiest examples of this kind, I may quote the common question of how many trains going from B to A a passenger from A to B would meet and pass on his way, assuming that the journey either way takes $4\frac{1}{2}$ hours and that the trains start from each end every hour. The answer is 9. Or again this: Take two pennies, face upwards on a table and edges in contact. Suppose that one is fixed and that the other rolls on it without slipping, making one complete revolution round it and returning to its initial position. How many revolutions round its own centre has the rolling coin made? The answer is 2.

---

[*] The proposition is put in this form in J. Richard's *Philosophie des mathématiques*, Paris, 1903, pp. 119–120.

Laws of Motion. I proceed next to make a few remarks on points connected with the laws of motion.

The first law of motion is often said to define *force*, but it is in only a qualified sense that this is true. Probably the meaning of the law is best expressed in Clifford's phrase, that force is "the description of a certain kind of motion"—in other words it is not an entity but merely a convenient way of stating, without circumlocution, that a certain kind of motion is observed.

It is not difficult to show that any other interpretation lands us in difficulties. Thus some authors use the law to justify a definition that force is that which moves a body or changes its motion; yet the same writers speak of a steam-engine moving a train. It would seem then that, according to them, a steam-engine is a force. That such statements are current may be fairly reckoned among mechanical paradoxes.

The idea of force is difficult to grasp. How many people, for instance, could predict correctly what would happen in a question as simple as the following? A rope (whose weight may be neglected) hangs over a smooth pulley; it has one end fastened to a weight of 10 stone, and the other end to a sailor of weight 10 stone, the sailor and the weight hanging in the air. The sailor begins to climb up the rope; will the weight move at all; and, if so, will it rise or fall?

It will be noted that in the first law of motion it is asserted that, unless acted on by an external force, a body in motion continues to move (i) with uniform velocity, and (ii) in a straight line.

The tendency of a body to continue in its state of rest or of uniform motion is called its *inertia*. This tendency may be used to explain various common phenomena and experiments. Thus, if a number of dominoes or draughts are arranged in a vertical pile, a sharp horizontal blow on one of those near the bottom will send it out of the pile, and those above will merely drop down to take its place—in fact they have not time to change their relative positions before there is sufficient space for them to drop vertically as if they were a solid body.

This also is the principle on which depends the successful playing of "Aunt Sally," and the performance of numerous tricks, described in collections of mathematical puzzles[*].

---

[*] See *Les récréations scientifiques* by G. Tissandier, where several ingenious illustrations of inertia are given.

The statement about inertia in the first law may be taken to imply that a body set in rotation about a principal axis passing through its centre of mass will continue to move with a uniform angular velocity and to keep its axis of rotation fixed in direction. The former of these statements is the assumption on which our measurement of time is based as mentioned below in chapter XIII. The latter assists us to explain the motion of a projectile in a resisting fluid. It affords the explanation of why the barrel of a rifle is grooved; and why, similarly, anyone who has to throw a flat body of irregular shape (such as a card) in a given direction usually gives it a rapid rotatory motion about a principal axis. Elegant illustrations of the fact just mentioned are afforded by a good many of the tricks of acrobats, though the full explanation of most of them also introduces other considerations. Thus when some few years ago the Japanese village at Knightsbridge was one of the shows of London, there were some acrobats there who tossed on to the top surface of an umbrella a penny so that it alighted on its edge, and then, by turning round the stick of the umbrella rapidly, the coin was caused to rotate, but as the umbrella moved away underneath it the coin remained apparently stationary and standing upright, while by diminishing or increasing the angular velocity of the umbrella the penny was caused to run forwards or backwards. This is not a difficult trick to execute.

The tendency of a body in motion to continue to move in a straight line is sometimes called its *centrifugal force*. Thus, if a train is running round a curve, it tends to move in a straight line, and is constrained only by the pressure of the rails to move in the required direction. Hence it presses on the outer rail of the curve. This pressure can be diminished to some extent both by raising the outer rail, and by putting a guard rail, parallel and close to the inner rail, against which the wheels on that side also will press.

An illustration of this fact occurred in a little known incident of the American civil war[*]. In the spring of 1862 a party of volunteers from the North made their way to the rear of the Southern armies and seized a train, intending to destroy, as they passed along it, the railway which was the main line of communication between various confederate corps and their base of operations. They were however detected and pursued. To save themselves, they stopped on a sharp curve and tore

---

[*] *Capturing a Locomotive* by W. Pittenger, London, 1882, p. 104.

up some rails so as to throw the engine which was following them off the line. Unluckily for themselves they were ignorant of dynamics and tore up the inner rails of the curve, an operation which did not incommode their pursuers.

The second law gives us the means of measuring mass, force, and therefore *work*. A given agent in a given time can do only a definite amount of work. This is illustrated by the fact that although, by means of a rigid lever and a fixed fulcrum, any force however small may be caused to move any mass however large, yet what is gained in power is lost in speed—as the popular phrase runs.

Montucla[*] inserted a striking illustration of this principle founded on the well-known story of Archimedes who is said to have declared to Hiero that, were he but given a fixed fulcrum, he could move the world. Montucla calculated the mass of the earth and, assuming that a man could work incessantly at the rate of 116 foot-lbs. per second, which is a very high estimate, he found that it would take over three billion centuries, *i.e.* 3×10^{14} years, before a mass equal to that of the earth was moved as much as one inch against gravity at the surface of the earth: to move it one inch along a horizonal plane would take about 74000 centuries.

Stability of Equilibrium. It is known to all those who have read the elements of mechanics that the centre of gravity of a body, which is resting in equilibrium under its own weight, must be vertically above its base: also, speaking generally, we may say that, if every small displacement has the effect of raising the centre of gravity, then the equilibrium is stable, that is, the body when left to itself will return to its original position; but, if a displacement has the effect of lowering the centre of gravity, then for that displacement the equilibrium is unstable; while, if every displacement does not alter the height above some fixed plane of the centre of gravity, then the equilibrium is neutral. In other words, if in order to cause a displacement work has to be done against the forces acting on the body, then for that displacement the equilibrium is stable, while if the forces do work the equilibrium is unstable.

A good many of the simpler mechanical toys and tricks afford illustrations of this principle.

---

[*] Ozanam, 1803 edition, vol. II, p. 18; 1840 edition, p. 202.

Magic Bottles[*]. Among the most common of such toys are the small bottles—trays of which may be seen any day in the streets of London—which keep always upright, and cannot be upset until their owner orders them to lie down. Such a bottle is made of thin glass or varnished paper fixed to the plane surface of a solid hemisphere or smaller segment of a sphere. Now the distance of the centre of gravity of a homogeneous hemisphere from the centre of the sphere is three-eighths of the radius, and the mass of the glass or varnished paper is so small compared with the mass of the lead base that the centre of gravity of the whole bottle is still within the hemisphere. Let us denote the centre of the hemisphere by C, and the centre of gravity of the bottle by G.

If such a bottle is placed with the hemisphere resting on a horizontal plane and GC vertical, any small displacement on the plane will tend to raise G, and thus the equilibrium is stable. This may be seen also from the fact that when slightly displaced there is brought into play a couple, of which one force is the reaction of the table passing through C and acting vertically upward, and the other the weight of the bottle acting vertically downward at G. If G is below C, this couple tends to restore the bottle to its original position.

If there is dropped into the bottle a shot or nail so heavy as to raise the centre of gravity of the whole above C, then the equilibrium is unstable, and, if any small displacement is given, the bottle falls over on to its side.

Montucla says that in his time it was not uncommon to see boxes of tin soldiers mounted on lead hemispheres, and when the lid of the box was taken off the whole regiment sprang to attention.

In a similar way we may explain how to balance a pencil in a vertical position, with its point resting on the top of one's finger, an experiment which is described in nearly every book of puzzles[†]. This is effected by taking a penknife, of which one blade is opened through an angle of (say) 120°, and sticking the blade in the pencil so that the handle of the penknife is below the finger. The centre of gravity is thus brought below the point of support, and a small displacement given to the pencil will raise the centre of gravity of the whole: thus the equilibrium is stable.

---

[*] Ozanam, 1803 edition, vol. II, p. 15; 1840 edition, p. 201.

[†] *Ex. gr.* Oughtred, *Mathematical Recreations*, p. 24; Ozanam, 1803 edition, vol. II, p. 14; 1840 edition, p. 200.

Other similar tricks are the suspension of a bucket over the edge of a table by a couple of sticks, and the balancing of a coin on the edge of a wine-glass by the aid of a couple of forks*—the sticks or forks being so placed that the centre of gravity of the whole is vertically below the point of support and its depth below it a maximum.

The toy representing a horseman, whose motion continually brings him over the edge of a table into a position which seems to ensure immediate destruction, is constructed in somewhat the same way. A wire has one end fixed to the feet of the rider; the wire is curved downwards and backwards, and at the other end is fixed a weight. When the horse is placed so that his hind legs are near the edge of the table and his forefeet over the edge, the weight is under his hind feet. Thus the whole toy forms a pendulum with a curved instead of a straight rod. Hence the farther it swings over the table, the higher is the centre of gravity raised, and thus the toy tends to return to its original position of equilibrium.

An elegant modification of the prancing horse was brought out at Paris in 1890 in the shape of a toy made of tin and in the figure of a man†. The legs are pivoted so as to be movable about the thighs, but with a wire check to prevent too long a step, and the hands are fastened to the top of a ∩-shaped wire weighted at its ends. If the figure is placed on a narrow sloping plank or strip of wood passing between the legs of the ∩, then owing to the ∩-shaped wire any lateral displacement of the figure will raise its centre of gravity, and thus for any such displacement the equilibrium is stable. Hence, if a slight lateral disturbance is given, the figure will oscillate and will rest alternately on each foot: when it is supported by one foot the other foot under its own weight moves forwards, and thus the figure will walk down the plank though with a slight reeling motion. Shortly after the publication of the third edition of this book an improved form of this toy, in the shape of a walking elephant made in heavy metal, was issued in England, and probably in that form it is now familiar to all who are interested in noticing street toys.

Columbus's Egg. The toy known as Columbus's egg depends on the same principle as the magic bottle, though it leads to the converse result. The shell of the egg is made of tin and cannot be opened. Inside

---

* Oughtred, p. 30; Ozanam, 1803 edition, vol. II, p. 12; 1840 edition, p. 199.
† *La Nature*, Paris, March, 1891.

it and fastened to its base is a hollow truncated tin cone, and there is also a loose marble inside the shell. If the egg is held properly, the marble runs inside the cone and the egg will stand on its base, but so long as the marble is outside the cone, the egg cannot be made to stand on its base.

Cones running up hill*. The experiment to make a double cone run up hill depends on the same principle as the toys above described; namely, on the tendency of a body to take a position so that its centre of gravity is as low as possible. In this case it produces the optical effect of a body moving by itself up a hill.

Usually the experiment is performed as follows. Arrange two sticks in the shape of a V, with the apex on a table and the two upper ends resting on the top edge of a book placed on the table. Take two equal cones fixed base to base, and place them with the curved surfaces resting on the sticks near the apex of the V, the common axis of the cones being horizontal and parallel to the edge of the book. Then, if properly arranged, the cones will run up the plane formed by the sticks.

The explanation is obvious. The centre of gravity of the cones moves in the vertical plane midway between the two sticks and it occupies a lower position as the points of contact on the sticks get farther apart. Hence as the cone rolls up the sticks its centre of gravity descends.

PERPETUAL MOTION. The idea of making a machine which once set going would continue to go for ever by itself has been the ignis fatuus of self-taught mechanicians in much the same way as the quadrature of the circle has been of self-taught geometricians.

Now the obvious meaning of the third law of motion is that a force is only one aspect of a stress, and that whenever a force is caused another equal and opposite one is brought also into existence—though it may act upon a different body, and thus be immaterial for the particular problem considered. The law however is capable of another interpretation†, namely, that the rate at which an agent does work (that is, its action) is equal to the rate at which work is done against it (that is, its reaction). If it is allowable to include in the reaction the rate at which kinetic energy is being produced, and if work is taken to include

---

* Ozanam, 1803 edition, vol. II, p. 49; 1840 edition, p. 216.
† Newton's *Principia*, last paragraph of the Scholium to the Laws of Motion.

that done against molecular forces, then it follows from this interpretation that the work done by an agent on a system is equivalent to the total increase of energy, that is, the power of doing work. Hence in an isolated system the total amount of energy is constant. If this is granted, then since friction and some molecular dissipation of energy cannot be wholly prevented, it must be impossible to construct in an isolated system a machine capable of perpetual motion.

I do not propose to describe in detail the various machines for producing perpetual motion which have been suggested, but I may add that a number of them are equivalent essentially to the one of which a section is represented in the accompanying figure.

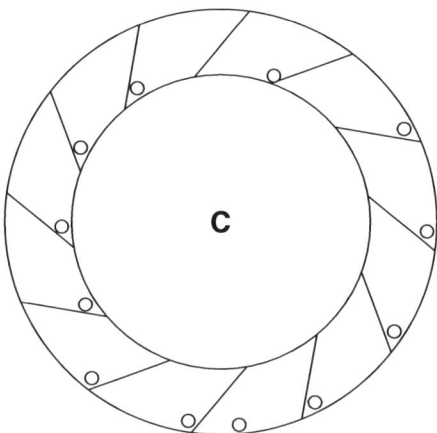

It consists of two concentric vertical wheels in the same plane, and mounted on a horizontal axle through their centre, C. The space between the wheels is divided into compartments by spokes inclined at a constant angle to the radii to the points whence they are drawn, and each compartment contains a heavy bullet. Apart from these bullets, the wheels would be in equilibrium. Each bullet tends to turn the wheels round their axle, and the moment which measures this tendency is the product of the weight of the bullet and its distance from the vertical through C.

The idea of the constructors of such machines was that, as the bullet in any compartment would roll under gravity to the lowest point of the compartment, the bullets on the right-hand side of the diagram would be farther from the vertical through C than those on the left. Hence the sum of the moments of the weights of the bullets on the right would be greater than the sum of the moments of those on the left. Thus the wheels would turn continually in the same direction as the hands of a watch. The fallacy in the argument is obvious.

Another large group of machines for producing perpetual motion depended on the use of a magnet to raise a mass which was then allowed to fall under gravity. Thus, if the bob of a simple pendulum was made of iron, it was thought that magnets fixed near the highest points which were reached by the bob in the swing of the pendulum would draw the bob up to the same height in each swing and thus give perpetual motion.

Of course it is only in isolated systems that the total amount of energy is constant, and, if a source of external energy can be obtained from which energy is continually introduced into the system, perpetual motion is, in a sense, possible; though even here materials would ultimately wear out. The solar heat and the tides are among the most obvious of such sources.

There was at Paris in the latter half of the eighteenth century a clock which was an ingenious illustration of such perpetual motion[*]. The energy which was stored up in it to maintain the motion of the pendulum was provided by the expansion of a silver rod. This expansion was caused by the daily rise of temperature, and by means of a train of levers it wound up the clock. There was a disconnecting apparatus, so that the contraction due to a fall of temperature produced no effect, and there was a similar arrangement to prevent overwinding. I believe that a rise of eight or nine degrees Fahrenheit was sufficient to wind up the clock for twenty-four hours.

I have in my possession a watch, which produces the same effect by somewhat different means. Inside the case is a steel weight, and if the watch is carried in a pocket this weight rises and falls at every step one takes, somewhat after the manner of a pedometer. The weight is raised by the action of the person who has it in his pocket in taking a step, and in falling it winds up the spring of the watch. On the face is a small dial showing the number of hours for which the watch is wound up. As soon as the hand of this dial points to fifty-six hours, the train of levers which winds up the watch disconnects automatically, so as to prevent overwinding the spring, and it reconnects again as soon as the watch has run down eight hours. The watch is an excellent timekeeper, and a walk of about a couple of miles is sufficient to wind it

---

[*] Ozanam, 1803 edition, vol. II, p. 105; 1840 edition, p. 238.

up for twenty-four hours.

MODELS. I may add here the observation, which is well known to mathematicians, but is a perpetual source of disappointment to ignorant inventors, that it frequently happens that an accurate model of a machine will work satisfactorily while the machine itself will not do so.

One reason for this is as follows. If all the parts of a model are magnified in the same proportion, say m, and if thereby a line in it is increased in the ratio $m : 1$, then the areas and volumes in it will be increased respectively in the ratios $m^2 : 1$ and $m^3 : 1$. For example, if the side of a cube is doubled then a face of it will be increased in the ratio $4 : 1$ and its volume will be increased in the ratio $8 : 1$.

Now if all the linear dimensions are increased m times, then some of the forces that act on a machine (such, for example, as the weight of part of it) will be increased m^3 times, while others which depend on area (such as the sustaining power of a beam) will be increased only m^2 times. Hence the forces that act on the machine and are brought into play by the various parts may be altered in different proportions, and thus the machine may be incapable of producing results similar to those which can be produced by the model.

The same argument has been adduced in the case of animal life to explain why very large specimens of any particular breed or species are usually weak. For example, if the linear dimensions of a bird were increased n times, the work necessary to give the power of flight would have to be increased no less than n^7 times[*]. Again, if the linear dimensions of a man of height 5 ft. 10 in. were increased by one-seventh his height would become 6 ft. 8 in., but his weight would be increased in the ratio $512 : 343$ (*i.e.* about half as much again), while the cross sections of his legs, which would have to bear this weight, would be increased only in the ratio $64 : 49$; thus in some respects he would be less efficient than before. Of course the increased dimensions, length of limb, or size of muscle might be of greater advantage than the relative loss of strength; hence the problem of what are the most efficient proportions is not simple, but the above argument will serve to illustrate the fact that the working of a machine may not be similar to that of a model of it.

---

[*] Helmholtz, *Gesammelte Abhandlungen*, Leipzig, 1881, vol. I, p. 165.

Leaving now these elementary considerations I pass on to some other mechanical questions.

SAILING QUICKER THAN THE WIND. As a kinematical paradox I may allude to the possibility of *sailing quicker than the wind blows*, a fact which strikes many people as curious.

The explanation[*] depends on the consideration of the velocity of the wind relative to the boat. Perhaps, however, a non-mathematician will find the solution simplified if I consider first the effect of the wind-pressure on the back of the sail which drives the boat forward, and second the resistance to motion caused by the sail being forced through the air.

When the wind is blowing against a plane sail the resultant pressure of the wind on the sail may be resolved into two components, one perpendicular to the sail (but which in general is not a function only of the component velocity in that direction, though it vanishes when that component vanishes) and the other parallel to its plane. The latter of these has no effect on the motion of the ship. The component perpendicular to the sail tends to move the ship in that direction. This pressure, normal to the sail, may be resolved again into two components, one in the direction of the keel of the boat, the other in the direction of the beam of the boat. The former component drives the boat forward, the latter to leeward. It is the object of a boat-builder to construct the boat on lines so that the resistance of the water to motion forward shall be as small as possible, and the resistance to motion in a perpendicular direction (*i.e.* to leeward) shall be as large as possible; and I will assume for the moment that the former of these resistances may be neglected, and that the latter is so large as to render motion in that direction impossible.

Now, as the boat moves forward, the pressure of the air on the front of the sail will tend to stop the motion. As long as its component normal to the sail is less than the pressure of the wind behind the sail and normal to it, the resultant of the two will be a force behind the sail and normal to it which tends to drive the boat forwards. But as the velocity of the boat increases, a time will arrive when the pressure of the wind is only just able to balance the resisting force which is caused by the sail moving through the air. The velocity of the boat will not

---

[*] Ozanam, 1803 edition, vol. III, pp. 359, 367; 1840 edition, pp. 540, 543.

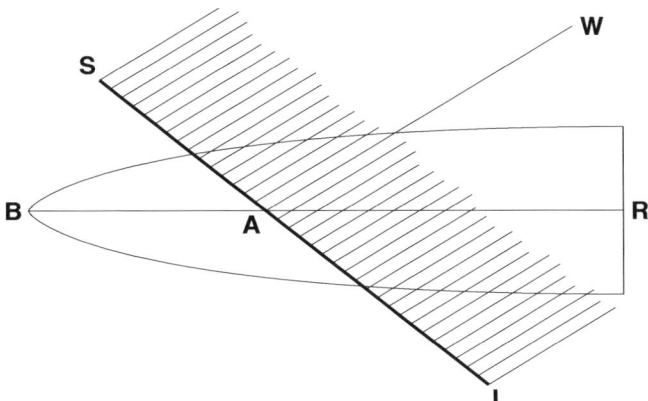

increase beyond this, and the motion will be then what mathematicians describe as "steady."

In the accompanying figure, let BAR represent the keel of a boat, B being the bow, and let SAL represent the sail. Suppose that the wind is blowing in the direction WA with a velocity u; and that this direction makes an angle θ with the keel, *i.e.* angle $WAR = \theta$. Suppose that the sail is set so as to make an angle α with the keel, *i.e.* angle $BAS = \alpha$, and therefore angle $WAL = \theta + \alpha$. Suppose finally that v is the velocity of the boat in the direction AB.

I have already shown that the solution of the problem depends on the relative directions and velocities of the wind and the boat; hence to find the result reduce the boat to rest by impressing on it a velocity v in the direction BA. The resultant velocity of v parallel to BA and of u parallel to WA will be parallel to SL, if $v \sin \alpha = u \sin(\theta + \alpha)$; and in this case the resultant pressure perpendicular to the sail vanishes.

Thus, for steady motion we have $v \sin \alpha = u \sin(\theta + \alpha)$. Hence, whenever $\sin(\theta + \alpha) > \sin \alpha$, we have $v > u$. Suppose, to take one instance, the sail to be fixed, that is, suppose α to be a constant. Then v is a maximum if $\theta + \alpha = \frac{1}{2}\pi$, that is, if θ is equal to the complement of α. In this case we have $v = u \operatorname{cosec} \alpha$, and therefore v is greater than u. Hence, if the wind makes the same angle α abaft the beam that the sail makes with the keel, the velocity of the boat will be greater than the velocity of the wind.

Next, suppose that the boat is running close to the wind, so that the wind is before the beam (see figure below), then in the same way as before we have $v \sin \alpha = u \sin(\theta + \alpha)$, or $v \sin \alpha = u \sin \varphi$, where

φ = angle $WAS = \pi - \theta - \alpha$. Hence $v = u \sin\varphi \csc\alpha$.

Let w be the component velocity of the boat in the teeth of the wind, that is, in the direction AW. Then we have $w = v\cos BAW = v\cos(\alpha + \varphi) = u\sin\varphi \csc\alpha \cos(\alpha + \varphi)$. If α is constant, this is a maximum when $\varphi = \frac{1}{4}\pi - \frac{1}{2}\alpha$; and, if φ has this value, then $w = \frac{1}{2}u(\csc\alpha - 1)$. This formula shows that w is greater than u, if $\sin\alpha < \frac{1}{3}$. Thus, if the sails can be set so that α is less than $\sin^{-1}\frac{1}{3}$, that is, rather less than $19°29'$, and if the wind has the direction above assigned, then the component velocity of the boat in the face of the wind is greater than the velocity of the wind.

The above theory is curious, but it must be remembered that in practice considerable allowance has to be made for the fact that no boat for use on water can be constructed in which the resistance to

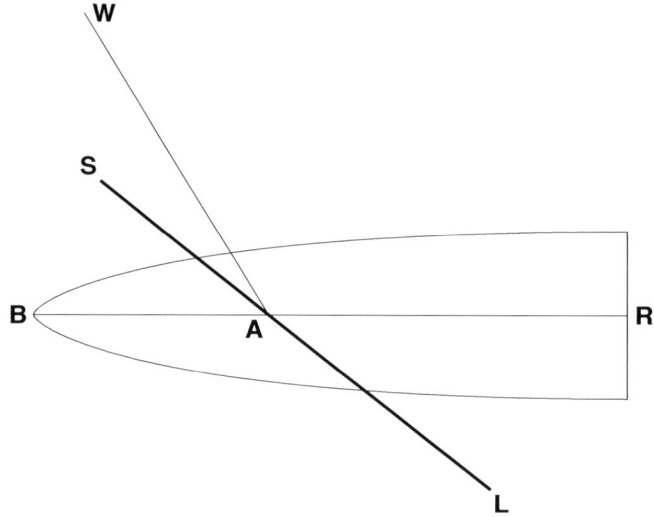

motion in the direction of the keel can be wholly neglected, or which would not drift slightly to leeward if the wind was not dead astern. Still this makes less difference than might be thought by a landsman. In the case of boats sailing on smooth ice the assumptions made are substantially correct, and the practical results are said to agree closely with the theory.

BOAT MOVED BY A ROPE. There is a form of boat-racing, occasionally used at regattas, which affords a somewhat curious illustration of certain mechanical principles. The only thing supplied to the crew is a coil of rope, and they have, without leaving the boat, to propel it from one point to another as rapidly as possible. The motion is given by tying one end of the rope to the after thwart, and giving the other end a series of violent jerks in a direction parallel to the keel. I am told that in still water a pace of two or three miles an hour can be thus attained.

The chief cause for this result seems to be that the friction between the boat and the water retards all relative motion, but is not great enough to materially affect motion caused by a sufficiently big impulse. Hence the usual movements of the crew in the boat do not sensibly move the centre of gravity of themselves and the boat, but this does not apply to an impulsive movement, and if the crew in making a jerk move their centre of gravity towards the bow n times more rapidly than it returns after the jerk, then the boat is impelled forwards at least n times more than backwards: hence on the whole the motion is forwards.

MOTION OF FLUIDS AND MOTION IN FLUIDS. The theories of *motion of fluids* and *motion in fluids* involve considerable difficulties. Here I will mention only one or two instances—mainly illustrations of Hauksbee's Law.

Hauksbee's Law. When a fluid is in rapid motion the pressure is less than when it is at rest[*]. Thus, if a current of air is moving in a tube, the pressure on the sides of the tube is less than when the air is at rest— and the quicker the air moves the smaller is the pressure. This fact was noticed by Hauksbee nearly two centuries ago. In an elastic perfect fluid in which the pressure is proportional to the density, the law connecting the pressure, p, and the steady velocity, v, is $p = \Pi \alpha^{-v^2}$ where Π and α are constants: the establishment of corresponding formula for gases where the pressure is proportional to a power of the density presents no difficulty.

This principle is illustrated by a twopenny toy, on sale in most toy-shops, called the *pneumatic mystery*. It consists of a tube, with a cup-shaped end in which rests a wooden ball. If the tube is held in a vertical position, with the mouthpiece at the upper end and the cup at

---

[*] See Besant, *Hydromechanics*, Cambridge, 1867, art. 149, where however it is assumed that the pressure is proportional to the density. Hauksbee was the earliest writer who called attention to the problem, but I do not know who first explained the phenomenon; some references to it are given by Willis, *Cambridge Philosophical Transactions*, 1830, vol. III, pp. 129–140.

the lower end, then, if anyone blows hard through the tube and places the ball against the cup, the ball will remain suspended there. The explanation is that the pressure of the air below the ball is so much greater than the pressure of the air in the cup that the ball is held up.

The same effect may be produced by fastening to one end of a tube a piece of cardboard having a small hole in it. If a piece of paper is placed over the hole and the experimenter blows through the tube, the paper will not be detached from the card but will bend so as to allow the egress of the air.

An exactly similar experiment, described in many text-books on hydromechanics, is made as follows. To one end of a straight tube a plane disc is fitted which is capable of sliding on wires projecting from the end of the tube. If the disc is placed at a small distance from the end, and anyone blows steadily into the tube, the disc will be drawn towards the tube instead of being blown off the wires, and will oscillate about a position near the end of the tube.

In the same way we may make a tube by placing two books on a table with their backs parallel and an inch or so apart and laying a sheet of newspaper over them. If anyone blows steadily through the tube so formed, the paper will be sucked in instead of being blown out.

The following experiment is explicable by the same argument. On the top of a vertical axis balance a thin horizontal rod. At each end of this rod fasten a small vertical square or sail of thin cardboard—the two sails being in the same plane. If anyone blows close to one of these squares and in a direction parallel to its plane, the square will move towards the side on which one is blowing, and the rod with the two sails will rotate about the axis.

The experiments above described can be performed so as to illustrate Hauksbee's Law; but unless care is taken other causes will be also introduced which affect the phenomena: it is however unnecessary for my purpose to go into these details.

Cut on a Tennis-Ball. Racquet and tennis players know that if a strong cut is given to a ball it can be made to rebound off a vertical wall and then (without striking the floor or any other wall) return and hit the wall again.

This affords another illustration of Hauksbee's Law. The explanation[*] is that the cut causes the ball to rotate rapidly about an axis

---

[*] See Magnus on '*Die Abweichung der Geschosse*' in the *Abhandlungen der*

through its centre of figure, and the friction of the surface of the ball on the air produces a sort of whirlpool. This rotation is in addition to its motion of translation. Suppose the ball to be spherical and rotating about an axis through its centre perpendicular to the plane of the paper in the direction of the arrow-head, and at the same time moving through still air from left to right parallel to PQ. Any motion of the ball perpendicular to PQ will be produced by the pressure of the air on the surface of the ball, and this pressure will, by Hauksbee's Law, be greatest where the velocity of the air relative to the ball is least, and vice versa. To find the velocity of the air relative to the ball we may reduce the centre of the ball to rest, and suppose a stream of air to impinge on the surface of the ball moving with a velocity equal and opposite to that of the centre of the ball. The air is not frictionless, and therefore the air in contact with the surface of the ball will be set in motion, by the rotation of the ball and will form a sort of whirlpool rotating in the direction of the arrow-head in the figure. To find the actual velocity of this air relative to the ball we must consider how the motion due to the whirlpool is affected by the motion of the stream of air parallel to QP. The air at A in the whirlpool is moving against the stream of air there, and therefore its velocity is retarded: the air at B in the whirlpool is moving in the same direction as the stream of air there, and therefore its velocity is increased. Hence the relative velocity of the air at A is less than that at B, and since the pressure of the air is greatest where the velocity is least, the pressure of the air on the surface of the ball at A is greater than on that at B, Hence the ball is forced by this pressure in the direction from the line PQ,

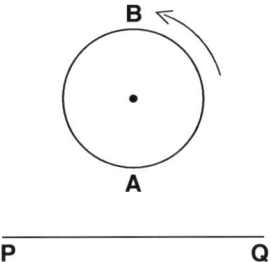

which we may suppose to represent the section of the vertical wall in

---

Akademie der Wissenschaften, Berlin, 1852, pp. 1–23; Lord Rayleigh, '*On the irregular flight of a tennis ball,*' *Messenger of Mathematics*, Cambridge, 1878, vol. VII, pp. 14–16.

a racquet-court. In other words, the ball tends to move at right angles to the line in which its centre is moving and in the direction in which the surface of the front of the ball is being carried by the rotation.

In the case of a lawn tennis-ball, the shape of the ball is altered by a strong cut, and this introduces additional complications.

Spin on a Cricket-Ball. The curl of a cricket-ball in its flight through the air, caused by a spin given by the bowler in delivering the ball, is explained by the same reasoning.

Thus suppose the ball is delivered in a direction lying in a vertical plane containing the two middle stumps of the wickets. A spin round a horizontal axis parallel to the crease in a direction which the bowler's umpire would describe as positive, namely, counter clock-wise, will, in consequence of the friction of the air, cause it to drop, and therefore decrease the length of the pitch. A spin in the opposite direction will cause it to rise, and therefore lengthen the pitch. A spin round a vertical axis in the positive direction, as viewed from above, will make it curl sideways in the air to the left, that is, from leg to off. A spin in the opposite direction will make it curl to the right. A spin given to the ball round the direction of motion of the centre of the ball will not sensibly affect the motion through the air, though it would cause the ball, on hitting the ground, to break. Of course these various kinds of spin can be combined.

The questions involving the application of Hauksbee's Law are easy as compared with many of the problems in fluid motion. The analysis required to attack most of these problems is beyond the scope of this book, but one of them may be worth mentioning even though no explanation is given.

The Theory of the Flight of Birds. A mechanical problem of great interest is the explanation of the means by which birds are enabled to fly for considerable distances with no (perceptible) motion of the wings. Albatrosses, to take an instance of special difficulty, have been known to follow for some days ships running at the rate of nine or ten knots, and sometimes for considerable periods there is no motion of the wings or body which can be detected, while even if the bird moved its wings it is not easy to understand how it has the muscular energy to propel itself so rapidly and for such a length of time. Of this phenomenon various explanations* have been suggested. Notable among these are Mr Maxim's of upward air-currents, Lord Rayleigh's of variations of the wind velocity at different heights above the ground, Dr S.P. Langley's of the incessant occurrence of gusts of wind separated by lulls, and Dr Bryan's of vortices in the atmosphere.

It now seems reasonably certain that the second and third of these sources of energy account for at least a portion of the observed phenomena. The effect of the third cause may be partially explained by noting that the centre of gravity of the bird with extended wings is slightly below the aeroplane or wing surface, so that the animal forms a sort of parachute. The effect of a sudden gust of wind upon such a body is that the aeroplane is set in motion more rapidly than the suspended mass, causing the structure to heel over so as to receive the wind on the under surface of the aeroplane, and this lifts the suspended mass giving it an upward velocity. When the wind falls the greater inertia of the mass carries it on upwards causing the aeroplane to again present its under side to the air; and if while the parachute is in this position the wind is still blowing from the side, the suspended mass is again lifted. Thus the more the bird is blown about, the more it rises in the air; actually birds in flight are carried up by a sudden side gust of wind as we should expect from this theory.

The fact that the bird is in motion tends also to keep it up, for it has been recently shown that a horizontal plane under the action of gravity falls to the ground more slowly if it is travelling through the air with horizontal velocity than it would do if allowed to fall vertically, hence the bird's forward motion causes it to fall through a smaller height between successive gusts of wind than it would do if it were at rest, Moreover it has been proved experimentally that the horsepower required to support a body in horizontal flight by means of an aeroplane is less for high than for low speeds: hence when a side-wind (that is, a wind at right angles to the bird's course) strikes the bird, the lift is increased in consequence of the bird's forward velocity.

CURIOSA PHYSICA. When I was writing the first edition of these "Recreations," I put together a chapter, following this one, on "Some Physical Questions," dealing with problems such as, in the Theory of

---

* See G.H. Bryan in the *Transactions of the British Association* for 1896, vol. LXVI, pp. 726-728.

Sound, the explanation of the fact that in some of Captain Parry's experiments the report of a cannon, when fired, travelled so much more rapidly than the sound of the human voice that observers heard the report of the cannon when fired before that of the order to fire it[*]: in the Kinetic Theory of Gases, the complications in our universe that might be produced by "Maxwell's demon"[†]: in the Theory of Optics, the explanation of the Japanese "magic mirrors,"[‡] which reflect the pattern on the back of the mirror (on which the light does not fall): to which I might add the theory of the "spectrum top," by means of which a white surface, on which some black lines are drawn, can be moved so as to give the impression[§] that the lines are coloured (red, green, blue, slate, or drab), and the curious fact that the colours change with the direction of rotation: it has also been recently shown that if two trains of waves, whose lengths are in the ratio $m - 1 : m + 1$, be superposed, then every mth wave in the system will be big—thus the current opinion that every ninth wave in the open sea is bigger than the other waves may receive scientific confirmation. There is no lack of interesting and curious phenomena in physics, and in some branches, notably in electricity and magnetism, the difficulty is rather one of selection, but I felt that the connection with mathematics was in general either too remote or too technical to justify the insertion of such a collection in a work on elementary mathematical recreations, and therefore I struck out the chapter. I mention the fact now partly to express the hope that some physicist will one day give us a collection of the kind, partly to suggest these questions to those who are interested in such matters.

---

[*] The fact is well authenticated. Mr Earnshaw (*Philosophical Transactions*, London, 1860, pp. 133–148) explained it by the acceleration of a wave caused by the formation of a kind of bore, a view accepted by Clerk Maxwell and most physicists, but Sir George Airy thought that the explanation was to be found in physiology; see Airy's *Sound*, second edition, London, 1871, pp. 141, 142.

[†] See *Theory of Heat*, by J. Clerk Maxwell, second edition, London, 1872, p. 308.

[‡] See a memoir by W.E. Ayrton and J. Perry, *Proceedings of the Royal Society of London*, part I, 1879, vol. XXVIII, pp. 127–148.

[§] See letters from Mr C.E. Benham and others in *Nature*, 1894–5; and a memoir by Prof. Liveing, Cambridge Philosophical Society, November 26, 1894.

CHAPTER IV.

SOME MISCELLANEOUS QUESTIONS.

I propose to discuss in this chapter the mathematical theory of certain of the more common mathematical amusements and games. Some of these might have been treated in the first two chapters, but, since most of them involve mixed geometry and algebra, it is rather more convenient to deal with them apart from the problems and puzzles which have been described already. This division, however, is by no means well defined, and the arrangement is based on convenience rather than on any logical distinction.

The majority of the questions here enumerated have no connection one with another, and I jot them down almost at random.

I shall discuss in succession the *Fifteen Puzzle*, the *Tower of Hanoï*, *Chinese Rings*, the *Eight Queens Problem*, the *Fifteen School-Girls Problem*, and some miscellaneous *Problems connected with a pack of cards*.

The Fifteen Puzzle[*]. Some years ago the so-called *fifteen puzzle* was on sale in all toy-shops. It consists of a shallow wooden box—one side being marked as the top—in the form of a square, and contains fifteen square blocks or counters numbered 1, 2, 3... up to 15. The box will hold just sixteen such counters, and, as it contains only fifteen, they can be moved about in the box relatively to one another. Initially they are put in the box in any order, but leaving the sixteenth

---

[*] There are two articles on the subject in the *American Journal of Mathematics* 1879, vol. II, by Professors Woolsey Johnson and Story; but the whole theory is deducible immediately from the proposition I give above in the text.

cell or small square empty; the puzzle is to move them so that finally they occupy the position shown in the first of the annexed figures.

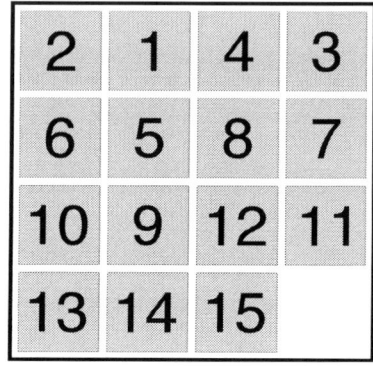

We may represent the various stages in the game by supposing that the blank space, occupying the sixteenth cell, is moved over the board, ending finally where it started.

The route pursued by the blank space may consist partly of tracks followed and again retraced, which have no effect on the arrangement, and partly of closed paths travelled round, which necessarily are cyclical permutations of an odd number of counters. No other motion is possible.

Now a cyclical permutation of n letters is equivalent to $n-1$ simple interchanges; accordingly an odd cyclical permutation is equivalent to an even number of simple interchanges. Hence, if we move the counters so as to bring the blank space back into the sixteenth cell, the new order must differ from the initial order by an even number of simple interchanges. If therefore the order we want to get can be obtained from this initial order only by an odd number of interchanges, the problem is incapable of solution; if it can be obtained by an even number, the problem is possible.

Thus the order in the second of the diagrams given on the current page is deducible from that in the first diagram by six interchanges; namely, by interchanging the counters 1 and 2, 3 and 4, 5 and 6, 7 and 8, 9 and 10, 11 and 12. Hence the one can be deduced from the other by moving the counters about in the box.

If however in the second diagram the order of the last three counters had been 13, 15, 14, then it would have required seven interchanges of counters to bring them into the order given in the first diagram. Hence in this case the problem would be insoluble.

The easiest way of finding the number of simple interchanges necessary in order to obtain one given arrangement from another is to make the transformation by a series of cycles. For example, suppose that we take the counters in the box in any definite order, such as taking the successive rows from left to right, and suppose the original order and the final order to be respectively

      1, 13, 2, 3, 5, 7, 12, 8, 15, 6, 9, 4, 11, 10, 14,
and 11, 2, 3, 4, 5, 6, 7, 1, 9, 10, 13, 12, 8, 14, 15.

We can deduce the second order from the first by 12 simple interchanges. The simplest way of seeing this is to arrange the process in three separate cycles as follows:—

| 1, 11, 8 ; | 13, 2, 3, 4, 12, 7, 6, 10, 14, 15, 9 ; | 5. |
| 11, 8, 1 ; | 2, 3, 4, 12, 7, 6, 10, 14, 15, 9, 13 ; | 5. |

Thus, if in the first row of figures 11 is substituted for 1, then 8 for 11, then 1 for 8, we have made a cyclical interchange of 3 numbers, which is equivalent to 2 simple interchanges (namely, interchanging 1 and 11, and then 1 and 8). Thus the whole process is equivalent to one cyclical interchange of 3 numbers, another of 11 numbers, and another of 1 number. Hence it is equivalent to $(2+10+0)$ simple interchanges. This is an even number, and thus one of these orders can be deduced from the other by moving the counters about in the box.

It is obvious that, if the initial order is the same as the required order except that the last three counters are in the order 15, 14, 13, it would require one interchange to put them in the order 13, 14, 15; hence the problem is insoluble.

If however the box is turned through a right angle, so as to make AD the top, this rotation will be equivalent to 13 simple interchanges. For, if we keep the sixteenth square always blank, then such a rotation would change any order such as

     1, 2, 3, 4, 5, 6, 7, 8, 9, 10, 11, 12, 13, 14, 15,
to 13, 9, 5, 1, 14, 10, 6, 2, 15, 11, 7, 3, 12, 8, 4,

which is equivalent to 13 simple interchanges. Hence it will change the arrangement from one in which a solution is impossible to one where it is possible, and vice versa.

Again, even if the initial order is one which makes a solution impossible, yet if the first cell and not the last is left blank it will be possible to arrange the fifteen counters in their natural order. For, if we represent the blank cell by b, this will be equivalent to changing the order

$$1,\ 2,\ 3,\ 4,\ 5,\ 6,\ 7,\ 8,\ 9,\ 10,\ 11,\ 12,\ 13,\ 14,\ 15,\ b,$$
$$b,\ 1,\ 2,\ 3,\ 4,\ 5,\ 6,\ 7,\ 8,\ 9,\ 10,\ 11,\ 12,\ 13,\ 14,\ 15:$$

this is a cyclical interchange of 16 things and therefore is equivalent to 15 simple interchanges. Hence it will change the arrangement from one in which a solution is impossible to one where it is possible, and vice versa.

It is evident that the above principles are applicable equally to a rectangular box containing mn cells or spaces and $mn - 1$ counters which are numbered. Of course m may be equal to n. If such a box is turned through a right angle, and m and n are both even, it will be equivalent to $mn - 3$ simple interchanges—and thus will change an impossible position to a possible one, and vice versa—but unless both m and n are even the rotation is equivalent to only an even number of interchanges. Similarly, if either m or n is even, and it is impossible to solve the problem when the last cell is left blank, then it will be possible to solve it by leaving the first cell blank.

The problem may be made more difficult by limiting the possible movements by fixing bars inside the box which will prevent the movement of a counter transverse to their directions. We can conceive also of a similar cubical puzzle, but we could not work it practically except by sections.

THE TOWER OF HANOÏ. I may mention next the ingenious puzzle known as the *Tower of Hanoï*. It was brought out in 1883 by M. Claus (Lucas).

It consists of three pegs fastened to a stand, and of eight circular discs of wood or cardboard each of which has a hole in the middle so that a peg can be put through it. These discs are of different radii, and initially they are placed all on one peg, so that the biggest is at the bottom, and the radii of the successive discs decrease as we ascend: thus the smallest disc is at the top. This arrangement is called the *Tower*. The problem is to shift the discs from one peg to another in such a way that a disc shall never rest on one smaller than itself, and finally to transfer the tower (*i.e.* all the discs in their proper order) from the peg on which they initially rested to one of the other pegs.

The method of effecting this is as follows. (i) If initially there are n discs on the peg A, the first operation is to transfer gradually the top $n-1$ discs from the peg A to the peg B, leaving the peg C vacant: suppose that this requires x separate transfers. (ii) Next, move the bottom disc to the peg C. (iii) Then, reversing the first process, transfer gradually the $n-1$ discs from B to C, which will necessitate x transfers. Hence, if it requires x transfers of simple discs to move a tower of $n-1$ discs, then it will require $2x+1$ separate transfers of single discs to move a tower of n discs. Now with 2 discs it requires 3 transfers, *i.e.* 2^2-1 transfers; hence with 3 discs the number of transfers required will be $2(2^2-1)+1$, that is, 2^3-1. Proceeding in this way we see that with a tower of n discs it will require 2^n-1 transfers of single discs to effect the complete transfer. Thus the eight discs of the puzzle will require 255 single transfers. The result can be also obtained by the theory of finite differences. It will be noticed that every alternate move consists of a transfer of the smallest disc from one peg to another, the pegs being taken in cyclical order.

M. De Parville gives an account of the origin of the toy which is a sufficiently pretty conceit to deserve repetition[*]. In the great temple at Benares, says he, beneath the dome which marks the centre of the world, rests a brass-plate in which are fixed three diamond needles, each a cubit high and as thick as the body of a bee. On one of these needles, at the creation, God placed sixty-four discs of pure gold, the largest disc resting on the brass plate, and the others getting smaller and smaller up to the top one. This is the Tower of Bramah. Day and night unceasingly the priests transfer the discs from one diamond needle to another according to the fixed and immutable laws of Bramah, which require that the priest must not move more than one disc at a time and that he must place this disc on a needle so that there is no smaller disc below it. When the sixty-four discs shall have been thus transferred from the needle on which at the creation God placed them to one of the other needles, tower, temple, and Brahmins alike will crumble into dust, and with a thunder-clap the world will vanish. Would that English writers were in the habit of inventing equally interesting origins for the puzzles they produce!

---

[*] *La Nature*, Paris, 1884, part I, pp. 285–286.

The number of separate transfers of single discs which the Brahmins must make to effect the transfer of the tower is $2^{64} - 1$, that is, is $18,446744,073709,551615$: a number which, even if the priests never made a mistake, would require many thousands of millions of years to carry out.

CHINESE RINGS[*]. A somewhat more elaborate toy, known as *Chinese Rings*, which is on sale in most English toy-shops, is represented in the accompanying figure. It consists of a number of rings

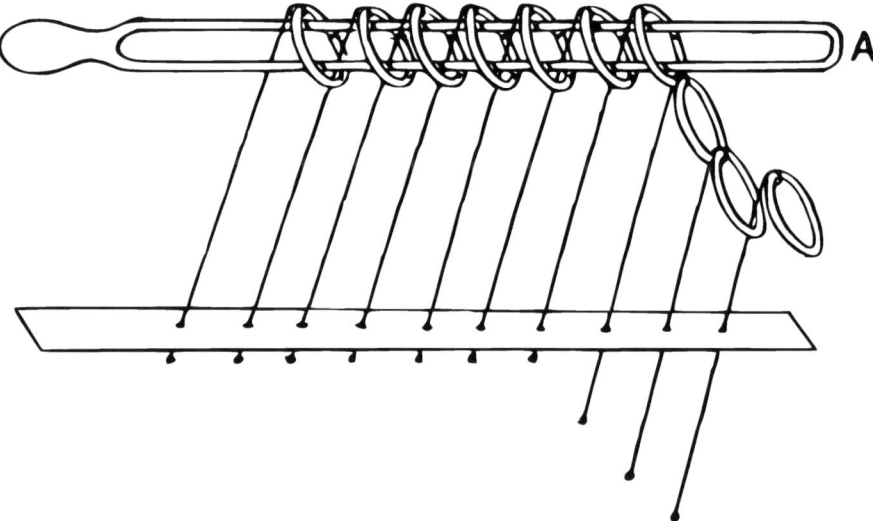

hung upon a bar in such a manner that the ring at one end (say A) can be taken off or put on the bar at pleasure; but any other ring can be taken off or put on only when the one next to it towards A is on, and all the rest towards A are off the bar. The order of the rings cannot be changed.

Only one ring can be taken off or put on at a time. [In the toy, as usually sold, the first two rings form an exception to the rule. Both these can be taken off or put on together. To simplify the discussion I shall assume at first that only one ring is taken off or put on at a time.] I proceed to show that, if there are n rings, then in order to

---

[*] It was described by Cardan in 1550 in his *De Subtilitate*, bk. XV, paragr. 2, ed. Sponius, vol. III, p. 587; by Wallis in his *Algebra*, second edition, 1693, *Opera*, vol. II, chap. 111, pp. 472–478; and allusion is made to it also in Ozanam's *Récréations*, 1723 edition, vol. IV, p. 439.

disconnect them from the bar, it will be necessary to take a ring off or to put a ring on either $\frac{1}{3}(2^{n+1} - 1)$ times or $\frac{1}{3}(2^{n+1} - 2)$ times according as n is odd or even.

Let the taking a ring off the bar or putting a ring on the bar be called a *step*. It is usual to number the rings from the free end A. Let us suppose that we commence with the first m rings off the bar and all the rest on the bar; and suppose that then it requires $x - 1$ steps to take off the next ring, that is, it requires $x - 1$ additional steps to arrange the rings so that the first $m + 1$ of them are off the bar and all the rest are on it. Before taking these steps we can take off the $(m + 2)$th ring and thus it will require x steps from our initial position to remove the $(m + 1)$th and $(m + 2)$th rings.

Suppose that these x steps have been made and that thus the first $m + 2$ rings are off the bar and the rest on it, and let us find how many additional steps are now necessary to take off the $(m + 3)$th and $(m + 4)$th rings. To take these off we begin by taking off the $(m + 4)$th ring: this requires 1 step. Before we can take off the $(m + 3)$th we must arrange the rings so that the $(m + 2)$th is on and the first $m + 1$ rings are off: to effect this, (i) we must get the $(m + 1)$th ring on and the first m rings off, which requires $x - 1$ steps, (ii) then we must put on the $(m + 2)$th ring, which requires 1 step, (iii) and lastly we must take the $(m + 1)$th ring off, which requires $x - 1$ steps: thus this series of movements requires in all $\{2(x - 1) + 1\}$ steps. Next we can take the $(m + 3)$th ring off, which requires 1 step; this leaves us with the first $m + 1$ rings off, the $(m + 2)$th on, the $(m + 3)$th off and all the rest on. Finally to take off the $(m + 2)$th ring, (i) we get the $(m + 1)$th ring on and the first m rings off, which requires $x - 1$ steps, (ii) we take off the $(m + 2)$th ring, which requires 1 step, (iii) we take $(m + 1)$th ring off, which requires $x - 1$ steps: thus this series of movements requires $\{2(x - 1) + 1\}$ steps.

Therefore, if when the first m rings are off it requires x steps to take off the $(m + 1)$th and $(m + 2)$th rings, then the number of additional steps required to take off the $(m + 3)$th and $(m + 4)$th rings is $1 + \{2(x - 1) + 1\} + 1 + \{2(x - 1) + 1\}$, that is, is $4x$.

To find the whole number of steps necessary to take off an odd number of rings we proceed as follows.

To take off the first ring requires 1 step;
∴ to take off the first 3 rings requires 4 additional steps;
∴   "   "   5   "   " 4^2 "   "   .

In this way we see that the number of steps required to take off the first $2n+1$ rings is $1+4+4^2+\cdots+4^n$, which is equal to $\frac{1}{3}(2^{2n+2}-1)$.

To find the number of steps necessary to take off an even number of rings we proceed in a similar manner.

To take off the first 2 rings requires 2 steps;
∴ to take off the first 4 rings requires 2×4 additional steps;
∴   ”   ”   6   ”   ” 2×4^2 ”   ”.

In this way we see that the number of steps required to take off the first $2n$ rings is $2+(2\times4)+(2\times4^2)+\cdots+(2\times4^{n-1})$, which is equal to $\frac{1}{3}(2^{2n+1}-2)$.

If we take off or put on the first two rings in one step instead of two separate steps, these results become respectively 2^{2n} and $2^{2n-1}-1$.

I give the above analysis because it is the direct solution of a problem attacked by Cardan in 1550 and by Wallis in 1693—in both cases unsuccessfully—and which at one time attracted some attention. I proceed next to give another solution, more elegant though rather artificial.

This solution, which is due to M. Gros[*], depends on a convention by which any position of the rings is denoted by a certain number expressed in the binary scale of notation in such a way that a step is indicated by the addition or subtraction of unity.

Let the rings be indicated by circles: if a ring is on the bar, it is represented by a circle drawn above the bar; if the ring is off the bar, it is represented by a circle below the bar. Thus figure i below represents a set of seven rings of which the first two are off the bar, the next three are on it, the sixth is off it, and the seventh is on it.

Denote the rings which are on the bar by the digits 1 or 0 alternately, reckoning from left to right, and denote a ring which is off the bar by the digit assigned to that ring on the bar which is nearest to it on the left of it, or by a 0 if there is no ring to the left of it.

Thus the three positions indicated below are denoted respectively by the numbers written below them. The position represented in figure ii is obtained from that in figure i by putting the first ring on to the bar, while the position represented in figure iii is obtained from that in figure i by taking the fourth ring off the bar.

It follows that every position of the rings is denoted by a number expressed in the binary scale: moreover, since in going from left to right

---

[*] *Théorie du Baguenodier*, by L. Gros, Lyons, 1872. I take the account of this from Lucas, vol. i, part 7.

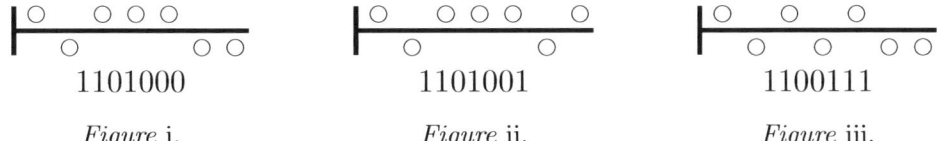

Figure i.　　　　Figure ii.　　　　Figure iii.

every ring on the bar gives a variation (that is, 1 to 0 or 0 to 1) and every ring off the bar gives a continuation, the effect of a step by which a ring is taken off or put on the bar is either to subtract unity from this number or to add unity to it. For example, the number denoting the position of the rings in figure ii is obtained from the number denoting that in figure i by adding unity to it. Similarly the number denoting the position of the rings in figure iii is obtained from the number denoting that in figure i by subtracting unity from it.

The position when all the seven rings are off the bar is denoted by the number 0000000: when all of them are on the bar by the number 1010101. Hence to change from one position to the other requires a number of steps equal to the difference between these two numbers in the binary scale. The first of these numbers is 0: the second is equal to $2^6 + 2^4 + 2^2 + 1$, that is, to 85. Therefore 85 steps are required. In a similar way we may show that to put on a set of $2n + 1$ rings requires $(1 + 2^1 + 2^2 + \ldots + 2^{2n})$ steps, that is, $\frac{1}{3}(2^{2n+2} - 1)$ steps; and to put on a set of $2n$ rings requires $(2 + 2^3 + \ldots + 2^{2n-1})$ steps, that is, $\frac{1}{3}(2^{n+1} - 2)$ steps.

I append a table indicating the steps necessary to take off the first four rings from a set of five rings. The diagrams in the middle column show the successive position of the rings after each step. The number following each diagram indicates that position, each number being obtained from the one above it by the addition of unity. The steps which are bracketed together can be made in one movement, and, if thus effected, the whole process is completed in 7 movements instead of 10 steps: this is in accordance with the formula given above.

M. Gros asserted that it is possible to take from 64 to 80 steps a minute, which in my experience is a rather high estimate. If we accept the lower of these numbers, it would be possible to take off 10 rings in less than 8 minutes; to take off 25 rings would require more than 582 days, each of ten hours work; and to take off 60 rings would

Initial position		⊢ ○ ○ ○ ○ ○	10101
After 1st step		⊢ ○ ○ ○ ○	10110 ⎫
" 2nd "		⊢ ○ ○ ○ ○ ○	10111 ⎭
" 3rd "		⊢ ○ ○ ○ ○	11000
" 4th "		⊢ ○ ○ ○ ○	11001 ⎫
" 5th "		⊢ ○ ○ ○ ○ ○	11010 ⎭
" 6th "		⊢ ○ ○ ○ ○	11011
" 7th "		⊢ ○ ○ ○ ○ ○	11100
" 8th "		⊢ ○ ○ ○ ○ ○	11101
" 9th "		⊢ ○ ○ ○ ○ ○	11110 ⎫
" 10th "		⊢ ○ ○ ○ ○ ○	11111 ⎭

necessitate no less than $768614,336404,564650$ steps, and would require nearly $55000,000000$ years work—assuming of course that no mistakes were made.

THE EIGHT QUEENS PROBLEM[*]. The determination of the number of ways in which eight queens can be placed on a chess-board—or more generally, in which n queens can be placed on a board of n^2 cells—so that no queen can take any other was proposed originally by Nauck in 1850.

In 1874 Dr S. Günther[†] suggested a method of solution by means of determinants. For, if each symbol represents the corresponding cell of the board, the possible solutions for a board of n^2 cells are given by

---

[*] On the history of this problem see W. Ahrens, *Mathematische Unterhaltungen und Spiele*, Leipzig, 1901, chap. IX—a work issued subsequent to the third edition of this book.

[†] Grunert's *Archiv der Mathematik und Physik*, 1874, vol. LVI pp. 281–292.

those terms, if any, of the determinant

$$\begin{vmatrix} a_1 & b_2 & c_3 & d_4 & \ldots\ldots\ldots\ldots \\ \beta_2 & a_3 & b_4 & c_5 & \ldots\ldots\ldots\ldots \\ \gamma_3 & \beta_4 & a_5 & b_6 & \ldots\ldots\ldots\ldots \\ \delta_4 & \gamma_5 & \beta_6 & a_7 & \ldots\ldots\ldots\ldots \\ \ldots & \ldots & \ldots & \ldots & \ldots\ldots\ldots\ldots \\ \ldots & \ldots & \ldots & \ldots & a_{2n-3} \quad b_{2n-2} \\ \ldots & \ldots & \ldots & \ldots & \beta_{2n-2} \quad a_{2n-1} \end{vmatrix}$$

in which no letter and no suffix appears more than once.

The reason is obvious. Every term in a determinant contains one and only one element out of every row and out of every column: hence any term will indicate a position on the board in which the queens cannot take one another by moves rook-wise. Again in the above determinant the letters and suffixes are so arranged that all the same letters and all the same suffixes lie along bishop's paths: hence, if we retain only those terms in each of which all the letters and all the suffixes are different, they will denote positions in which the queens cannot take one another by moves bishop-wise. It is clear that the signs of the terms are immaterial.

In the case of an ordinary chess-board the determinant is of the 8th order, and therefore contains 8!, that is, 40320 terms, so that it would be out of the question to use this method for the usual chess-board of 64 cells or for a board of larger size unless some way of picking out the required terms could be discovered.

A way of effecting this was suggested by Dr J.W.L. Glaisher[*] in 1874, and as far as I am aware the theory remains as he left it. He showed that if all the solutions of n queens on a board of n^2 cells were known, then all the solutions of a certain type for $n+1$ queens on a board of $(n+1)^2$ cells could be deduced, and that all the other solutions of $n+1$ queens on a board of $(n+1)^2$ cells could be obtained without difficulty. The method will be sufficiently illustrated by one instance of its application.

It is easily seen that there are no solutions when $n = 2$ and $n = 3$. If $n = 4$ there are two terms in the determinant which give solutions, namely, $b_2 c_5 \gamma_3 \beta_6$ and $c_3 \beta_2 b_6 \gamma_5$. To find the solutions when $n = 5$,

---

[*] *Philosophical Magazine*, London, December, 1874, series 4, vol. XLVIII, pp. 457–467.

Glaisher proceeded thus. In this case Günther's determinant is

$$\begin{vmatrix} a_1 & b_2 & c_3 & d_4 & e_5 \\ \beta_2 & a_3 & b_4 & c_5 & d_6 \\ \gamma_3 & \beta_4 & a_5 & b_6 & c_7 \\ \delta_4 & \gamma_5 & \beta_6 & a_7 & b_8 \\ \varepsilon_5 & \delta_6 & \gamma_7 & \beta_8 & a_9 \end{vmatrix}$$

To obtain those solutions (if any) which involve a_9 it is sufficient to append a_9 to such of the solutions for a board of 16 cells as do not involve a. As neither of those given above involves an a we thus get two solutions, namely, $b_2 c_5 \gamma_3 \beta_6 a_9$ and $c_3 \beta_2 b_6 \gamma_5 a_9$. The solutions which involve a_1, e_5 and ε_5 can be written down by symmetry. The eight solutions thus obtained are all distinct; we may call them of the first type.

The above are the only solutions which can involve elements in the corner squares of the determinant. Hence the remaining solutions are obtainable from the determinant

$$\begin{vmatrix} 0 & b_2 & c_3 & d_4 & 0 \\ \beta_2 & a_3 & b_4 & c_5 & d_6 \\ \gamma_3 & \beta_4 & a_5 & b_6 & c_7 \\ \delta_4 & \gamma_5 & \beta_6 & a_7 & b_8 \\ 0 & \delta_6 & \gamma_7 & \beta_8 & 0 \end{vmatrix}$$

If, in this, we take the minor of b_2 and in it replace by zero every term involving the letter b or the suffix 2 we shall get all solutions involving b_2. But in this case the minor at once reduces to $d_6 a_5 \delta_4 \beta_8$. We thus get one solution, namely, $b_2 d_6 a_5 \delta_4 \beta_8$. The solutions which involve β_2, δ_4, δ_6, β_8, b_8, d_6, and d_4 can be obtained by symmetry. Of these eight solutions it is easily seen that only two are distinct: these may be called solutions of the second type.

Similarly the remaining solutions must be obtained from the determinant

$$\begin{vmatrix} 0 & 0 & c_3 & 0 & 0 \\ 0 & a_3 & b_4 & c_5 & 0 \\ \gamma_3 & \beta_4 & a_5 & b_6 & c_7 \\ 0 & \gamma_5 & \beta_6 & a_7 & 0 \\ 0 & 0 & \gamma_7 & 0 & 0 \end{vmatrix}$$

If, in this, we take the minor of c_3, and in it replace by zero every term involving the letter c or the suffix 3, we shall get all the solutions which involve c_3. But in this case the minor vanishes. Hence there is no solution involving c_3, and therefore by symmetry no solutions which involve γ_3, γ_7, or c_3. Had there been any solutions involving the third element in the first or last row or column of the determinant we should have described them as of the third type.

Thus in all there are ten and only ten solutions, namely, eight of the first type, two of the second type, and none of the third type.

Similarly, if $n = 6$, we obtain no solutions of the first type, four solutions of the second type, and no solutions of the third type; that is, four solutions in all. If $n = 7$, we obtain sixteen solutions of the first type, twenty-four solutions of the second type, no solutions of the third type, and no solutions of the fourth type; that is, forty solutions in all. If $n = 8$, we obtain sixteen solutions of the first type, fifty-six solutions of the second type, and twenty solutions of the third type, that is, ninety-two solutions in all.

It will be noticed that all the solutions of one type are not always distinct. In general, from any solution seven others can be obtained at once. Of these eight solutions, four consist of the initial or fundamental solution and the three similar ones obtained by turning the board through one, two, or three right angles; the other four are the reflexions of these in a mirror: but in any particular case it may happen that the reflexions reproduce the originals, or that a rotation through one or two right angles makes no difference. Thus on boards of 4^2, 5^2, 6^2, 7^2, 8^2, 9^2, 10^2 cells there are respectively 1, 2, 1, 6, 12, 46, 92 fundamental solutions; while altogether there are respectively 2, 10, 4, 40, 92, 352, 724 solutions.

The following collection of fundamental solutions may interest the reader. The positions on the board of the queens are indicated by digits: the first digit represents the number of the cell occupied by the queen in the first column reckoned from one end of the column, the second digit the number in the second column, and so on. Thus on a board of 4^2 cells the solution 3142 means that one queen is on the 3rd square of the first column, one on the 1st square of the second column, one on the 4th square of the third column, and one on the 2nd square of the fourth column. If a fundamental solution gives rise to only four solutions the number which indicates it is placed in curved brackets, ( ); if it gives rise to only two solutions the number which indicates it is placed in square brackets, [ ]; the other fundamental solutions give rise to eight solutions each.

On a board of 4^2 cells there is 1 fundamental solution: namely, [3142].

On a board of 5^2 cells there are 2 fundamental solutions: namely, 14253, [25314].

On a board of 6^2 cells there is 1 fundamental solution: namely, (246135).

On a board of 7^2 cells there are 6 fundamental solutions: namely, 1357246, 3572461, (5724613), 4613572, 3162574, (2574136).

On a board of 8^2 cells there are 12 fundamental solutions: namely, 25713864, 57138642, 71386425, 82417536, 68241753, 36824175, 64713528, 36814752, 36815724, 72418536, 26831475, (64718253). The arrangement in this order is due to Mr Oram. It will be noticed that the 10th, 11th, and 12th solutions somewhat resemble the 4th, 6th, and 7th respectively. The 6th solution is the only one in which no three queens are in a straight line.

On a board of 9^2 cells there are 46 fundamental solutions; one of them is 248396157. On a board of 10^2 cells there are 92 fundamental solutions; these were given by Dr A. Pein[*]; one of them is $2468t13579$, where t stands for ten. On a board of 11^2 cells there are 341 fundamental solutions; these have been given by Dr T.B. Sprague[†]: one of them is $15926t37e48$. I may add that for a board of n^2 cells there is always a symmetrical solution of the form $246\ldots n135\ldots(n-1)$, when $n = 6m$ or $n = 6m + 4$, Also Mr Oram has shown that for a board of n^2 cells, when n is a prime, cyclical arrangements of the n natural numbers, other than in their natural order, will give solutions; see, for instance, the solution quoted above.

The puzzle in the form of a board of 36 squares is sold in the streets of London for a penny, a small wooden board being ruled in the manner shown in the diagram and having holes drilled in it at the points marked by dots. The object is to put six pins into the holes so that no two are connected by a straight line.

---

[*] *Aufstellung von n Königinnen auf einem Schachbrett von n^2 Feldern*, Leipzig, 1889.

[†] *Proceedings of the Edinburgh Mathematical Society*, vol. XVII, 1898–9, pp. 43–68.

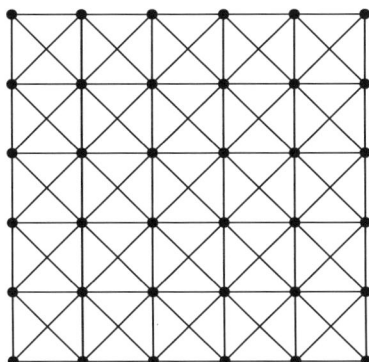

Other Problems with Queens. Captain Turton called my attention to two other problems of a somewhat analogous character, neither of which, as far as I know, has been published elsewhere, or solved otherwise than empirically.

The first of these is to place eight queens on a chess-board so as to command the fewest possible squares. Thus, if queens are placed on cells 1 and 2 of the second column, on cell 2 of the sixth column, on cells 1, 3, and 7 of the seventh column, and on cells 2 and 7 of the eighth column, eleven cells on the board will not be in check; the same number can be obtained by other arrangements. Is it possible to place the eight queens so as to leave more than eleven cells out of check? I have never succeeded in doing so, nor in showing that it is impossible to do it.

The other problem is to place m queens (m being less than 5) on a chess-board so as to command as many cells as possible. For instance, four queens can be placed in several ways on the board so as to command 58 cells besides those on which the queens stand, thus leaving only 2 cells which are not commanded: *ex. gr.* this is effected if the queens are placed on cell 5 of the third column, cell 1 of the fourth column, cell 6 of the seventh column, and cell 2 of the eighth column; or on cell 1 of the first column, cell 7 of the third column, cell 3 of the fifth column, and cell 5 of the seventh column. A similar problem is to determine the minimum number and the position of queens which can be placed on a board of n^2 cells so as to occupy or command every cell. It would seem that, even with the additional restriction that no queen shall be able to take any other queen, there are no less than ninety-one typical solutions in which five queens can be placed on a chess-board so as to command every cell*.

Extension to other Chess Pieces. Analogous problems may be proposed with other chess-pieces. For instance, questions as to the maximum number of knights which can be placed on a board of n^2 cells so that no knight can take any other, and the minimum number of knights which can be placed on it so as to occupy or command every cell have been propounded†.

Similar problems have also been proposed for k kings placed on a chess-board of n^2 cells‡. It has been asserted that, if $k = 2$, the number of ways in which two kings can be placed on a board so that they may not occupy adjacent squares is $\frac{1}{2}(n-1)(n-2)(n^2+3n-2)$. Similarly, if $k = 3$, the number of ways in which three kings can be placed on a board so that no two of them occupy adjacent squares is said to be $\frac{1}{6}(n-1)(n-2)(n^4+3n^3-20n^2-30n+132)$.

THE FIFTEEN SCHOOL-GIRLS PROBLEM. This problem—which was first enunciated by Mr T.P. Kirkman, and is sometimes known as *Kirkman's problem*§—consists in arranging fifteen things in different sets of triplets. It is usually presented in the form that a school-mistress was in the habit of taking her girls for a daily walk.

---

* *L'Intermédiaire des mathématiciens*, Paris, 1901, vol. VIII, p. 88.
† *Ibid.*, March, 1896, vol. III, p. 58; 1897, vol. IV, p. 15, 254; and 1898, vol. V, p. 87.
‡ *Ibid.*, June, 1901, p. 140.
§ It was published first in the *Lady's and Gentleman's Diary* for 1850, p. 48, and has been the subject of numerous memoirs. Among these I may single out the papers by A. Cayley in the *Philosophical Magazine*, July, 1850, series 3, vol. XXXVII, pp. 50–53; by T.P. Kirkman in the *Cambridge and Dublin Mathematical Journal*, 1850, vol. V, p. 260; by R.R. Anstice, *Ibid.*, 1852, vol. VII, pp. 279–292; by B. Pierce, *Gould's Journal*, Cambridge, U.S., 1860, vol. VI, pp. 169–174; by T.P. Kirkman, *Philosophical Magazine*, March, 1862, series 4, vol. XXIII, pp. 198–204; by W.S.B. Woolhouse in the *Lady's Diary* for 1862, pp. 84–88, and for 1863, pp. 79–90, and in the *Educational Times Reprints*, 1867, vol. VIII, pp. 76–83; by J. Power in the *Quarterly Journal of Mathematics*, 1867, vol. VIII, pp. 236–251; by A.H. Frost, *Ibid.*, 1871, vol. XI, pp. 26–37; by E. Carpmael in the *Proceedings of the London Mathematical Society*, 1881, vol. XII, pp. 148–156; by Lucas in his *Récréations*, vol. II, part vi; by A.C. Dixon in the *Messenger of Mathematics*, Cambridge, October, 1893, vol. XXIII, pp. 88–89; and by W. Burnside, *Ibid.*, 1894, vol. XXIII, pp. 137–143. It has also, since the issue of my third edition, been discussed by W. Ahrens in his *Mathematische Unterhaltungen und Spiele*, Leipzig, 1901, chapter xiv.

The girls were fifteen in number, and were arranged in five rows of three each so that each girl might have two companions. The problem is to dispose them so that for seven consecutive days no girl will walk with any of her school-fellows more than once. More generally we may require to arrange $3m$ girls in triplets to walk out for $\frac{1}{2}(3m-1)$ days, so that no girl will walk with any of her school-fellows more than once.

The theory of the formation of all such possible triplets in the case of fifteen girls is not difficult, but the extension to $3m$ girls is, as yet, unsolved. I proceed to describe three methods of solution: these methods are analytical, but I may add that the problem can be also attacked by geometrical methods.

Frost's Method. The first of these solutions is due to Mr Frost. A full exposition of it would occupy a good deal of space, but I hope that the following sketch will make the process intelligible.

Denote one of the girls by k. Her companions on each day are different: suppose that on Sunday they are a_1 and a_2, on Monday b_1 and b_2, and so on, and finally on Saturday g_1 and g_2. Hence for each day we have one triplet, and we have to find four others, but in each of the latter no two like letters can occur together, that is, the three letters in any of them must be all different.

Let a stand for a_1, or a_2, b for b_1 or b_2, and so on. The suffixes 1 and 2 are called complementary. Then, since the three letters in each of the triplets we are trying to find must be different, we must make some arrangement such as putting a with bc, de, and fg; and, if so, b may be associated with df and eg; and c with dg and ef. Thus there are seven possible triads, such as abc, ade, afg, bdf, beg, cdg, and cef. Moreover each of these may stand for any one of four triplets: for instance, the triad bdf may stand for any of the triplets $b_1d_1f_1$, $b_1d_2f_2$, $b_2d_1f_2$, $b_2d_2f_1$.

The four triads which do not involve a must be placed in the Sunday column, the four which do not involve b in the Monday column, and so on. Thus each triad will occur four times.

It only remains to insert the proper suffixes. This is done as follows. Take one triad, such as bdf, and insert a different set of suffixes each time that it occurs; for instance, the four sets given above. Next, the other like letters (b, d, or f as the case may be) in these four columns must have the complementary suffixes attached.

After this is done, the next triplet in the Sunday column will be b_2eg. The triad beg occurs in four columns and includes four possible triplets, such as $b_2e_1g_1$, $b_2e_2g_2$, $b_1e_1g_2$, $b_1e_2g_1$. Insert these, and then give the complementary suffixes to the other like letters in these four columns.

In this way the arrangement is constructed gradually, by taking one triad at a time, inserting the proper suffixes to the four triplets included in it, and then the complementary suffixes in the other like letter in the same columns.

One final arrangement, thus obtained, is as follows:

Sunday	Monday	Tuesday	Wednesday	Thursday	Friday	Saturday
ka_1a_2	kb_1b_2	kc_1c_2	kd_1d_2	ke_1e_2	kf_1f_2	kg_1g_2
$b_1d_1f_1$	$a_1d_2e_2$	$a_1d_1e_1$	$a_2b_2c_2$	$a_2b_1c_1$	$a_1b_2c_1$	$a_1b_1c_2$
$b_2e_1g_1$	$a_2f_2g_2$	$a_2f_1g_1$	$a_1f_2g_1$	$a_1f_1g_2$	$a_2d_2e_1$	$a_2d_1e_2$
$c_1d_2g_2$	$c_1d_1g_1$	$b_1d_2f_2$	$b_1e_1g_2$	$b_2d_1f_2$	$b_1e_2g_1$	$b_2d_2f_1$
$c_2e_2f_2$	$c_2e_1f_1$	$b_2e_2g_2$	$c_1e_2f_1$	$c_2d_2g_1$	$c_2d_1g_2$	$c_1e_1f_2$

We might obtain other solutions by selecting other seven triads or by choosing other arrangements of the suffixes in each triad (or by merely interchanging letters and suffixes in the above order). By these means Mr Power showed that there are no less than $15567,552000$ different solutions; but, since the total number of ways in which the school can walk out for a week in triplets is $(455)^7$, the probability that any chance way satisfies Kirkman's condition is very small.

Frost's method is applicable to the case of $2^{2n}-1$ girls walking out for $2^{2n-1}-1$ days in triplets. The detailed solution for 63 girls walking out for 31 days, which corresponds to $n=3$, have been given.

Anstice's Method. Another method of attacking the problem is due to Mr Anstice; it is illustrated by the following elegant solution, by which from the order on Sunday we can obtain the order on the following six days by a cyclical permutation. Let the girls be denoted respectively by the letters k, a_1, a_2, a_3, a_4, a_5, a_6, a_7, b_1, b_2, b_3, b_4, b_5, b_6, b_7; and suppose the order on Sunday to be

$$ka_1b_1,\ a_2a_3a_5,\ a_4b_3b_6,\ a_6b_2b_7,\ a_7b_4b_5\ .$$

Then, if the suffixes are permuted cyclically, we obtain six other arrangements which satisfy the conditions of the problem: the reason being that in the above arrangement the difference of the suffixes of every pair of like letters—such as either the "a"s or the "b"s—in a triplet is different for each triplet, as also is the difference of the suffixes of every pair of unlike letters which are in a triplet.

Two other arrangements for Sunday, from which those for the remaining days are obtainable by cyclical permutations can be formed. These are $ka_1b_1, a_2a_3a_5, a_4b_5b_7, a_6b_3b_4, a_7b_2b_6$; and $ka_1b_1, a_2a_3a_5, a_4b_2b_6, a_6b_5b_7, a_7b_3b_4$.

Anstice's method is applicable to the case of $2p+1$ girls walking out for p days in triplets so that no pair may walk together more than once, provided p is a prime of the form $12m+7$. In such a case he showed how to construct a fundamental arrangement for one day from which the arrangements for the remaining $p-1$ days can be obtained by cyclical permutations of suffixes. The number of such fundamental arrangements is $3(2m+1)(3m+1)$.

The problem of 15 girls corresponds to $m=0$, and the three fundamental Anstician arrangements are given above. If $m=1$ we have the problem of 39 girls. One Anstician arrangement in this case is as follows: $ka_1b_1, a_2a_8a_{12}, a_5a_7a_{10}, a_6a_{17}a_{18}, a_3b_{10}b_{15}, a_4b_3b_5, a_9b_{18}b_{19}, a_{11}b_8b_{14}, a_{13}b_9b_{17}, a_{14}b_{12}b_{16}, a_{15}b_4b_7, a_{16}b_2b_{11}, a_{19}b_6b_{13}$. If $m=2$ we have the problem of 63 girls, of which Frost has given one solution; and so on.

Gill's Method. Another method of attacking the problem has been suggested to me by Mr T.H. Gill. Representing the girls by $a_1, a_2, a_3, \ldots, a_{3m}$ he (i) forms one triplet of the type $a_1a_{m+1}a_{2m+1}$, from which, by cyclical permutation of the suffixes $1, 2, \ldots, 3m$ he obtains m triplets which constitute an arrangement for one day, and (ii) forms $\frac{1}{2}(m-1)$ other triplets such that the three differences of the suffixes are different, from which, by cyclical permutations of the suffixes, the arrangements for the remaining $\frac{3}{2}(m-1)$ other days can be obtained. Thus in the case of 15 girls, the triplet $a_1a_6a_{11}$ gives, by cyclical permutations of the suffixes, an arrangement for the first day and two triplets such as $a_1a_2a_5, a_1a_3a_9$ enable us to form 30 triplets from which an arrangement for the other six days can be found. Here is a solution thus determined.

First Day:	1. 6.11;	2. 7.12;	3. 8.13;	4. 9.14;	5.10.15.
Second Day:	1. 2. 5;	3. 4. 7;	8. 9.12;	10.11.14;	13.15. 6.
Third Day:	2. 3. 6;	4. 5. 8;	9.10.13;	11.12.15;	14. 1. 7.
Fourth Day:	5. 6. 9;	7. 8.11;	12.13. 1;	14.15. 3;	2. 4.10.
Fifth Day:	7. 9.15;	8.10. 1;	3. 5.11;	4. 6.12;	13.14. 2.
Sixth Day:	9.11. 2;	10.12. 3;	5. 7.13;	6. 8.14;	15. 1. 4.
Seventh Day:	11.13. 4;	12.14. 5;	15. 2. 8;	1. 3. 9;	6. 7.10.

THE FIFTEEN SCHOOL-GIRLS PROBLEM.

But, although this method gives triplets with which the problem can be solved, the final arrangement is empirical.

A solution of the problem of 21 girls for 10 days can be got by the same method: $a_1 a_8 a_{15}$ giving 7 triplets which constitute an arrangement for one day; and $a_1 a_2 a_6$, $a_1 a_3 a_{11}$, $a_1 a_4 a_{10}$ giving 63 triplets from which an arrangement for the other nine days can be formed. Here is the solution thus determined.

First Day:	1. 8.15;	2. 9.16;	3.10.17;	4.11.18;	5.12.19;	6.13.20;	7.14.21.
Second Day:	1. 2. 6;	4. 5. 9;	7. 8.12;	10.11.15;	13.14.18;	16.17.21;	19.20. 3.
Third Day:	7.10.16;	8.11.17;	12.15.21;	18.19. 2;	20. 1. 9;	3. 5.13;	4. 6.14.
Fourth Day:	13.16. 1;	14.17. 2;	18.21. 6;	3. 4. 8;	5. 7.15;	9.11.19;	10.12.20.
Fifth Day:	4. 7.13;	5. 8.14;	9.12.18;	15.16.20;	17.19. 6;	21. 2.10;	1. 3.11.
Sixth Day:	1. 4.10;	2. 5.11;	6. 9.15;	12.13.17;	14.16. 3;	18.20. 7;	19.21. 8.
Seventh Day:	2. 3. 7;	5. 6.10;	8. 9.13;	11.12.16;	14.15.19;	17.18. 1;	20.21. 4.
Eighth Day:	10.13.19;	11.14.20;	15.18. 3;	21. 1. 5;	2. 4.12;	6. 8.16;	7. 9.17.
Ninth Day:	16.19. 4;	17.20. 5;	21. 3. 9;	6. 7.11;	8.10.18;	12.14. 1;	13.15. 2.
Tenth Day:	19. 1. 7;	20. 2. 8;	3. 6.12;	9.10.14;	11.13.21;	15.17. 4;	16.18. 5.

I should be interested if any of my readers could give me a similar solution of the analogous arrangement of 33 girls for 16 days formed from typical triplet suffixes like 1, 12, 23; 1, 2, 10; 1, 3, 16; 1, 4, 18; 1, 5, 11; 1, 6, 13; or from other sets of triplets formed in a similar way so that (except in the first triplet) the differences of the suffixes are all different.

Walecki's Theorem. Lastly, Walecki—quoted by Lucas—has shown that, if a solution for the case of n girls walking out in triplets for $\frac{1}{2}(n-1)$ days is known, then a solution for $3n$ girls walking out for $\frac{1}{2}(3n-1)$ days can be deduced.

For if an arrangement of the n girls, $a_1, a_2, \ldots, a_n$ for $\frac{1}{2}(n-1)$ days is known; and also one of the n girls, $b_1, b_2, \ldots, b_n$; and also one of the n girls $c_1, c_2, \ldots, c_n$; then an arrangement of these $3n$ girls for $\frac{1}{2}(n-1)$ days is known. A set of n triplets for another day will be given by $a_m b_{m+k} c_{m+2k}$ where m is put equal to $1, 2, \ldots, n$ successively. Here k may have any of the n values, $0, 1, 2, \ldots, (n-1)$; but, wherever a suffix is greater than n, it is to be divided by n and only the remainder retained. Hence altogether we have an arrangement for $n + \frac{1}{2}(n-1)$ days, *i.e.* for $\frac{1}{2}(3n-1)$ days.

The arrangement of 3 girls for one day is obvious. Hence, by Walecki's theorem, we can deduce at once an arrangement of 3^m girls for $\frac{1}{2}(3^m - 1)$ days. And, generally, as I have given solutions of the problem in the case of $3n$ girls when $n = 1, 3, 5, 7, 9, 13, 15$, it follows that for the same values of n, we can solve the analogous arrangements of $3^m \times n$ girls.

To the original theorem J.J. Sylvester* added the corollary that the school of 15 girls could walk out in triplets on 13×7 days until every possible triplet had walked abreast once.

The generalized problem of finding the greatest number of ways in which x girls walking in rows of a abreast can be arranged so that every possible combination of b of them may walk abreast once and only once has been solved for various cases. Suppose that this greatest number of ways is y. It is obvious that, if all the x girls are to walk out each day in rows of a abreast, then x must be an exact multiple of a and the number of rows formed each day is x/a. If such an arrangement can be made for z days, then we have a solution of the problem to arrange x girls to walk out in rows of a abreast for z days so that they all go out each day and so that every possible combination of b girls may walk together once, and only once. In the corresponding generalization of Kirkman's problem no companionship of girls which has occurred once may occur again, but it does not follow necessarily that every possible companionship must occur once.

An example where the solution is obvious is if $x = 2n$, $a = 2$, $b = 2$, in which case $y = n(2n - 1)$, $z = 2n - 1$.

If we take the case $x = 15$, $a = 3$, $b = 2$, we find $y = 35$; and it happens that these 35 rows can be divided into 7 sets, each of which contains all the symbols; hence $z = 7$. More generally, if $x = 5 \times 3^m$, $a = 3$, $b = 2$, we find $y = \frac{3}{2}(x-1)/x$, $z = \frac{1}{2}(x-1)$. It will be noticed that in the solutions of the original fifteen school-girls problem and of Walecki's extension of it given above every possible pair of girls walk together once; hence we might infer that in these cases we could determine z as well as y.

The results of the last paragraph were given by Kirkman† in 1850. In the same memoir he also proved that, if p is a prime, and if $x = p^m$, $a = p$, $b = 2$, then $y = (p^m - 1)/(p - 1)$; if $x = (p^2 + p + 1)(p + 1)$ where $p^2 + p + 1$ has no divisor less than $p + 1$, $a = p + 1$, $b = 2$, then $y = x(x-1)/p(p+1)$; if $x = p^3 + p + 1$, $a = p + 1$, $b = 2$, then $y = x$;

---

* *Philosophical Magazine*, July, 1850, series 3, vol. XXXVII, p. 52; a solution by Sylvester is given in the *Philosophical Magazine*, May, 1861, series 4, vol. XXI, p. 371.

† *Cambridge and Dublin Mathematical Journal*, 1850, vol. V, pp. 255–262.

and Sylvester's result that if $x = 15$, $a = 3$, $b = 3$, $y = 455$, $z = 91$. Three years later Kirkman[*] solved the problem when $x = 2^n$, $a = 4$, $b = 3$. Lastly, in 1893, Sylvester[†] published the solution when $x = 9$, $a = 3$, $b = 3$, in which case $y = 84$, $z = 28$; and stated that a similar method was applicable when $x = 3^m$, $a = 3$, $b = 3$: thus 9 girls can be arranged to walk out 28 times (say 4 times a day for a week) so that in any day the same pair never are together more than once and so that at the end of the week each girl has been associated with every possible pair of her schoolfellows.

In 1867 Mr S. Bills[‡] showed that if $x = 7$, $a = 3$, $b = 2$, then $y = 7$: if $x = 15$, $a = 3$, $b = 2$, then $y = 35$: if $x = 31$, $a = 3$, $b = 2$, then $y = 155$: and the method by which these results are proved will give the value of y, if $x = 2^n - 1$, $a = 3$, $b = 2$. Shortly afterwards Mr W. Lea[§] showed that if $x = 11$, $a = 5$, $b = 4$, then $y = 66$; also that if $x = 16$, $a = 4$, $b = 3$, then $y = 140$; the latter result is a particular case of Kirkman's theorems. It will be noticed that these writers did not confine their discussion to cases where x is an exact multiple of a.

PROBLEMS CONNECTED WITH A PACK OF CARDS. I mentioned in chapter I that an ordinary pack of playing cards could be used to illustrate many tricks depending on simple properties of numbers. Most of these involve the relative position of the cards. The principle of solution generally consists in re-arranging the pack in a particular manner so as to bring the card into some definite position. Any such rearrangement is a species of shuffling.

I shall treat in succession of problems connected with *shuffling a pack*, *arrangements by rows and columns*, the *determination of a pair out of $\frac{1}{2}n(n+1)$ pairs*, *Gergonne's pile problem*, and the game known as *the mouse trap*.

SHUFFLING A PACK. Any system of *shuffling a pack* of cards, if carried out consistently, leads to an arrangement which can be calculated; but tricks that depend on it generally require considerable

---
[*] *Ibid.*, 1853, vol. VIII, pp. 38–42.
[†] *Messenger of Mathematics*, February, 1893, vol. XXII, pp. 159–160.
[‡] *Educational Times Reprints*, London, 1867, vol. VIII, pp. 32–33.
[§] *Ibid.*, 1868, vol. IX, pp. 35–36; and 1874, vol. XXII, pp. 74–76; see also the volume for 1869, vol. XI, p. 97.

technical skill.

Suppose for instance that a pack of n cards is shuffled, as is not unusual, by placing the second card on the first, the third below these, the fourth above them, and so on. The theory of this system of shuffling is due to Monge*. The following are some of the results and are not difficult to prove directly.

If the pack contains $6p + 4$ cards, the $(2p + 2)$th card will occupy the same position in the shuffled pack. For instance, if a complete pack of 52 cards is shuffled as described above, the 18th card will remain the 18th card.

Again, if a pack of $10p+2$ cards is shuffled in this way, the $(2p+1)$th and the $(6p \times 2)$th cards will interchange places. For instance, if an écarté pack of 32 cards is shuffled as described above, the 7th and the 20th cards will change places.

More generally, one shuffle of a pack of $2p$ cards will move the card which was in the x_0th place to the x_1th place, where $x_1 = \frac{1}{2}(2p+x_0+1)$ if x_0 is odd, and $x_1 = \frac{1}{2}(2p - x_0 + 2)$ if x_0 is even, from which the above results can be deduced. By repeated applications of the above formulae we can show that the effect of m such shuffles is to move the card which was initially in the x_0th place to the x_mth place where $2^{m+1}x_m = (4p+1)(2^{m-1} \pm 2^{m-2} \pm \cdots \pm 2 \pm 1) \pm 2x_0 + 2^m \pm 1$, the sign $\pm$ representing an ambiguity of sign.

Again, in any pack of n cards after a certain number of shufflings, not greater than n, the cards will return to their primitive order. This will always be the case as soon as the original top card occupies that position again. To determine the number of shuffles required for a pack of $2p$ cards, it is sufficient to put $x_m = x_0$ and find the smallest value of m which satisfies the resulting equation for all values of x_0 from 1 to $2p$. It follows that, if m is the least number which makes $4^m - 1$ divisible by $4p + 1$, then m shuffles will be required if either $2^m + 1$ or $2^m - 1$ is divisible by $4p+1$, otherwise $2m$ shuffles will be required. The number

---

* Monge's investigations are printed in the *Mémoires de l'Académie des Sciences*, Paris, 1773, pp. 390–412. Among those who have studied the subject afresh I may in particular mention V. Bouniakowski, *Bulletin physico-mathématique de St Pétersbourg*, 1857, vol. XV, pp. 202–205, summarised in the *Nouvelles annales de mathématiques*, 1858, pp. 66–67; T. de St Laurent, *Mémoires de l'Académie de Gard*, 1865; L. Tanner, *Educational Times Reprints*, 1880, vol. XXXIII, pp. 73–75; and M.J. Bourget, *Liouville's Journal*, 1882, pp. 413–434. The solutions given by Prof. Tanner are simple and concise.

for a pack of $2p+1$ cards is the same as that for a pack of $2p$ cards. With an écarté pack of 32 cards, six shuffles are sufficient; with a pack of 2^n cards, $n+1$ shuffles are sufficient; with a full pack of 52 cards, twelve shuffles are sufficient; with a pack of 13 cards ten shuffles are sufficient; while with a pack of 50 cards fifty shuffles are required; and so on.

Mr W.H.H. Hudson[*] has also shown that, whatever is the law of shuffling, yet if it is repeated again and again on a pack of n cards, the cards will ultimately fall into their initial positions after a number of shufflings not greater than the greatest possible L.C.M. of all numbers whose sum is n.

For suppose that any particular position is occupied after the 1st, 2nd, ..., pth shuffles by the cards $A_1, A_2, \ldots, A_p$ respectively, and that initially the position is occupied by the card A_0. Suppose further that after the pth shuffle A_0 returns to its initial position, therefore $A_0 = A_p$. Then at the second shuffling A_2 succeeds A_1 by the same law by which A_1 succeeded A_0 at the first; hence it follows that previous to the second shuffling A_2 must have been in the place occupied by A_1 previous to the first. Thus the cards which after the successive shuffles take the place initially occupied by A_1 are $A_2, A_3, \ldots, A_p, A_1$; that is, after the pth shuffle A_1 has returned to the place initially occupied by it: and so for all the other cards $A_2, A_3, \ldots, A_{p-1}$.

Hence the cards $A_1, A_2, \ldots, A_p$ form a cycle of p cards, one or other of which is always in one or other of p positions in the pack, and which go through all their changes in p shufflings. Let the number n of the pack be divided into $p, q, r, \ldots$ such cycles, whose sum is n; then the L.C.M. of $p, q, r, \ldots$ is the utmost number of shufflings necessary before all the cards will be brought back to their original places.

In the case of a pack of 52 cards, the greatest L.C.M. of numbers whose sum is 52 will be found by trial to be 180180.

ARRANGEMENTS BY ROWS AND COLUMNS. A not uncommon trick, which rests on a species of shuffling, depends on the obvious fact that if n^2 cards are arranged in the form of a square of n rows, each containing n cards, then any card will be defined if the row and the column in which it lies are mentioned.

This information is generally elicited by first asking in which row the selected card lies, and noting the extreme left-hand card of that

---

[*] *Educational Times Reprints*, London, 1865, vol. II, p. 105.

row. The cards in each column are then taken up, face upwards, one at a time beginning with the lowest card of each column and taking the columns in their order from right to left—each card taken up being placed on the top of those previously taken up. The cards are then dealt out again in rows, from left to right, beginning with the top left-hand corner, and a question is put as to which row contains the card. The selected card will be that card in the row mentioned which is in the same vertical column as the card which was originally noted.

The above is the form in which the trick is usually presented, but it is greatly improved by allowing the pack to be cut as often as is liked before the cards are re-dealt, and then giving one cut at the end so as to make the top card in the pack one of those originally in the top row.

The explanation is obvious. For, if 16 cards are taken, the first

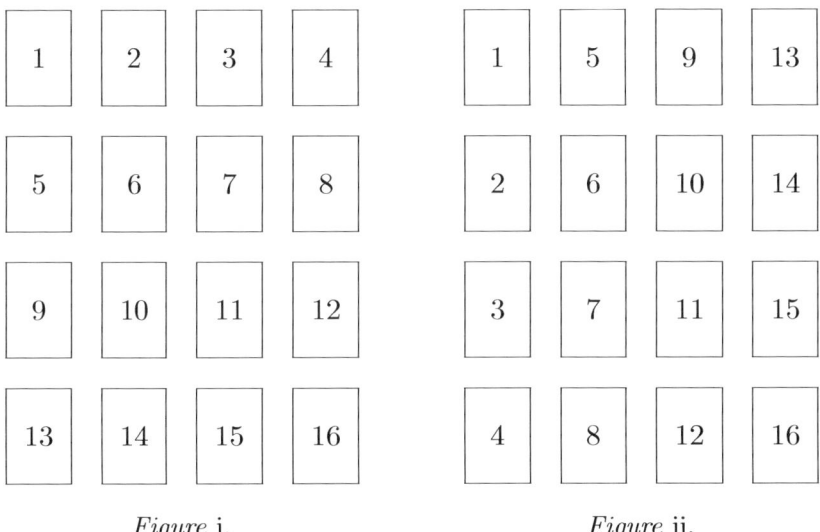

Figure i.   *Figure* ii.

and second arrangements may be represented by figures i and ii. For example, if we are told that in figure i the card is in the third row, it must be either 9, 10, 11, or 12: hence, if we know in which row of figure ii it lies, it is determined. If we allow the pack to be cut between the deals, we must secure somehow that the top card is either 1, 2, 3, or 4, since that will leave the cards in each row of figure ii unaltered though the positions of the rows will be changed.

DETERMINATION OF A SELECTED PAIR OF CARDS OUT OF $\frac{1}{2}n(n+1)$ GIVEN PAIRS*. Another common trick is to throw twenty cards on to a table in ten couples, and ask someone to select one couple. The cards are then taken up, and dealt out in a certain manner into four rows each containing five cards. If the rows which contain the given cards are indicated, the cards selected are known at once.

This depends on the fact that the number of homogeneous products of two dimensions which can be formed out of four things is 10. Hence the homogeneous products of two dimensions formed out of four things can be used to define ten things.

Suppose that ten pairs of cards are placed on a table and someone selects one couple. Take up the cards in their couples. Then the first

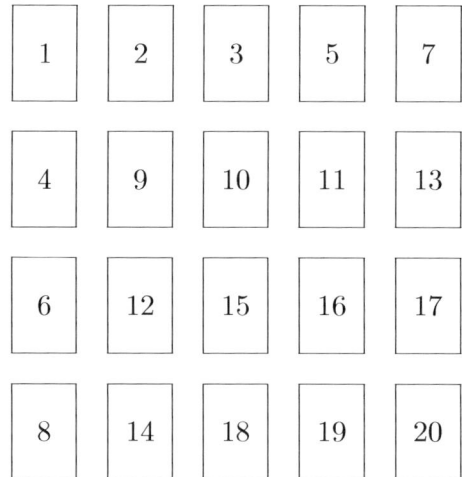

two cards form the first couple, the next two the second couple, and so on. Deal them out in four rows each containing five cards according to the scheme shown in the diagram.

The first couple (1 and 2) are in the first row. Of the next couple (3 and 4), put one in the first row and one in the second. Of the next couple (5 and 6), put one in the first row and one in the third, and so on, as indicated in the diagram. After filling up the first row proceed similarly with the second row, and so on.

Enquire in which rows the two selected cards appear. If only one line, the mth, is mentioned as containing the cards then the required pair of cards are the mth and $(m+1)$th cards in that line. These

---

* Bachet, problem XVII, avertissement, p. 146 *et seq.*

occupy the clue squares of that line. Next, if two lines are mentioned, then proceed as follows. Let the two lines be the pth and the qth and suppose $q > p$. Then that one of the required cards which is in the qth line will be the $(q-p)$th card which is below the first of the clue squares in the pth line. The other of the required cards is in the pth line and is the $(q-p)$th card to the right of the second of the clue squares.

Bachet's rule, in the form in which I have given it, is applicable to a pack of $n(n+1)$ cards divided into couples, and dealt in n rows each containing $n+1$ cards; for there are $\frac{1}{2}n(n+1)$ such couples, also there are $\frac{1}{2}n(n+1)$ homogeneous products of two dimensions which can be formed out of n things. Bachet gave the diagrams for the cases of 20, 30, and 42 cards: these the reader will have no difficulty in constructing for himself, and I have enunciated the rule for 20 cards in a form which covers all the cases.

I have seen the same trick performed by means of a sentence and not by numbers. If we take the case of ten couples, then after collecting the pairs the cards must be dealt in four rows each containing five cards, in the order indicated by the sentence *Matas dedit nomen Cocis*. This sentence must be imagined as written on the table, each word forming one line, The first card is dealt on the M. The next card (which is the pair of the first) is placed on the second m in the sentence, that is, third in the third row. The third card is placed on the a. The fourth card (which is the pair of the third) is placed on the second a, that is, fourth in the first row. Each of the next two cards is placed on a t, and so on. Enquire in which rows the two selected cards appear. If two rows are mentioned, the two cards are on the letters common to the words that make these rows. If only one row is mentioned, the cards are on the two letters common to that row.

The reason is obvious: let us denote each of the first pair by an a, and similarly each of any of the other pairs by an e, i, o, c, d, m, n, s, or t respectively. Now the sentence *Matas dedit nomen Cocis* contains four words each of five letters; ten letters are used, and each letter is repeated only twice. Hence, if two of the words are mentioned, they will have one letter in common, or, if one word is mentioned, it will have two like letters.

To perform the same trick with any other number of cards we should require a different sentence.

The number of homogeneous products of three dimensions which can be formed out of four things is 20, and of these the number consisting of products in which three things are alike and those in which three things are different is 8. This leads to a trick with 8 trios of things, which is similar to that last given—the cards being arranged in the order indicated by the sentence *Lanata levete livini novoto*.

I believe that these arrangements by sentences are known, but I am not aware who invented them.

GERGONNE'S PILE PROBLEM. Before discussing Gergonne's theorem I will describe the familiar three pile problem, the theory of which is included in his results.

The Three Pile Problem[*]. This trick is usually performed as follows. Take 27 cards and deal them into three piles, face upwards. By "dealing" is to be understood that the top card is placed as the bottom card of the first pile, the second card in the pack as the bottom card of the second pile, the third card as the bottom card of the third pile, the fourth card on the top of the first one, and so on: moreover I assume that throughout the problem the cards are held in the hand face upwards. The result can be modified to cover any other way of dealing.

Request a spectator to note a card, and remember in which pile it is. After finishing the deal, ask in which pile the card is. Take up the three piles, placing that pile between the other two. Deal again as before, and repeat the question as to which pile contains the given card. Take up the three packs again, placing the pile which now contains the selected card between the other two. Deal again as before, but in dealing note the middle card of each pile. Ask again for the third time in which pile the card lies, and you will know that the card was the one which you noted as being the middle card of that pile. The trick can be finished then in any way that you like. The usual method—but a very clumsy one—is to take up the three piles once more, placing the named pile between the other two as before, when the selected card will be the middle one in the pack, that is, if 27 cards are used it will be the fourteenth.

The trick is often performed with 15 cards or with 21 cards, in either of which cases the same rule holds.

---

[*] The trick is mentioned by Bachet, problem XVIII, p. 143, but his analysis of it is insufficient.

Gergonne's Generalization. The general theory for a pack of m^m cards was given by M. Gergonne[*]. Suppose the pack is arranged in m piles, each containing m^{m-1} cards, and that, after the first deal, the pile indicated as containing the selected card is taken up ath; after the second deal, is taken up bth; and so on, and finally after the mth deal, the pile containing the card is taken up kth. Then when the cards are collected after the mth deal the selected card will be nth from the top where

if m is even, $\quad n = km^{m-1} - jm^{m-2} + \cdots + bm - a + 1$,
if m is odd, $\quad n = km^{m-1} - jm^{m-2} + \cdots - bm + a$.

For example, if a pack of 256 cards (*i.e.* $m = 4$) was given, and anyone selected a card out of it, the card could be determined by making four successive deals into four piles of 64 cards each, and after each deal asking in which pile the selected card lay. The reason is that after the first deal you know it is one of sixty-four cards. In the next deal these sixty-four cards are distributed equally over the four piles, and therefore, if you know in which pile it is, you will know that it is one of sixteen cards. After the third deal you know it is one of four cards. After the fourth deal you know which card it is.

Moreover, if the pack of 256 cards is used, it is immaterial in what order the pile containing the selected card is taken up after a deal. For, if after the first deal it is taken up ath, after the second bth, after the third cth, and after the fourth dth, the card will be the $(64d - 16c + 4b - a + 1)$th from the top of the pack, and thus will be known. We need not take up the cards after the fourth deal, for the same argument will show that it is the $(64 - 16c + 4b - a + 1)$th in the pile then indicated as containing it. Thus if $a = 3$, $b = 4$, $c = 1$, $d = 2$, it will be the 62nd card in the pile indicated after the fourth deal as containing it and will be the 126th card in the pack as then collected.

In exactly the same way a pack of twenty-seven cards may be used, and three successive deals, each into three piles of nine cards, will suffice to determine the card. If after the deals the pile indicated as containing the given card is taken up ath, bth, and cth respectively, then the card will be the $(9c - 3b + a)$th in the pack or will be the $(9 - 3b + a)$th card in the pile indicated after the third deal as containing it.

---

[*] Gergonne's *Annales de Mathématiques*, Nismes, 1813–4, vol. IV, pp. 276–283.

The method of proof will be illustrated sufficiently by considering the usual case of a pack of twenty-seven cards, for which $m = 3$, which are dealt into three piles each of nine cards.

Suppose that, after the first deal, the pile containing the selected card is taken up ath: then (i) at the top of the pack there are $a - 1$ piles each containing nine cards; (ii) next there are 9 cards, of which one is the selected card; and (iii) lastly there are the remaining cards of the pack. The cards are dealt out now for the second time: in each pile the bottom $3(a - 1)$ cards will be taken from (i), the next 3 cards from (ii), and the remaining $9 - 3a$ cards from (iii).

Suppose that the pile now indicated as containing the selected card is taken up bth: then (i) at the top of the pack are $9(b - 1)$ cards; (ii) next are $9 - 3a$ cards; (iii) next are 3 cards, of which one is the selected card; and (iv) lastly are the remaining cards of the pack. The cards are dealt out now for the third time: in each pile the bottom $3(b - 1)$ cards will be taken from (i), the next $3 - a$ cards will be taken from (ii), the next card will be one of the three cards in (iii), and the remaining $8 - 3b + a$ cards are from (iv).

Hence, after this deal, as soon as the pile is indicated, it is known that the card is the $(9 - 3b + a)$th from the top of that pile. If the process is continued by taking up this pile as cth, then the selected card will come out in the place $9(c - 1) + (8 - 3b + a) + 1$ from the top, that is, will come out as the $(9c - 3b + a)$th card.

Since, after the third deal, the position of the card in the pile then indicated is known, it is easy to notice the card, in which case the trick can be finished in some way more effective than dealing again.

If we put the pile indicated always in the middle of the pack we have $a = 2$, $b = 2$, $c = 2$, hence $n = 9c - 3b + a = 14$, which is the form in which the trick is usually presented, as was explained above on page 115.

I have shown that if a, b, c are known, then n is determined. We may modify the rule so as to make the selected card come out in any assigned position, say the nth. In this case we have to find values of a, b, c which will satisfy the equation $n = 9c - 3b + a$, where a, b, c can have only the values 1, 2, or 3.

Hence, if we divide n by 3 and the remainder is 1 or 2, this remainder will be a; but, if the remainder is 0, we must decrease the quotient by unity so that the remainder is 3, and this remainder will be a. In other words a is the smallest positive number (exclusive of zero) which must be subtracted from n to make the difference a multiple of 3.

Next let p be this multiple, *i.e.* p is the next lowest integer to $n/3$: then $3p = 9c - 3b$, therefore $p = 3c - b$. Hence b is the smallest positive number (exclusive of zero) which must be added to p to make the sum a multiple of 3, and c is that multiple.

A couple of illustrations will make this clear. Suppose we wish the card to come out 22nd from the top, therefore $22 = 9c - 3b + a$. The smallest number which must be subtracted from 22 to leave a multiple of 3 is 1, therefore $a = 1$. Hence $22 = 9c - 3b + 1$, therefore $7 = 3c - b$. The smallest number which must be added to 7 to make a multiple of 3 is 2, therefore $b = 2$. Hence $7 = 3c - 2$, therefore $c = 3$. Thus $a = 1$, $b = 2$, $c = 3$.

Again, suppose the card is to come out 21st. Hence $21 = 9c - 3b + a$. Therefore a is the smallest number which subtracted from 21 makes a multiple of 3, therefore $a = 3$. Hence $6 = 3c - b$. Therefore b is the smallest number which added to 6 makes a multiple of 3, therefore $b = 3$. Hence $9 = 3c$, therefore $c = 3$. Thus $a = 3$, $b = 3$, $c = 3$.

If any difficulty is experienced in this work, we can proceed thus. Let $a = x + 1$, $b = 3 - y$, $c = z + 1$; then x, y, z may have only the values 0, 1, or 2. In this case Gergonne's equation takes the form $9z + 3y + x = n - 1$. Hence, if $n - 1$ is expressed in the ternary scale of notation, x, y, z will be determined, and therefore a, b, c will be known.

The rule in the case of a pack of m^m cards is exactly similar. We want to make the card come out in a given place. Hence, in Gergonne's formula, we are given n and we have to find $a, b, \ldots, k$. We can effect this by dividing n continually by m, with the convention that the remainder are to be alternately positive and negative and that their numerical values are to be not greater than m or less than unity.

An analogous theorem with a pack of lm cards can be constructed. C.T. Hudson and L.E. Dickson[*] have discussed the general case where such a pack is dealt n times, each time into l piles of m cards; and they have shown how the piles must be taken up in order that after the nth deal the selected card may be rth from the top.

The principle will be sufficiently illustrated by one example treated in a manner analogous to the cases already discussed. For instance,

---

[*] *Educational Times Reprints*, 1868, vol. IX, pp. 89–91; and *Bulletin* of the American Mathematical Society, New York, April, 1895, vol. I, pp. 184–186.

suppose that an écarté pack of 32 cards is dealt into 4 piles each of 8 cards, and that the pile which contains some selected card is picked up ath. Suppose that on dealing again into four piles, one pile is indicated as containing the selected card, the selected card cannot be one of the bottom $2(a-1)$ cards, or of the top $8-2a$ cards, but must be one of the intermediate 2 cards, and the trick can be finished in any way, as for instance by the common conjuring ambiguity of asking someone to choose one of them, leaving it doubtful whether the one he takes is to be rejected or retained.

THE MOUSE TRAP. I will conclude this chapter with the bare mention of another game of cards, known as the *mouse trap*, the discussion of which involves some rather difficult algebraic analysis.

It is played as follows. A set of cards, marked with the numbers $1, 2, 3, \ldots, n$, is dealt in any order, face upwards, in the form of a circle. The player begins at any card and counts round the circle always in the same direction. If the kth card has the number k on it—which event is called a *hit*—the player takes up the card and begins counting afresh. According to Cayley, the player wins if he thus takes up all the cards, and the cards win if at any time the player counts up to n without being able to take up a card.

For example, if a pack of only four cards is used and these cards come in the order, 3214, then the player would obtain the second card 2 as a hit, next he would obtain 1 as a hit, but if he went on for ever he would not obtain another hit. On the other hand, if the cards in the pack were initially in the order 1423, the player would obtain successively all four cards in the order 1, 2, 3, 4.

The problem may be stated as the determination of what hits and how many hits can be made with a given number of cards; and what permutations will give a certain number of hits in a certain order.

Cayley[*] showed that there are 9 arrangements of a pack of four cards in which no hit will be made, 7 arrangements in which only one hit will be made, 3 arrangements in which only two hits will be made, and 5 arrangements in which four hits will be made.

Prof. Steen[†] has investigated the general theory for a pack of n cards. He has shown how to determine the number of arrangements in

---

[*] *Quarterly Journal of Mathematics*, 1878, vol. XV, pp. 8–10.
[†] *Ibid.*, vol. XV, pp. 230–241.

which x is the first hit [Arts. 3–5]; the number of arrangements in which 1 is the first hit and x is the second hit [Art. 6]; and the number of arrangements in which 2 is the first hit and x the second hit [Arts. 7–8]; but beyond this point the theory has not been carried. It is obvious that, if there are $n-1$ hits, the nth hit will necessarily follow.

TREIZE. The French game of *treize* is very similar. It is played with a full pack of fifty-two cards (knave, queen, and king counting as 11, 12, and 13 respectively). The dealer calls out $1, 2, 3, \ldots, 13$, as he deals the 1st, 2nd, 3rd, ..., 13th cards respectively. At the beginning of a deal the dealer offers to lay or take certain odds that he will make a hit in the thirteen cards next dealt.

One of the innumerable forms of *patience* is played in a similar way.

CHAPTER V.

MAGIC SQUARES.

A *magic square* consists of a number of integers arranged in the form of a square, so that the sum of the numbers in every row, in every column, and in each diagonal is the same. If the integers are the consecutive numbers from 1 to n^2 the square is said to be of the nth order, and it is easily seen that in this case the sum of the numbers in any row, column, or diagonal is equal to $\frac{1}{2}n(n^2+1)$: this number may be denoted by N. I confine my account to such magic squares, that is, to squares formed with consecutive integers, from 1 upwards.

Thus the first 16 integers, arranged in either of the forms given in figures i and ii below, form a magic square of the fourth order, the sum

1	15	14	4
12	6	7	9
8	10	11	5
13	3	2	16

Figure i.

15	10	3	6
4	5	16	9
14	11	2	7
1	8	13	12

Figure ii.

of the numbers in any row, column, or diagonal being 34. Similarly figures iii and v on page 124, figure viii on page 126, and figures xii and xiii on page 136, show magic squares of the fifth order; and figure xi on page 133 shows a magic square of the sixth order; and figures xiv and xv on pages 137, 138, show magic squares of the eighth order.

The formation of these squares is an old amusement, and in times when mystical philosophical ideas were associated with particular numbers it was natural that such arrangements should be deemed to possess magical properties. Magic squares of an odd order were constructed in India before the Christian era according to a law of formation which is explained hereafter. Their introduction into Europe appears to have been due to Moschopulus, who lived at Constantinople in the early part of the fifteenth century, and enunciated two methods for making such squares. The majority of the medieval astrologers and physicians were much impressed by such arrangements. In particular the famous Cornelius Agrippa (1486–1535) constructed magic squares of the orders 3, 4, 5, 6, 7, 8, 9, which were associated respectively with the seven astrological "planets": namely, Saturn, Jupiter, Mars, the Sun, Venus, Mercury, and the Moon. He taught that a square of one cell, in which unity was inserted, represented the unity and eternity of God; while the fact that a square of the second order could not be constructed illustrated the imperfection of the four elements, air, earth, fire, and water; and later writers added that it was symbolic of original sin. A magic square engraved on a silver plate was sometimes prescribed as a charm against the plague, and one, namely, that represented in figure i on page 121, is drawn in the picture of Melancholy, painted about the year 1500 by Albert Dürer. Such charms are still worn in the East.

The development of the theory has been due mainly to French mathematicians. Bachet gave a rule for the construction of any square of an odd order in a form substantially equivalent to one of the rules given by Moschopulus. The formation of magic squares, especially of even squares, was considered by Frénicle and Fermat. The theory was continued by Poignard, De la Hire, Sauveur, D'Ons-en-bray, and Des Ourmes. Ozanam included in his work an essay on magic squares which was amplified by Montucla. From this and from De la Hire's memoirs the larger part of the materials for this chapter are derived. Like most algebraical problems, the construction of magic squares attracted the attention of Euler, but he did not advance the general theory. In 1837 an elaborate work on the subject was compiled by B. Violle, which is useful as containing numerous illustrations. I give the references in a footnote[*].

---

[*] Bachet, *Problèmes plaisans*, Lyons, 1624, problem XXI, p. 161; Frénicle, *Divers Ouvrages de Mathématique par Messieurs de l'Académie des Sciences*, Paris,

I shall confine myself to establishing rules for the construction of squares subject to no conditions beyond those given in the definition. Rules sufficient for this purpose are contained in the works to which I have just referred and on which I have based this sketch; some extensions and developments will be found in the memoirs mentioned below*. I shall commence by giving rules for the construction of a square of an odd order, and then shall proceed to similar rules for one of an even order.

It will be convenient to use the following terms. The spaces or small squares occupied by the numbers are called *cells*. The diagonal from the top left-hand cell to the bottom right-hand cell is called the *leading diagonal* or *left diagonal*. The diagonal from the top right-hand cell to the bottom left-hand cell is called the *right diagonal*.

MAGIC SQUARES OF AN ODD ORDER.   I proceed to give methods for constructing *odd magic squares*, but for simplicity I shall apply them to the formation of squares of the fifth order, though exactly similar

---

1693, pp. 423–483; with an appendix (pp. 484–507), containing diagrams of all the possible magic squares of the fourth order, 880 in number: Fermat, *Opera Mathematica*, Toulouse, 1679, pp. 173–178; or Brassinne's *Précis*, Paris, 1853, pp. 146–149: Poignard, *Traité des Quarrés Sublimes*, Brussels, 1704: De la Hire, *Mémoires de l'Académie des Sciences* for 1705, Paris, 1706, part I, pp. 127–171; part II, pp. 364–382: Sauveur, *Construction des Quarrés Magiques*, Paris, 1710: D'Ons-en-bray, *Mémoires de l'Académie des Sciences* for 1750, Paris, 1754, pp. 241–271: Des Ourmes, *Mémoires de Mathématique et de Physique* (French Academy), Paris, 1763, vol. IV, pp. 196–241: Ozanam and Montucla, *Récréations*, part I, chapter XII: Euler, *Commentationes Arithmeticae Collectae*, St Petersburg, 1849, vol. II, pp. 593–602: Violle, *Traité Complet des Carrés Magiques*, 3 vols, Paris, 1837–8. A sketch of the history of the subject is given in chap. iv of S. Günther's *Geschichte der mathematischen Wissenschaften*, Leipzig, 1876. See also W. Ahrens, *Mathematische Unterhaltungen und Spiele*, Leipzig, 1901, chapter xii.

* In England the subject has been studied by R. Moon, *Cambridge Mathematical Journal*, 1845, vol. IV, pp. 209–214; H. Holditch, *Quarterly Journal of Mathematics*, London, 1864, vol. VI, pp. 181–189; W.H. Thompson, *Ibid.*, 1870, vol. X, pp. 186–202; J. Horner, *Ibid.*, 1871, vol. XI, pp. 57–65, 123–132, 213–224; S.M. Drach, *Messenger of Mathematics*, Cambridge, 1873, vol. II, pp. 169–174, 187; A.H. Frost, *Quarterly Journal of Mathematics*, London, 1878, vol. XV, pp. 34–49, 93–123, 366–368, in which the results of previous memoirs are included: there are also some pamphlets and articles on it of a more popular character. Of recent Continental works on the subject I have no complete bibliography, and probably it is better to omit all rather than give an imperfect list.

proofs will apply equally to any odd square.

De la Loubère's Method[*]. If the reader will look at figure iii he will see one way in which such a square containing 25 cells can be constructed. The middle cell in the top row is occupied by 1. The successive numbers are placed in their natural order in a diagonal line

17	24	1	8	15
23	5	7	14	16
4	6	13	20	22
10	12	19	21	3
11	18	25	2	9

15+2	20+4	0+1	5+3	10+5
20+3	0+5	5+2	10+4	15+1
0+4	5+1	10+3	15+5	20+2
5+5	10+2	15+4	20+1	0+3
10+1	15+3	20+5	0+2	5+4

23	6	19	2	15
10	18	1	14	22
17	5	13	21	9
4	12	25	8	16
11	24	7	20	3

De la Loubère's Method.   De la Loubère's Method.   Bachet's Method.
Figure iii.   Figure iv.   Figure v.

which slopes upwards to the right, except that (i) when the top row is reached the next number is written in the bottom row as if it came immediately above the top row; (ii) when the right-hand column is reached, the next number is written in the left-hand column, as if it immediately succeeded the right-hand column; and (iii) when a cell which has been filled up already, or when the top right-hand square is reached, the path of the series drops to the row vertically below it and then continues to mount again. Probably a glance at the diagram in figure iii will make this clear.

The reason why such a square is magic can be explained best by expressing the numbers in the scale of notation whose radix is 5 (or n, if the magic square is of the order n), except that 5 is allowed to appear as a unit-digit and 0 is not allowed to appear as a unit-digit. The result is shown in figure iv. From that figure it will be seen that the method of construction ensures that every row and every column shall contain one and only one of each of the unit-digits 1, 2, 3, 4, 5, the sum of which is 15; and also one and only one of each of the radix-digits 0, 5, 10, 15, 20, the sum of which is 50. Hence, as far as rows and columns are concerned, the square is magic. Moreover if the square is odd, each of the diagonals will contain one and only one of each of the unit-digits

---

[*] De la Loubère, *Du Royaume de Siam* (Eng. Trans.), London, 1693, vol. II, pp. 227–247. De la Loubère was the envoy of Louis XIV to Siam in 1687–8, and there learnt this method.

1, 2, 3, 4, 5. Also the leading diagonal will contain one and only one of the radix-digits 0, 5, 10, 15, 20, the sum of which is 50; and if, as is the case in the square drawn above, the number 10 is the radix-digit to be added to the unit-digits in the right diagonal, then the sum of the radix-digits in that diagonal is also 50. Hence the two diagonals also possess the magical property.

And generally if a magic square of an odd order n is constructed by De la Loubère's method, every row and every column must contain one and only one of each of the unit-digits $1, 2, 3 \ldots, n$; and also one and only one of each of the radix-digits $0, n, 2n, \ldots, n(n-1)$. Hence, as far as rows and columns are concerned, the square is magic. Moreover each diagonal will either contain one and only one of the unit-digits or will contain n unit-digits each equal to $\frac{1}{2}(n+1)$. It will also either contain one and only one of the radix-digits or will contain n radix-digits each equal to $\frac{1}{2}n(n-1)$. Hence the two diagonals will also possess the magical property. Thus the square will be magic.

I may notice here that, if we place 1 in any cell and fill up the square by De la Loubère's rule, we shall obtain a square that is magic in rows and in columns, but it will not in general be magic in its diagonals.

It is evident that other squares can be derived from De la Loubère's square by permuting the symbols properly. For instance, in figure iv, we may permute the symbols 1, 2, 3, 4, 5 in 5! ways, and we may permute the symbols 0, 5, 15, 20 in 4! ways. Any one of these 5! arrangements combined with any one of these 4! arrangements will give a magic square. Hence we can obtain 2880 magic squares of the fifth order of this kind, though only 720 of them are really distinct. Other squares can however be deduced, for it may be noted that from any magic square, whether even or odd, other magic squares of the same order can be formed by the mere interchange of the row and the column which intersect in a cell on a diagonal with the row and the column which intersect in the complementary cell of the same diagonal.

Bachet's Method[*]. Another method, very similar to that of De la Loubère, for constructing an odd magic square is as follows. We begin by placing 1 in the cell above the middle one (that is, in a square of the fifth order in the cell occupied by the number 7 in figure iii), and then we write the successive numbers in a diagonal line sloping upwards to the right, subject to the condition that when the cases (i)

---

[*] Bachet, Problem XXI, p. 161.

and (ii) mentioned under De la Loubère's method occur the rules there given are followed, but when the case (iii) occurs the path of the series rises *two* rows, *i.e.* it is continued from one cell to the cell next but one vertically above it, if this cell is above the top row the path continues from the corresponding cell in one of the bottom two rows following the analogy of rule (i) in De la Loubère's method. Such a square is delineated in figure v on page 124. Bachet's method leads ultimately to this arrangement; except that the rules are altered so as to make the line slope downwards. This method also gives 720 magic squares of the fifth order.

De la Hire's Method[*]. I shall now give another rule for the formation of odd magic squares. To form an odd magic square of the order n by this method, we begin by constructing two subsidiary squares, one of the unit-digits, $1, 2, \ldots, n$, and the other of multiples of the radix, namely, $0, n, 2n, \ldots, (n-1)n$. We then form the magic square by adding together the numbers in the corresponding cells in the two subsidiary squares.

De la Hire gave several ways of constructing such subsidiary squares. I select the following method (props. x and xiv of his memoir) as being the simplest, but I shall apply it to form a square of only the fifth order. It leads to the same results as the second of the two rules given by Moschopulus.

The first of the subsidiary squares (figure vi, below), is constructed thus. First, 3 is put in the top left-hand corner, and then the numbers 1, 2, 4, 5 are written in the other cells of the top line (in any order). Next, the number in each cell of the top line is repeated in all the cells

3	4	1	5	2
2	3	4	1	5
5	2	3	4	1
1	5	2	3	4
4	1	5	2	3

First Subsidiary Square
Figure vi.

15	0	20	5	10
0	20	5	10	15
20	5	10	15	0
5	10	15	0	20
10	15	0	20	5

Second Subsidiary Square
Figure vii.

18	4	21	10	12
2	23	9	11	20
25	7	13	19	1
6	15	17	3	24
14	16	5	22	8

Resulting Magic Square
Figure viii.

which lie in a diagonal line sloping downwards to the right (see figure vi)

---

[*] *Mémoires de l'Académie des Sciences* for 1705, part I, pp. 127–171.

according to the rule (ii) in De la Loubère's method. The cells filled by the same number form a *broken diagonal*. It follows that every row and every column contains one and only one 1, one and only one 2, and so on. Hence the sum of the numbers in every row and in every column is equal to 15; also, since we placed 3, which is the average of these numbers, in the top left-hand corner, the sum of the numbers in the left diagonal is 15; and, since the right diagonal contains one and only one of each of the numbers 1, 2, 3, 4, and 5, the sum of the numbers in that diagonal also is 15.

The second of the subsidiary squares (figure vii) is constructed in a similar way with the numbers 0, 5, 10, 15, and 20, except that the mean number 10 is placed in the top right-hand corner; and the broken diagonals formed of the same numbers all slope downwards to the left. It follows that every row and every column in figure vii contains one and only one 0, one and only one 5, and so on; hence the sum of the numbers in every row and every column is equal to 50. Also the sum of the numbers in each diagonal is equal to 50.

If now we add together the numbers in the corresponding cells of these two squares, we shall obtain 25 numbers such that the sum of the numbers in every row, every column, and each diagonal is equal to $15 + 50$, that is, to 65. This is represented in figure viii. Moreover, no two cells in that figure contain the same number. For instance, the numbers 21 to 25 can occur only in those five cells which in figure vii are occupied by the number 20, but the corresponding cells in figure vi contain respectively the numbers 1, 2, 3, 4, and 5; and thus in figure viii each of the numbers from 21 to 25 occurs once and only once. De la Hire preferred to have the cells in the subsidiary squares which are filled by the same number connected by a knight's move and not by a bishop's move; and usually his rule is enunciated in that form.

By permuting the numbers 1, 2, 4, 5 in figure vi we get 4! other arrangements, each of which combined with that in figure vii would give a magic square. Similarly by permuting the numbers 0, 5, 15, 20 in figure vii we obtain 4! other squares, each of which might be combined with any of the 4! arrangements deduced from figure vi. Hence altogether we can obtain in this way 576 magic squares of the fifth order.

There is yet another method of constructing odd squares which is due to Poignard, and was improved by De la Hire in the memoir already cited. I shall not discuss it, because, though for certain assigned values of n it is simpler than the methods which I have given, it depends on the form of n, and particularly on the number of prime factors of n. In the case of a square of the fifth order, this gives an even larger number of magic squares than the methods of De la Loubère, Bachet, and De la Hire. I may also add that it has been shown that magic squares whose order is a prime number can be constructed by a rule similar to De la Loubère's, except that we begin by placing 1 in the bottom left-hand cell, and the subsequent consecutive numbers fill cells forming a knight's path on the square and not a bishop's path. A square of the fifth order of this kind is given in figure xiii on page 136. There are 2880 magic squares of the fifth order of this kind.

De la Hire showed that, apart from mere inversions, there were 57600 magic squares of the fifth order which could be formed by the methods he enumerated. Taking account of other methods, it would seem that the total number of magic squares of the fifth order is very large, perhaps exceeding 500000.

MAGIC SQUARES OF AN EVEN ORDER. The above methods are inapplicable to squares of an even order. I proceed to give two methods for constructing any *even magic square* of an order higher than two.

It will be convenient to use the following terms. Two rows which are equidistant, the one from the top, the other from the bottom, are said to be *complementary*. Two columns which are equidistant, the one from the left-hand side, the other from the right-hand side, are said to be *complementary*. Two cells in the same row, but in complementary columns, are said to be *horizontally related*. Two cells in the same column, but in complementary rows, are said to be *vertically related*. Two cells in complementary rows and columns are said to be *skewly related*; thus, if the cell b is horizontally related to the cell a, and the cell d is vertically related to the cell a, then the cells b and d are skewly related; in such a case if the cell c is vertically related to the cell b, it will be horizontally related to the cell d, and the cells a and c are skewly related: the cells a, b, c, d constitute an *associated group*, and if the square is divided into four equal quarters, one cell of an associated group is in each quarter.

A *horizontal interchange* consists in the interchange of the numbers in two horizontally related cells. A *vertical interchange* consists in the interchange of the numbers in two vertically related cells. A *skew interchange* consists in the interchange of the numbers in two skewly related cells. A *cross interchange* consists in the change of the numbers in any cell and in its horizontally related cell with the numbers in the cells skewly related to them; hence, it is equivalent to two vertical interchanges and two horizontal interchanges.

First Method[*]. This method is the simplest with which I am acquainted, and I believe, at any rate as far as concerns singly-even squares, was published for the first time in 1893.

Begin by filling the cells of the square with the numbers $1, 2, \ldots, n^2$ in their natural order commencing (say) with the top left-hand corner, writing the numbers in each row from left to right, and taking the rows in succession from the top. I will commence by proving that a certain number of horizontal and vertical interchanges in such a square must make it magic, and will then give a rule by which the cells whose numbers are to be interchanged can be at once picked out.

First, we may notice that the sum of the numbers in each diagonal is equal to N, where $N = \frac{1}{2}n(n^2 + 1)$; hence the diagonals are already magic, and will remain so if the numbers therein are not altered.

Next, consider the rows. The sum of the numbers in the xth row from the top is $N - \frac{1}{2}n^2(n - 2x + 1)$. The sum of the numbers in the complementary row, that is, the xth row from the bottom, is $N + \frac{1}{2}n^2(n - 2x + 1)$. Also the number in any cell in the xth row is less than the number in the cell vertically related to it by $n(n - 2x + 1)$. Hence, if in these two rows we make $\frac{1}{2}n$ interchanges of the numbers which are situated in vertically related cells, then we increase the sum of the numbers in the xth row by $\frac{1}{2}n \times n(n - 2x + 1)$, and therefore make that row magic; while we decrease the sum of the numbers in the complementary row by the same number, and therefore make that row magic. Hence, if in every pair of complementary rows we make $\frac{1}{2}n$ interchanges of the numbers situated in vertically related cells, the square will be made magic in rows. But, in order that the diagonals may remain magic, either we must leave both the diagonal numbers in any row unaltered, or we must change both of them with those in the cells vertically related to them.

The square is now magic in diagonals and in rows, and it remains to make it magic in columns. Taking the original arrangement of the numbers (in their natural order) we might have made the square magic

---

[*] See an article in the *Messenger of Mathematics*, Cambridge, September, 1893, vol. XXIII, pp. 65–69.

in columns in a similar way to that in which we made it magic in rows. The sum of the numbers originally in the yth column from the left-hand side is $N - \frac{1}{2}n(n - 2y + 1)$. The sum of the numbers originally in the complementary column, that is, the yth column from the right-hand side, is $N + \frac{1}{2}n(n - 2y + 1)$. Also the number originally in any cell in the yth column was less than the number in the cell horizontally related to it by $n - 2y + 1$. Hence, if in these two columns we had made $\frac{1}{2}n$ interchanges of the numbers situated in horizontally related cells, we should have made the sum of the numbers in each column equal to N. If we had done this in succession for every pair of complementary columns, we should have made the square magic in columns. But, as before, in order that the diagonals might remain magic, either we must have left both the diagonal numbers in any column unaltered, or we must have changed both of them with those in the cells horizontally related to them.

It remains to show that the vertical and horizontal interchanges, which have been considered in the last two paragraphs, can be made independently, that is, that we can make these interchanges of the numbers in complementary columns in such a manner as will not affect the numbers already interchanged in complementary rows. This will require that in every column there shall be exactly $\frac{1}{2}n$ interchanges of the numbers in vertically related cells, and that in every row there shall be exactly $\frac{1}{2}n$ interchanges of the numbers in horizontally related cells. I proceed to show how we can always ensure this, if n is greater than 2. I continue to suppose that the cells are initially filled with the numbers $1, 2, \ldots, n^2$ in their natural order, and that we work from that arrangement.

A *doubly-even square* is one where n is of the form $4m$. If the square is divided into four equal quarters, the first quarter will contain $2m$ columns and $2m$ rows. In each of these columns take m cells so arranged that there are also m cells in each row, and change the numbers in these $2m^2$ cells and the $6m^2$ cells associated with them by a cross interchange. The result is equivalent to $2m$ interchanges in every row and in every column, and therefore renders the square magic.

One way of selecting the $2m^2$ cells in the first quarter is to divide the whole square into sixteen subsidiary squares each containing m^2 cells, which we may represent by the diagram below, and then we may take either the cells in the a squares or those in the b squares; thus, if

a	b	b	a
b	a	a	b
b	a	a	b
a	b	b	a

every number in the eight a squares is interchanged with the number skewly related to it the resulting square is magic. A magic square of the eighth order, constructed in this way, is shown in figure xv on page 138.

Another way of selecting the $2m^2$ cells in the first quarter would be to take the first m cells in the first column, the cells 2 to $m+1$ in the second column, and so on, the cells $m+1$ to $2m$ in the $(m+1)$th column, the cells $m+2$ to $2m$ and the first cell in the $(m+2)$th column, and so on, and finally the $2m$th cell and the cells 1 to $m-1$ in the $2m$th column.

A *singly-even square* is one where n is of the form $2(2m+1)$. If the square is divided into four equal quarters, the first quarter will contain $2m+1$ columns and $2m+1$ rows. In each of these columns take m cells so arranged that there are also m cells in each row: as, for instance, by taking the first m cells in the first column, the cells 2 to $m+1$ in the second column, and so on, the cells $m+2$ to $2m+1$ in the $(m+2)$th column, the cells $m+3$ to $2m+1$ and the first cell in the $(m+3)$th column, and so on, and finally the $(2m+1)$th cell and the cells 1 to $m-1$ in the $(2m+1)$th column. Next change the numbers in these $m(2m+1)$ cells and the $3m(2m+1)$ cells associated with them by cross interchanges. The result is equivalent to $2m$ interchanges in every row and in every column. In order to make the square magic we must have $\frac{1}{2}n$, that is, $2m+1$ such interchanges in every row and in every column, that is, we must have one more interchange in every row and in every column. This presents no difficulty, for instance, in the arrangement indicated above the numbers in the $(2m+1)$th cell of the first column, in the first cell of the second column, in the second cell of the third column, and so on, to the $2m$th cell in the $(2m+1)$th column may be interchanged with the numbers in their vertically related cells; this will make all the rows magic. Next, the numbers in the $2m$th cell of the first column, in the $(2m+1)$th cell of the second column, in the first cell of the third column, in the second cell of the fourth column, and so on, to the $(2m-1)$th cell of the $(2m+1)$th column may be interchanged with those in the cells horizontally related to them; and this will make the columns magic without affecting the magical properties of the rows.

It will be observed that we have implicitly assumed that m is not zero, that is, that n is greater than 2; also it would seem that, if $m = 1$ and therefore $n = 6$, then the numbers in the diagonal cells must be included in those to which the cross interchange is applied, but, if $n > 6$, this is not necessary, though it may be convenient.

The construction of odd magic squares and of doubly-even magic squares is very easy. But though the rule given above for singly-even squares is not difficult, it is tedious of application. It is unfortunate that no more obvious rule—such, for instance, as one for bordering a doubly-even square—can be suggested for writing down instantly and without thought singly-even magic squares.

De la Hire's Method[*]. I now proceed to give another way due to De la Hire, of constructing any even magic square of an order higher than two.

In the same manner as in his rule for making odd magic squares, we begin by constructing two subsidiary squares, one of the unit-digits, $1, 2, 3, \ldots, n$, and the other of the radix-digits $0, n, 2n, \ldots, (n-1)n$. We then form the magic square by adding together the numbers in the corresponding cells in the two subsidiary squares. Following the analogy of the notation used above, two numbers which are equidistant from the ends of the series $1, 2, 3, \ldots, n$ are said to be *complementary*. Similarly numbers which are equidistant from the ends of the series $0, n, 2n, \ldots, (n-1)n$ are said to be *complementary*.

For simplicity I shall apply this method to construct a magic square of only the sixth order, though an exactly similar method will apply to any even square of an order higher than the second.

The first of the subsidiary squares (figure ix) is constructed as follows. First, the cells in the leading diagonal are filled with the numbers 1, 2, 3, 4, 5, 6 placed in any order whatever that puts complementary numbers in complementary positions (*ex. gr.* in the order 2, 6, 3, 4, 1, 5, or in their natural order 1, 2, 3, 4, 5, 6). Second, the cells vertically related to these are filled respectively with the same numbers.

---

[*] The rule is due to De la Hire (part 2 of his memoir) and is given by Montucla in his edition of Ozanam's work: I have used the modified enunciation of it inserted in Labosne's edition of Bachet's *Problèmes*, as it saves the introduction of a third subsidiary square. I do not know to whom the modification is due.

1	5	4	3	2	6
6	2	4	3	5	1
6	5	3	4	2	1
1	5	3	4	2	6
6	2	3	4	5	1
1	2	4	3	5	6

First Subsidiary Square
Figure ix.

0	30	30	0	30	0
24	6	24	24	6	6
18	18	12	12	12	18
12	12	18	18	18	12
6	24	6	6	24	24
30	0	0	30	0	30

Second Subsidiary Square
Figure x.

1	35	34	3	32	6
30	8	28	27	11	7
24	23	15	16	14	19
13	17	21	22	20	18
12	26	9	10	29	25
31	2	4	33	5	36

Resulting Magic Square
Figure xi.

Third, each of the remaining cells in the first vertical column is filled either with the same number as that already in two of them or with the complementary number (*ex. gr.* in figure ix with a "1" or a "6") in any way, provided that there are an equal number of each of these numbers in the column, and subject also to the provisoes mentioned in the next paragraph but one. Fourth, the cells horizontally related to those in the first column are filled with the complementary numbers. Fifth, the remaining cells in the second and third columns are filled in an analogous way to that in which those in the first column were filled: and then the cells horizontally related to them are filled with the complementary numbers. The square so formed is necessarily magic in rows, columns, and diagonals.

The second of the subsidiary squares (figure x) is constructed as follows. First, the cells in the leading diagonal are filled with the numbers 0, 6, 12, 18, 24, 30 placed in any order whatever that puts complementary numbers in complementary positions. Second, the cells horizontally related to them are filled respectively with the same numbers. Third, each of the remaining cells in the first horizontal row is filled either with the same number as that already in two of them or with the complementary number (*ex. gr.* in figure x with a "0" or a "30") in any way, provided (i) that there are an equal number of each of these numbers in the row, and (ii) that if any cell in the first row of figure ix and its vertically related cell are filled with complementary numbers, then the corresponding cell in the first row of figure x and its horizontally related cell must be occupied by the same number[*]. Fourth, the cells vertically related to those in the first row are filled with the

---

[*] The insertion of this step evades the necessity of constructing (as Montucla did) a third subsidiary square.

complementary numbers. Fifth, the remaining cells in the second and the third rows are filled in an analogous way to that in which those in the first row were filled: and then the cells vertically related to them are filled with the complementary numbers. The square so formed is necessarily magic in rows, columns, and diagonals.

It remains to show that proviso (ii) in the third step described in the last paragraph can be satisfied always. In a doubly even square, that is, one in which n is divisible by 4, we need not have any complementary numbers in vertically related cells in the first subsidiary square unless we please, but even if we like to insert them they will not interfere with the satisfaction of this proviso. In the case of a singly even square, that is, one in which n is divisible by 2, but not by 4, we cannot satisfy the proviso if any horizontal row in the first square has all its vertically related squares, other than the two squares in the diagonals, filled with complementary numbers. Thus in the case of a singly even square it will be necessary in constructing the first square to take care in the third step that in every row at least one cell which is not in a diagonal shall have its vertically related cell filled with the same number as itself: this is always possible if n is greater than 2.

The required magic square will be constructed if in each cell we place the sum of the numbers in the corresponding cells of the subsidiary squares, figures ix and x. The result of this is given in figure xi. The square is evidently magic. Also every number from 1 to 36 occurs once and only once, for the numbers from 1 to 6 and from 31 to 36 can occur only in the top or the bottom rows, and the method of construction ensures that the same number cannot occur twice. Similarly the numbers from 7 to 12 and from 25 to 30 occupy two other rows, and no number can occur twice; and so on. The square in figure i on page 121 may be constructed by the above rules; and the reader will have no difficulty in applying them to any other even square.

OTHER METHODS FOR CONSTRUCTING ANY MAGIC SQUARE. The above methods appear to me to be the simplest which have been proposed. There are however *two other methods*, of less generality, to which I will allude briefly in passing. Both depend on the principle that, if every number in a magic square is multiplied by some constant, and a constant is added to the product, the square will remain magic.

The *first method* applies only to such squares as can be divided into smaller magic squares of some order higher than two. It depends on the fact that, if we know how to construct magic squares of the mth and nth orders, we can construct one of the mnth order. For example, a square of 81 cells may be considered as composed of 9 smaller squares each containing 9 cells, and by filling the cells in each of these small squares in the same relative order and taking the small squares themselves in the same order, the square can be constructed easily. Such squares are called *composite magic squares*.

The *second method*, which was introduced by Frénicle, consists in surrounding a magic square with a *border*. Thus in figure xii on the following page the inner square is magic, and it is surrounded with a border in such a way that the whole square is also magic. In this manner from the magic square of the 3rd order we can build up successively squares of the orders 5, 7, 9, &c., that is, any odd magic square. Similarly from the magic square of the 4th order we can build up successively any higher even magic square.

If we construct a magic square of the first n^2 numbers by bordering a magic square of $(n-2)^2$ numbers, the usual process is to reserve for the $4(n-1)$ numbers in the border the first $2(n-1)$ natural numbers and the last $2(n-1)$ numbers. Now the sum of the numbers in each line of a square of the order $(n-2)$ is $\frac{1}{2}(n-2)\{(n-2)^2+1\}$, and the average is $\frac{1}{2}\{(n-2)^2+1\}$. Similarly the average number in a square of the nth order is $\frac{1}{2}(n^2+1)$. The difference of these is $2(n-1)$. We begin then by taking any magic square of the order $(n-2)$, and we add to every number in it $2(n-1)$; this makes the average number $\frac{1}{2}(n^2+1)$.

The numbers reserved for the border occur in pairs, n^2 and 1, n^2-1 and 2, n^2-2 and 3, &c., such that the average of each pair is $\frac{1}{2}(n^2+1)$, and they must be bordered on the square so that these numbers are opposite to one another. Thus the bordered square will be necessarily magic, provided that the sum of the numbers in two adjacent sides of the external border is correct. The arrangement of the numbers in the borders will be somewhat facilitated if the number n^2+1-p (which has to be placed opposite to the number p) is denoted by $\bar{p}$, but it is not worth while going into further details here.

It will illustrate sufficiently the general method if I explain how the square in figure xii is constructed. A magic square of the third order is formed by De la Loubère's rule, and to every number in it 8 is added: the result is the inner square in figure xii. The numbers not used are 25 and 1, 24 and 2, 23 and 3, 22 and 4, 21 and 5, 20 and 6, 19 and

7, 18 and 8. The sum of each pair is 26, and obviously they must be placed at opposite ends of any line.

I believe that with a little patience a magic square of any order can be thus built up, and of course it will have the property that, if each border is successively stripped off, the square will still remain magic. Some examples are given by Violle. This is the method of construction commonly adopted by self-taught mathematicians, many

1	2	19	20	23
18	16	9	14	8
21	11	13	15	5
22	12	17	10	4
3	24	7	6	25

Bordered Magic Square.
Figure xii.

7	20	3	11	24
13	21	9	17	5
19	2	15	23	6
25	8	16	4	12
1	14	22	10	18

Nasik Magic Square.
Figure xiii.

of whom seem to think that the empirical formation of such squares is a valuable discovery.

There are magic circles, rectangles, crosses, diamonds, stars, and other figures: also magic cubes, cylinders, and spheres. The theory of the construction of such figures is of no value, and I cannot spare the space to describe rules for forming them.

HYPER-MAGIC SQUARES. In recent times attention has been mainly concentrated on the formation of magic squares with the imposition of additional conditions; some of the resulting problems involve mathematical difficulties of a high order.

Nasik Squares. In one species of hyper-magic squares the squares are formed so that the sums of the numbers along all the rows and columns, both diagonals, and all the broken diagonals are the same. In England these are called *nasik squares* or *pan-diagonal magic squares*: in France *carrés diaboliques* or *carrés magiquement magiques*. These squares were mentioned by De la Hire, Sauveur, and Euler; but the theory is mainly due to Mr A.H. Frost, who has expounded it in the memoirs mentioned in the footnote on page 123, and to M. Frolow, who treated it in two memoirs, St Petersburg, 1884, and Paris, 1886. Of course a nasik square can be divided by a vertical or horizontal cut and the pieces interchanged without affecting the magical property. By one vertical and one horizontal transposition of this kind any number can be moved to any specified cell.

A nasik square of the fourth order is represented in figure ii on page 121, and one of the fifth order is represented in figure xiii on the preceding page. Nasik squares of the order $6n\pm1$ can be constructed by rules analogous to those given by De la Loubère, except that a knight's and not a bishop's move must be used in connecting cells filled by consecutive numbers and that for orders higher than five special rules for going from the cell occupied by the number kn to that occupied by the number $kn+1$ have to be laid down.

Doubly-Magic Squares. In another species of hyper-magic squares the problem is to construct a magic square of the nth order in such a way that if the number in each cell is replaced by its mth power the

5	31	35	60	57	34	8	30
19	9	53	46	47	56	18	12
16	22	42	39	52	61	27	1
63	37	25	24	3	14	44	50
26	4	64	49	38	43	13	23
41	51	15	2	21	28	62	40
54	48	20	11	10	17	55	45
36	58	6	29	32	7	33	59

A Doubly-Magic Square.
Figure xiv.

resulting square shall also be magic. Here for example (see figure xiv) is a magic square[*] of the eighth order, the sum of the numbers in each line being equal to 260, so constructed that if the number in each cell is replaced by its square the resulting square is also magic (the sum of the numbers in each line being equal to 11180).

MAGIC PENCILS. Hitherto I have concerned myself with numbers arranged in lines. By reciprocating the figures composed of the

---

[*] See M. Coccoz in *L'Illustration*, May 29, 1897.

points on which the numbers are placed we obtain a collection of lines forming pencils, and, if these lines be numbered to correspond with the points, the pencils will be magic*. Thus, in a magic square of the nth order, we arrange n^2 consecutive numbers to form $2n + 2$ lines, each containing n numbers so that the sum of the numbers in each line is the same. Reciprocally we can arrange n^2 lines, numbered consecutively to form $2n + 2$ pencils, each containing n lines, so that in each pencil the sum of the numbers designating the lines is the same.

For instance, figure xv represents a magic square of 64 consecutive

1	2	62	61	60	59	7	8
9	10	54	53	52	51	15	16
48	47	19	20	21	22	42	41
40	39	27	28	29	30	34	33
32	31	35	36	37	38	26	25
24	23	43	44	45	46	18	17
49	50	14	13	12	11	55	56
57	58	6	5	4	3	63	64

Figure xv.

numbers arranged to form 18 lines, each of 8 numbers. Reciprocally, figure xvi represents 64 lines arranged to form 18 pencils, each of 8 lines. The method of construction is fairly obvious. The eight-rayed pencil, vertex O, is cut by two parallels perpendicular to the axis of the pencil, and all the points of intersection are joined cross-wise. This gives 8 pencils, with vertices $A, B, \ldots, H$; 8 pencils, with vertices $A', \ldots H'$; one pencil with its vertex at O; and one pencil with its vertex on the axis of the last-named pencil.

The sum of the numbers in each of the 18 lines in figure xv is the same. To make figure xvi correspond to this we must number the lines in the pencil A from left to right, $1, 9, \ldots, 57$, following the order of the numbers in the first column of the square: the lines in pencil B must be numbered similarly to correspond to the numbers in the second column of the square, and so on. To prevent confusion in the figure I

---

* See *Magic-Reciprocals* by G. Frankenstein, Cincinnati, 1875.

CH. V] MAGIC PENCILS. 139

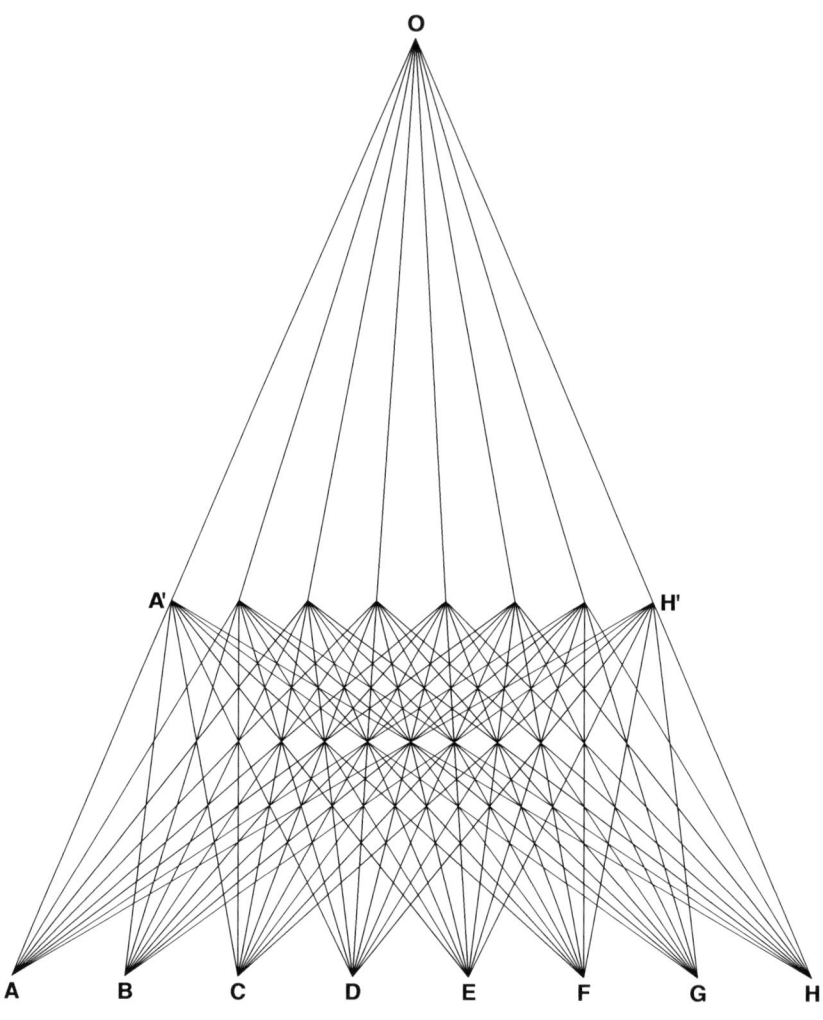

Figure xvi.

have not inserted the numbers, but it will be seen that the method of construction ensures that the sum of the 8 numbers which designate the lines in each of these 18 pencils is the same.

We can proceed a step further, if the resulting figure is cut by two other parallel lines perpendicular to the axis, and if the points of their intersection with the cross-joins be joined cross-wise, these new cross-joins will intersect on the axis of the original pencil or on lines perpendicular to it. The whole figure will now give 8^3 lines, arranged in 244 pencils each of 8 rays, and will be the reciprocal of a magic cube of the 8th order. If we reciprocate back again we obtain a representation in a plane of a magic cube.

MAGIC SQUARE PUZZLES. Many empirical problems, closely related to magic squares, will suggest themselves; but most of them are more correctly described as ingenious puzzles than as mathematical recreations. The following will serve as specimens.

Magic Card Square*. The first of these is the familiar problem of placing the sixteen court cards (taken out of a pack) in the form of a square so that no row, no column, and neither of the diagonals shall contain more than one card of each suit and one card of each rank. The solution presents no difficulty, and is indicated in figure xviii on the next page.

Euler's Officers Problem†. A similar problem, proposed by Euler in 1779, consists in arranging, if it be possible, thirty-six officers taken from six regiments—the officers being in six groups, each consisting of six officers of equal rank, one drawn from each regiment; say officers of rank a, b, c, d, e, f, drawn from the 1st, 2nd, 3rd, 4th, 5th, and 6th regiments—in a solid square formation of six by six, so that each row and each file shall contain one and only one officer of each rank and one and only one officer from each regiment. The problem is insoluble.

Extension of Euler's Problem. More generally we may investigate the arrangement on a chess-board, containing n^2 cells, of n^2 counters (the counters being divided into n groups, each group consisting of n counters of the same colour numbered consecutively $1, 2, \ldots, n$) so that each row and each column shall contain no two counters of the same colour or marked with the same number.

For instance, if $n = 3$, with three red counters a_1, a_2, a_3, three white counters b_1, b_2, b_3, and three black counters c_1, c_2, c_3, we can satisfy the conditions by arranging them as in figure xvii on the facing page. If $n = 4$, then with counters a_1, a_2, a_3, a_4; b_1, b_2, b_3, b_4; c_1, c_2,

---

* Ozanam, 1723 edition, vol. IV, p. 434.

† Euler's *Commentationes Arithmeticae*, St Petersburg, 1849, vol II, pp. 302–361. See also a paper by G. Tarry in the *Comptes rendus* of the French Association for the Advancement of Science, Paris, 1900, vol. II, pp. 170–203; and various notes in *L'Intermédiaire des mathématiciens*, Paris, vol. III, 1896, pp. 17, 90; vol. V, 1898, pp. 83, 176, 252, vol. VI, 1899, p. 251; vol. VII, 1900, pp. 14, 311.

c_3, c_4; d_1, d_2, d_3, d_4, we can arrange them as in figure xviii below. A solution when $n = 5$ is indicated in figure xix.

a_1	b_2	c_3
b_2	c_1	a_2
c_3	a_3	b_1

Figure xvii.

a_1	b_2	c_3	d_4
c_4	d_3	a_2	b_1
d_2	c_1	b_4	a_3
b_3	a_4	d_1	c_2

Figure xviii.

a_1	b_2	c_3	d_4	e_5
b_5	c_1	d_2	e_3	a_4
c_4	d_5	e_1	a_2	b_3
d_3	e_4	a_5	b_1	c_2
e_2	a_3	b_4	c_5	d_1

Figure xix.

The problem is soluble if n is odd or if n is of the form $4m$. If solutions when $n = a$ and when $n = b$ are known, a solution when $n = ab$ can be written down at once. The theory is closely connected with that of magic squares and need not be here discussed further.

Magic Domino Squares. Analogous problems can be made with dominoes. An ordinary set of dominoes, ranging from double zero to double six, contains 28 dominoes. Each domino is a rectangle formed by fixing two small square blocks together side by side: of these 56 blocks, eight are blank, on each of eight of them is one pip, on each of another eight of them are two pips, and so on. It is required to arrange the dominoes so that the 56 blocks form a square of 7 by 7 bordered by one line of 7 blank squares and so that the sum of the pips in each row, each column, and in the two diagonals of the square is equal to 24. A solution[*] is given on the following page.

Similarly, a set of dominoes, ranging from double zero to double n, contains $\frac{1}{2}(n+1)(n+2)$ dominoes and therefore $(n+1)(n+2)$ blocks. Can these dominoes be arranged in the form of a square of $(n+1)^2$ cells, bordered by a row of blanks, so that the sum of the pips in each row, each column, and in the two diagonals of the square is equal to $\frac{1}{2}n(n+2)$?

Magic Coin Squares[†]*.* There are somewhat similar questions concerned with coins. Here is one applicable to a square of the third order divided into nine cells, as in figure xvii above. If a five-shilling piece is placed in the middle cell c_1 and a florin in the cell below it, namely, in a_3 it is required to place the fewest possible current English coins in

---

[*] See *L'Illustration*, July 10, 1897.
[†] See *The Strand Magazine*, London, December, 1896, pp. 720, 721.

142   MAGIC SQUARES.   [CH. V

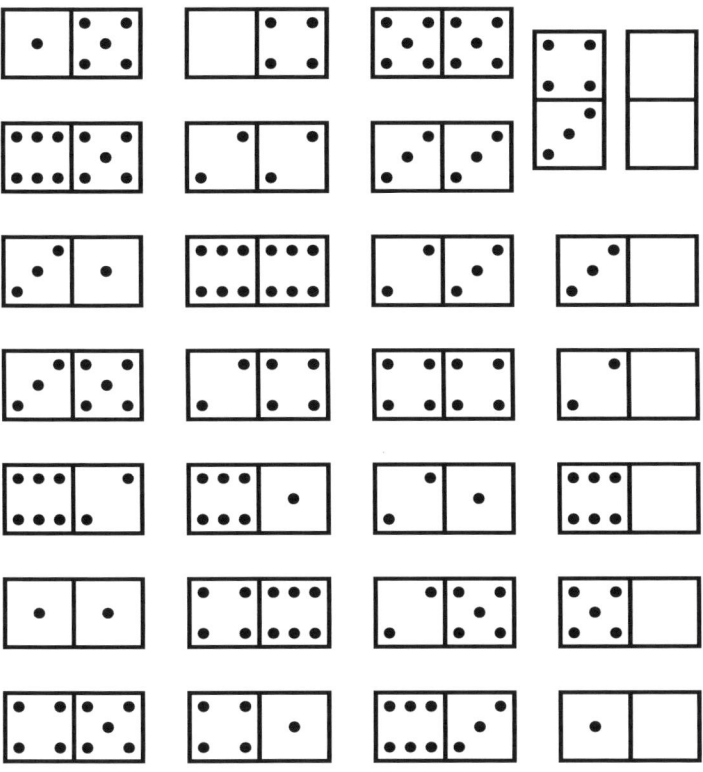

Magic Domino Square.

the remaining seven cells so that in each cell there is at least one coin, so that the total value of the coins in every cell is different, and so that the sum of the values of the coins in each row, column, and diagonal is fifteen shillings: it will be found that thirteen additional coins will suffice. A similar problem is to place ten current English postage stamps, all but two being different, in the nine cells so that the sum of the values of the stamps in each row, column, and diagonal is ninepence.

CHAPTER VI.

UNICURSAL PROBLEMS.

I propose to consider in this chapter some problems which arise out of the theory of unicursal curves. I shall commence with *Euler's Problem and Theorems*, and shall apply the results briefly to the theories of *Mazes* and *Geometrical Trees*. The reciprocal unicursal problems of the *Hamilton Game* and the *Knight's Path on a Chess-board* will be discussed in the latter half of the chapter.

Euler's Problem.   Euler's problem has its origin in a memoir[*] presented by him in 1736 to the St Petersburg Academy, in which he solved a question then under discussion as to whether it was possible to take a walk in the town of Königsberg in such a way as to cross every bridge in it once and only once.

The town is built near the mouth of the river Pregel, which there takes the form indicated on the following page and includes the island of Kneiphof. In 1759 there were (and according to Baedeker there are still) seven bridges in the positions shown in the diagram, and it is easily seen that with such an arrangement the problem is insoluble. Euler however did not confine himself to the case of Königsberg, but discussed the general problem of any number of islands connected in any way by bridges. It is evident that the question will not be affected if

---

[*] *Solutio problematis ad Geometriam situs pertinentis*, *Commentarii Academiae Scientiarum Petropolitanae* for 1736, St Petersburg, 1741, vol. VIII, pp. 128–140. This has been translated into French by M. Ch. Henry; see Lucas, vol. I, part 2, pp. 21–33.

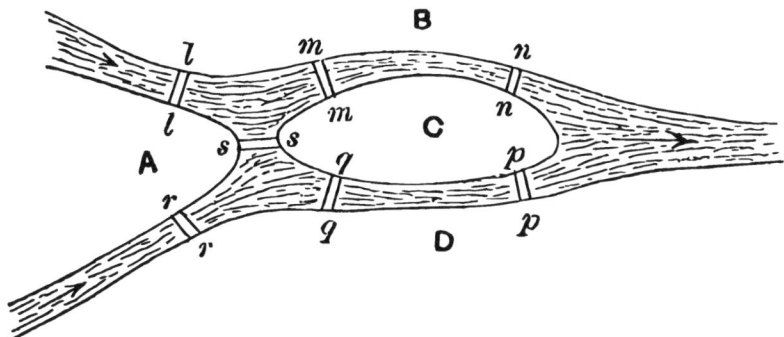

we suppose the islands to diminish to points and the bridges to lengthen out. In this way we ultimately obtain a geometrical figure or network. In the Königsberg problem this figure is of the shape indicated below, the areas being represented by the points A, B, C, D, and the bridges being represented by the lines l, m, n, p, q, r, s.

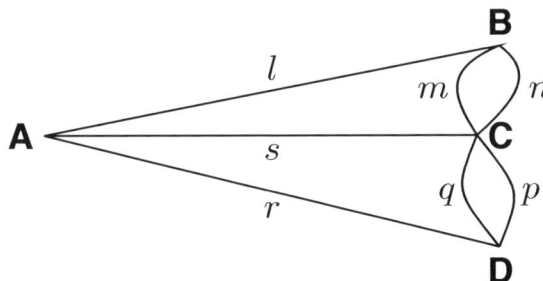

Euler's problem consists therefore in finding whether a given geometrical figure can be described by a point moving so as to traverse every line in it once and only once. A more general question is to determine how many strokes are necessary to describe such a figure so that no line is traversed twice: this is covered by the rules hereafter given. The figure may be either in three or in two dimensions, and it may be represented by lines, straight, curved, or tortuous, joining a number of given points, or a model may be constructed by taking a number of rods or pieces of string furnished at each end with a hook so as to allow of any number of them being connected together at one point.

The theory of such figures is included as a particular case in the propositions proved by Listing in his *Topologie*[*]. I shall, however, adopt here the methods of Euler, and I shall begin by giving some definitions, as it will enable me to put the argument in a more concise form.

A *node* (or isle) is a point to or from which lines are drawn. A *branch* (or bridge or path) is a line connecting two consecutive nodes. An *end* (or hook) is the point at each termination of a branch. The *order* of a node is the number of branches which meet at it. A node to which only one branch is drawn is a *free* node or a free end. A node at which an even number of branches meet is an *even* node: evidently the presence of a node of the second order is immaterial. A node at which an odd number of branches meet is an *odd* node. A figure is closed if it has no free end: such a figure is often called a closed network.

A *route* consists of a number of branches taken in consecutive order and so that no branch is traversed twice. A closed route terminates at the point from which it started. A figure is described *unicursally* when the whole of it is traversed in one route.

The following are Euler's results. (i) In a closed network the number of odd nodes is even. (ii) A figure which has no odd node can be described unicursally, in a re-entrant route, by a moving point which starts from any point on it. (iii) A figure which has two and only two odd notes can be described unicursally by a moving point which starts from one of the odd nodes and finishes at the other. (iv) A figure which has more than two odd nodes cannot be described completely in one route; to which Listing added the corollary that a figure which has $2n$ odd nodes, and no more, can be described completely in n separate routes. I now proceed to prove these theorems.

First. *The number of odd nodes in a closed network is even.*

Suppose the number of branches to be b. Therefore the number of hooks is $2b$. Let k_n be the number of nodes of the nth order. Since a node of the nth order is one at which n branches meet, there are n hooks there. Also since the figure is closed, n cannot be less than 2.

$$\therefore 2k_2 + 3k_3 + 4k_4 + \cdots + nk_n + \cdots = 2b.$$
Hence $\quad 3k_3 + 5k_5 + \cdots \quad$ is even.
$$\therefore k_3 + k_5 + \cdots \quad \text{is even.}$$

---

[*] *Die Studien*, Göttingen, 1847, part x. See also Tait on *Listing's Topologie*, *Philosophical Magazine*, London, January, 1884, series 5, vol. XVII, pp. 30–46.

Second. A figure which has no odd node can be described unicursally in a re-entrant route.

Since the route is to be re-entrant it will make no difference where it commences. Suppose that we start from a node A. Every time our route takes us through a node we use up one hook in entering it and one in leaving it. There are no odd nodes, therefore the number of hooks at every node is even: hence, if we reach any node except A, we shall always find a hook which will take us into a branch previously untraversed. Hence the route will take us finally to the node A from which we started. If there are more than two hooks at A, we can continue the route over one of the branches from A previously untraversed, but in the same way as before we shall finally come back to A.

It remains to show that we can arrange our route so as to make it cover all the branches. Suppose each branch of the network to be represented by a string with a hook at each end, and that at each node all the hooks there are fastened together. The number of hooks at each node is even, and if they are unfastened they can be re-coupled together in pairs, the arrangement of the pairs being immaterial. The whole network will then form one or more closed curves, since now each node consists merely of two ends hooked together.

If this random coupling gives us one single curve then the proposition is proved; for starting at any point we shall go along every branch and come back to the initial point. But if this random coupling produces anywhere an isolated loop, L, then where it touches some other loop, M, say at the node P, unfasten the four hooks there (viz. two of the loop L and two of the loop M) and re-couple them in any other order: then the loop L will become a part of the loop M. In this way, by altering the couplings, we can transform gradually all the separate loops into parts of only one loop.

For example, take the case of three isles, A, B, C, each connected with both the others by two bridges. The most unfavourable way of re-coupling the ends at A, B, C would be to make ABA, ACA, and BCB separate loops. The loops ABA and ACA are separate and touch at A; hence we should re-couple the hooks at A so as to combine ABA and ACA into one loop $ABACA$. Similarly, by re-arranging the couplings of the four hooks at B, we can combine the loop BCB with $ABACA$ and thus make only one loop.

I infer from Euler's language that he had attempted to solve the

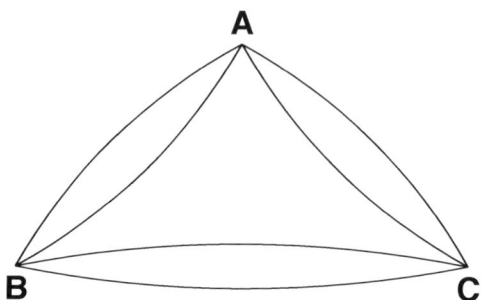

problem of giving a practical rule which would enable one to describe such a figure unicursally without knowledge of its form, but that in this he was unsuccessful. He however added that any geometrical figure can be described completely in a single route provided each part of it is described twice and only twice, for, if we suppose that every branch is duplicated, there will be no odd nodes and the figure is unicursal. In this case any figure can be described completely without knowing its form: rules to effect this are given below.

Third. A figure which has two and only two odd nodes can be described unicursally by a point which starts from one of the odd nodes and finishes at the other odd node.

This at once reduces to the second theorem. Let A and Z be the two odd nodes. First, suppose that Z is not a free end. We can, of course, take a route from A to Z; if we imagine the branches in this route to be eliminated, it will remove one hook from A and make it even, will remove two hooks from every node intermediate between A and Z and therefore leave each of them even, and will remove one hook from Z and therefore will make it even. All the remaining network is now even: hence, by Euler's second proposition, it can be described unicursally, and, if the route begins at Z, it will end at Z. Hence, if these two routes are taken in succession, the whole figure will be described unicursally, beginning at A and ending at Z. Second, if Z is a free end, then we must travel from Z to some node, Y, at which more than two branches meet. Then a route from A to Y which covers the whole figure exclusive of the path from Y to Z can be determined as before and must be finished by travelling from Y to Z.

Fourth. A figure having $2n$ odd nodes, and no more, can be described completely in n separate routes.

If any route starts at an odd node, and if it is continued until it reaches a node where no fresh path is open to it, this latter node must be an odd one. For every time we enter an even node there is necessarily a way out of it; and similarly every time we go through an odd node we use up one hook in entering and one hook in leaving, but whenever we reach it as the end of our route we use only one hook. If this route is suppressed there will remain a figure with $2n - 2$ odd nodes. Hence n such routes will leave one or more networks with only even nodes. But each of these must have some node common to one of the routes already taken and therefore can be described as a part of that route. Hence the complete passage will require n and not more than n routes. It follows, as stated by Euler, that, if there are more than two odd nodes, the figure cannot be traversed completely in one route.

The Königsberg bridges lead to a network with four odd nodes; hence, by Euler's fourth proposition, it cannot be described unicursally in a single journey, though it can be traversed completely in two separate routes.

The first and second diagrams figured below contain only even nodes, and therefore each of them can be described unicursally. The

 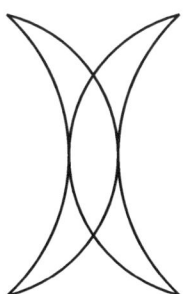 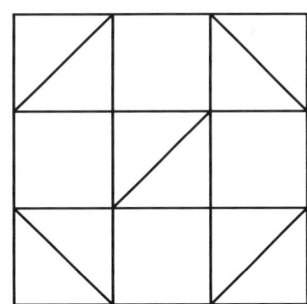

first of these—a re-entrant pentagon—was one of the Pythagorean symbols. The other is the so-called sign-manual of Mohammed, said to have been originally traced in the sand by the point of his scimetar without taking the scimetar off the ground or retracing any part of the figure—which, as it contains only even nodes, is possible. The third diagram is taken from Tait's article: it contains only two odd nodes, and therefore can be described unicursally if we start from one of them and finish at the other.

As other examples I may note that the geometrical figure formed by taking a $(2n + 1)$gon and joining every angular point with every other angular point is unicursal. On the other hand a chess-board, divided as usual by straight lines into 64 cells, has 28 odd nodes and 53 even nodes: hence it would require 14 separate pen-strokes to trace out all the boundaries without going over any more than once. Again, the diagram on page 102 has 20 odd nodes and therefore would require 10 separate pen-strokes to trace it out.

It is well known that a curve which has as many nodes as is consistent with its degree is unicursal.

Mazes. Everyone has read of the labyrinth of Minos in Crete and of Rosamund's Bower. A few modern mazes exist here and there—notably one, which is a very poor specimen of its kind, at Hampton Court—and in one of these, or at any rate on a drawing of one, most of us have threaded our way to the interior. I proceed now to consider the manner in which any such construction may be completely traversed even by one who is ignorant of its plan.

The theory of the description of mazes is included in Euler's theorems given above. The paths in the maze are what previously we have termed branches, and the places where two or more paths meet are nodes. The entrance to the maze, the end of a blind alley, and the centre of the maze are free ends and therefore odd nodes.

If the only odd nodes are the entrance to the maze and the centre of it–which will necessitate the absence of all blind alleys–the maze can be described unicursally. This follows from Euler's third proposition. Again, no matter how many odd nodes there may be in a maze, we can always find a route which will take us from the entrance to the centre without retracing our steps, though such a route will take us through only a part of the maze. But in neither of the cases mentioned in this paragraph can the route be determined without a plan of the maze.

A plan is not necessary, however, if we make use of Euler's suggestion, and suppose that every path in the maze is duplicated. In this case we can give definite rules for the complete description of the whole of any maze, even if we are entirely ignorant of its plan. Of course to walk twice over every path in a labyrinth is not the shortest way of arriving at the centre, but, if it is performed correctly, the whole maze is traversed, the arrival at the centre at some point in the course of the route is certain, and it is impossible to lose one's way.

I need hardly explain why the complete description of such a duplicated maze is possible, for now every node is even, and hence, by Euler's second proposition, if we begin at the entrance we can traverse the whole maze; in so doing we shall at some point arrive at the centre, and finally shall emerge at the point from which we started. This description will require us to go over every path in the maze twice, and as a matter of fact the two passages along any path will be always made in opposite directions.

If a maze is traced on paper, the way to the centre is generally obvious, but in an actual labyrinth it is not so easy to find the correct route unless the plan is known. In order to make sure of describing a maze without knowing its plan it is necessary to have some means of marking the paths which we traverse and the direction in which we have traversed them—for example, by drawing an arrow at the entrance and end of every path traversed, or better perhaps by marking the wall on the right-hand side, in which case a path may not be entered when there is a mark on each side of it. If we can do this, and if when a node is reached, we take, if it be possible, some path not previously used, or, if no other path is available, we enter on a path already traversed once only, we shall completely traverse any maze in two dimensions[*]. Of course a path must not be traversed twice in the same direction, a path already traversed twice (namely, once in each direction) must not be entered, and at the end of a blind alley it is necessary to turn back along the path by which it was reached.

I think most people would understand by a maze a series of interlacing paths through which some route can be obtained leading to a space or building at the centre of the maze. I believe that few, if any, mazes of this type existed in classical or medieval times.

One class of what the ancients called mazes or labyrinths seems to have comprised any complicated building with numerous vaults and passages[†]. Such a building might be termed a labyrinth, but it is not

---

[*] See *Le problème des labyrinthes* by G. Tarry, *Nouvelles Annales de mathématiques*, May, 1895, series 3, vol. XIV.

[†] For instance, see the descriptions of the labyrinth at Lake Moeris given by Herodotus, bk. ii, c. 148; Strabo, bk. xvii, c. 1, art. 37; Diodorus, bk. i, cc. 61, 66; and Pliny, *Hist. Nat.*, bk. xxxvi, c. 13, arts. 84–89. On these and other references see A. Wiedemann, *Herodots zweites Buch*, Leipzig, 1890, p. 522 *et seq.* See also Virgil, *Aeneid*, bk. v, c. v, 588; Ovid, *Met.*, bk. viii, c. 5, 159; Strabo, bk. viii, c. 6.

what is usually understood by the word. The above rules would enable anyone to traverse the whole of any structure of this kind. I do not know if there are any accounts or descriptions of Rosamund's Bower other than those by Drayton, Bromton, and Knyghton: in the opinion of some, these imply that the bower was merely a house, the passages in which were confusing and ill-arranged.

Another class of ancient mazes consisted of a tortuous path confined to a small area of ground and leading to a place or shrine in the centre*. This is a maze in which there is no chance of taking a wrong turning; but, as the whole area can be occupied by the windings of one path, the distance to be traversed from the entrance to the centre may be considerable, even though the piece of ground covered by the maze is but small.

The traditional form of the labyrinth constructed for the Minotaur is a specimen of this class. It was delineated on the reverses of the coins of Cnossus, specimens of which are not uncommon; one form of it is indicated in the accompanying diagram (figure i). The design really is the same as that drawn in figure ii, as can be easily seen by bending round a circle the rectangular figure there given.

Mr Inwards has suggested[†] that this design on the coins of Cnossus may be a survival from that on a token given by the priests as a clue to

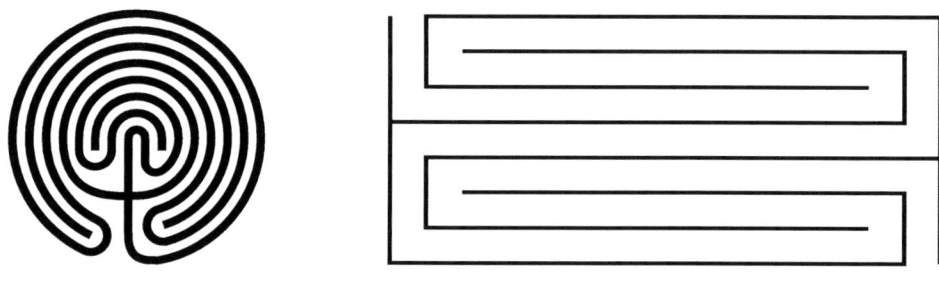

Figure i.   *Figure* ii.

the right path in the labyrinth there. Taking the circular form of the design shown above he supposed each circular wall to be replaced by two equidistant walls separated by a path, and thus obtained a maze

---

* On ancient and medieval labyrinths—particularly of this kind—see an article by Mr E. Trollope in *The Archaeological Journal*, 1858, vol. XV, pp. 216–235, from which much of the historical information given above is derived

† *Knowledge*, London, October, 1892.

to which the original design would serve as the key. The route thus indicated may be at once obtained by noticing that when a node is reached (*i.e.* a point where there is a choice of paths) the path to be taken is that which is next but one to that by which the node was approached. This maze may be also threaded by the simple rule of always following the wall on the right-hand side or always that on the left-hand side. The labyrinth may be somewhat improved by erecting a few additional barriers, without affecting the applicability of the above rules, but it cannot be made really difficult. This makes a pretty toy, but though the conjecture on which it is founded is ingenious it must be regarded as exceedingly improbable. Another suggestion is that the curved line on the reverse of the coins indicated the form of the rope held by those taking part in some rhythmic dance; while others consider that the form was gradually evolved from the widely prevalent svastika.

Copies of the maze of Cnossus were frequently engraved on Greek and Roman gems; similar but more elaborate designs are found in numerous Roman mosaic pavements[*]. A copy of the Cretan labyrinth was embroidered on many of the state robes of the later Emperors, and, apparently thence, was copied on to the walls and floors of various churches[†]. At a later time in Italy and in France these mural and pavement decorations were developed into scrolls of great complexity, but consisting, as far as I know, always of a single line. Some of the best specimens now extant are on the walls of the cathedrals at Lucca, Aix in Provence, and Poitiers; and on the floors of the churches of Santa Maria in Trastevere at Rome, San Vitale at Ravenna, Notre Dame at St Omer, and the cathedral at Chartres. It is possible that they were used to represent the journey through life as a kind of pilgrim's progress.

In England these mazes were usually, perhaps always, cut in the turf adjacent to some religious house or hermitage: and there are some slight reasons for thinking that, when traversed as a religious exercise, a *pater* or *ave* had to be repeated at every turning. After the Renaissance, such labyrinths were frequently termed Troy-towns or Julian's bowers. Some of the best specimens, which are still extant, are those at Rockliff Marshes, Cumberland; Asenby, Yorkshire; Alkborough, Lincolnshire; Wing, Rutlandshire; Boughton-Green, Northamptonshire; Comberton, Cambridgeshire; Saffron Walden, Essex; and Chilcombe,

---

[*] See *ex. gr.* Breton's *Pompeia*, p. 303.
[†] Ozanam, *Graphia aureae urbis Romae*, pp. 92, 178.

near Winchester.

The modern maze seems to have been introduced—probably from Italy—during the Renaissance, and many of the palaces and large houses built in England during the Tudor and the Stuart periods had labyrinths attached to them. Those adjoining the royal palaces at Southwark, Greenwich, and Hampton Court were particularly well known from their vicinity to the capital. The last of these was designed by London and Wise in 1690, for William III, who had a fancy for such conceits: a plan of it is given in various guide-books. For the majority

Maze at Hampton Court.

of the sight-seers who enter, it is sufficiently elaborate; but it is an indifferent construction, for it can be described completely by always following the hedge on one side (either the right hand or the left hand), and no node is of an order higher than three.

Unless at some point the route to the centre forks and subsequently the two forks reunite, forming a loop in which the centre of the maze is situated, the centre can be reached by the rule just given, namely, by following the wall on one side—either on the right hand or on the left hand. No labyrinth is worthy of the name of a puzzle which can be threaded in this way. Assuming that the path forks as described above, the more numerous the nodes and the higher their order the more difficult will be the maze, and the difficulty might be increased considerably by using bridges and tunnels so as to construct a labyrinth in three dimensions. In an ordinary garden and on a small piece of ground, often of an inconvenient shape, it is not easy to make a maze which fulfils these conditions. Here on the following page is a plan of one which I put up in my own garden on a plot of ground which would

not allow of more than 36 by 23 paths, but it will be noticed that none of the nodes are of a high order.

GEOMETRICAL TREES. Euler's original investigations were confined to a closed network. In the problem of the maze it was assumed that there might be any number of blind alleys in it, the ends of which formed free nodes. We may now progress one step farther, and suppose that the network or closed part of the figure diminishes to a point. This last arrangement is known as a *tree*. The number of unicursal descriptions necessary to completely describe a tree is called the *base* of the ramification.

We can illustrate the possible form of these trees by rods, having a hook at each end. Starting with one such rod, we can attach at either end one or more similar rods. Again, on any free hook we can attach one or more similar rods, and so on. Every free hook, and also every point where two or more rods meet, are what hitherto we have called nodes. The rods are what hitherto we have termed branches or paths.

The theory of trees—which already plays a somewhat important part in certain branches of modern analysis, and possibly may contain the key to certain chemical and biological theories—originated in a memoir by Cayley[*], written in 1856. The discussion of the theory has

---

[*] *Philosophical Magazine*, March, 1857, series 4, vol. XIII, pp. 172–176; or *Collected*

been analytical rather than geometrical. I content myself with noting the following results.

The number of trees with n given nodes is n^{n-2}. If A_n is the number of trees with n branches, and B_n the number of trees with n free branches which are bifurcations at least, then

$$(1-x)^{-1}(1-x^2)^{-A_1}(1-x^3)^{-A_2}\cdots = 1 + A_1 x + A_2 x^2 + A_3 x^3 + \cdots,$$
$$(1-x)^{-1}(1-x^2)^{-B_2}(1-x^3)^{-B_3}\cdots = 1 + x + 2B_2 x^2 + 2B_3 x^3 + \cdots.$$

Using these formulae we can find successively the values of $A_1, A_2, \ldots$, and $B_1, B_2, \ldots$. The values of A_n when $n = 2, 3, 4, 5, 6, 7$, are 2, 4, 9, 20, 48, 115; and of B_n are 1, 2, 5, 12, 33, 90.

I turn next to consider some problems where it is desired to find a route which will pass once and only once through each node of a given geometrical figure. This is the reciprocal of the problem treated in the first part of this chapter, and is a far more difficult question. I am not aware that the general theory has been considered by mathematicians, though two special cases—namely, the *Hamiltonian* (or Icosian) *Game* and the *Knight's Path on a Chess-Board*—have been treated in some detail; and I confine myself to a discussion of these.

THE HAMILTONIAN GAME. The Hamiltonian Game consists in the determination of a route along the edges of a regular dodecahedron which will pass once and only once through every angular point. Sir William Hamilton[*], who invented this game—if game is the right term for it—denoted the twenty angular points on the solid by letters which stand for various towns. The thirty edges constitute the only possible

---

Works, Cambridge, 1890, vol. III, no. 203, pp. 242–346: see also the paper on double partitions, *Philosophical Magazine*, November, 1860, series 4, vol. XX, pp. 337–341. On the number of trees with a given number of nodes, see the *Quarterly Journal of Mathematics*, London, 1889, vol. XXIII, pp. 376–378. The connection with chemistry was first pointed out in Cayley's paper on isomers, *Philosophical Magazine*, June, 1874, series 4, vol. XLVII, pp. 444–447, and was treated more fully in his report on trees to the British Association in 1875, *Reports*, pp. 257–305.

[*] See *Quarterly Journal of Mathematics*, London, 1862, vol. V, p. 305; or *Philosophical Magazine*, January, 1884, series 5, vol. XVII, p. 42; also Lucas, vol. II, part vii.

paths. The inconvenience of using a solid is considerable, and the dodecahedron may be represented conveniently in perspective by a flat board marked as shown in the first of the annexed diagrams. The second and third diagrams will answer our purpose equally well and are easier to draw.

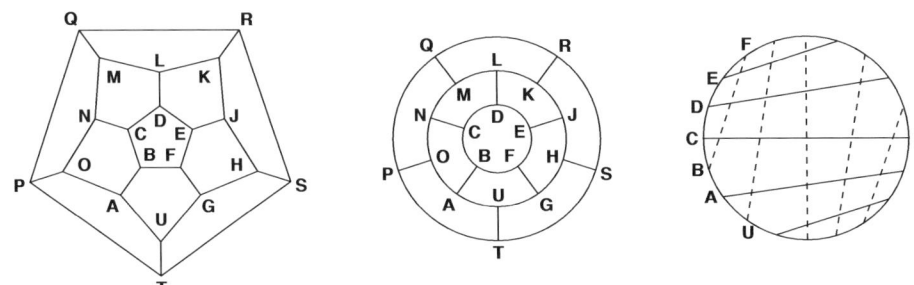

The first problem is go "all round the world," that is, starting from any town, to go to every other town once and only once and to return to the initial town; the order of the n towns to be first visited being assigned, where n is not greater than five.

Hamilton's rule for effecting this was given at the meeting in 1857 of the British Association at Dublin. At each angular point there are three and only three edges. Hence, if we approach a point by one edge, the only routes open to us are one to the right, denoted by r, and one to the left, denoted by l. It will be found that the operations indicated on opposite sides of the following equalities are equivalent,

$$lr^2l = rlr, \quad rl^2r = lrl, \quad lr^3l = r^2, \quad rl^3r = l^2.$$

Also the operation l^5 or r^5 brings us back to the initial point: we may represent this by the equations

$$l^5 = 1, \quad r^5 = 1.$$

To solve the problem for a figure having twenty angular points we must deduce a relation involving twenty successive operations, the total effect of which is equal to unity. By repeated use of the relation

$l^2 = rl^3r$ we see that

$$1 = l^5 = l^2 l^3 = (rl^3 r)l^3 = \{rl^3\}^2 = \{r(rl^3 r)l\}^2$$
$$= \{r^2 l^3 rl\}^2 = \{r^2(rl^3 r)lrl\}^2 = \{r^3 l^3 rlrl\}^2.$$

Therefore $\quad\quad\quad \{r^3 l^3 (rl)^2\}^2 = 1 \quad\quad\quad$ .... (i),

and similarly $\quad\quad \{l^3 r^3 (lr)^2\}^2 = 1 \quad\quad\quad$ .... (ii).

Hence on a dodecahedron either of the operations

$$r\ r\ r\ l\ l\ l\ r\ l\ r\ l\ r\ r\ r\ l\ l\ l\ r\ l\ r\ l \quad\quad \text{....... (i)},$$
$$l\ l\ l\ r\ r\ r\ l\ r\ l\ r\ l\ l\ l\ r\ r\ r\ l\ r\ l\ r \quad\quad \text{....... (ii)},$$

indicates a route which takes the traveller through every town, The arrangement is cyclical, and the route can be commenced at any point in the series of operations by transferring the proper number of letters from one end to the other. The point at which we begin is determined by the order of certain towns which is given initially.

Thus, suppose that we are told that we start from F and then successively go to B, A, U, and T, and we want to find a route from T through all the remaining towns which will end at F. If we think of ourselves as coming into F from G, the path FB would be indicated by l, but if we think of ourselves as coming into F from E, the path FB would be indicated by r. The path from B to A is indicated by l, and so on. Hence our first paths are indicated either by $l\ l\ l\ r$ or by $r\ l\ l\ r$. The latter operation does not occur either in (i) or in (ii), and therefore does not fall within our solutions. The former operation may be regarded either as the 1st, 2nd, 3rd, and 4th steps of (ii), or as the 4th, 5th, 6th, and 7th steps of (i). Each of these leads to a route which satisfies the problem. These routes are

$$F\ B\ A\ U\ T\ P\ O\ N\ C\ D\ E\ J\ K\ L\ M\ Q\ R\ S\ H\ G\ F,$$
and $\quad F\ B\ A\ U\ T\ S\ R\ K\ L\ M\ Q\ P\ O\ N\ C\ D\ E\ J\ H\ G\ F.$

It is convenient to make a mark or to put down a counter at each corner as soon as it is reached, and this will prevent our passing through the same town twice.

A similar game may be played with other solids provided that at each angular point three and only three edges meet. Of such solids a tetrahedron and a cube are the simplest instances, but the reader can make for himself any number of plane figures representing such solids similar to those drawn on page 156. Some of these were indicated by Hamilton. In all such cases we must obtain from the formulae analogous to those given above cyclical relations like (i) or (ii) there given. The solution will then follow the lines indicated above. This method may be used to form a rule for describing any maze in which no node is of an order higher than three.

For solids having angular points where more than three edges meet—such as the octahedron where at each angular point four edges meet, or the icosahedron where at each angular point five edges meet—we should at each point have more than two routes open to us; hence (unless we suppress some of the edges) the symbolical notation would have to be extended before it could be applied to these solids. I offer the suggestion to anyone who is desirous of inventing a new game.

Another and a very elegant solution of the Hamiltonian dodecahedron problem has been given by M. Hermary. It consists in unfolding the dodecahedron into its twelve pentagons, each of which is attached to the preceding one by only one of its sides; but the solution is geometrical, and not directly applicable to more complicated solids.

Hamilton suggested as another problem to start from any town, to go to certain specified towns in an assigned order, then to go to every other town once and only once, and to end the journey at some given town. He also suggested the consideration of the way in which a certain number of towns should be blocked so that there was no passage through them, in order to produce certain effects. These problems have not, so far as I know, been subjected to mathematical analysis.

KNIGHT'S PATH ON A CHESS-BOARD. Another geometrical problem on which a great deal of ingenuity has been expended, and of a kind somewhat similar to the Hamiltonian game, consists in moving a knight on a chess-board in such a manner that it shall move successively on to every cell[*] once and only once. The literature on this subject is so extensive[†] that I make no pretence to give a full account of the various methods for solving the problem, and I shall content myself

---

[*] The 64 small squares into which a chess-board is divided are termed *cells*.

[†] For a bibliography see A. van der Linde, *Geschichte und Literatur des Schachspiels*, Berlin, 1874, vol. II, pp. 101–111. On the problem see a memoir by P. Volpicelli in *Atti della Reale Accademia dei Lincei*, Rome, 1872, vol. XXV, pp. 87–162: also *Applications de l'Analyse Mathématique au Jeu des échecs*,

by putting together a few notes on some of the solutions I have come across, particularly on those due to De Moivre, Euler, Vandermonde, Warnsdorff, and Roget.

On a board containing an even number of cells the path may or may not be re-entrant, but on a board containing an odd number of cells it cannot be re-entrant. For, if a knight begins on a white cell, its first move must take it to a black cell, the next to a white cell, and so on. Hence, if its path passes through all the cells, then on a board of an odd number of cells the last move must leave it on a cell of the same colour as that on which it started, and therefore these cells cannot be connected by one move.

The earliest solutions of which I have any knowledge are those given about the end of the seventeenth century by De Montmort and De Moivre[*]. They apply to the ordinary chess-board of 64 cells, and depend on dividing (mentally) the board into an inner square containing sixteen cells surrounded by an outer ring of cells two deep. If initially the knight is placed on a cell in the outer ring, it moves round that ring always in the same direction so as to fill it up completely—only going into the inner square when absolutely necessary. When the outer ring is filled up the order of the moves required for filling the remaining cells presents but little difficulty. If initially the knight is placed on the inner square the process must be reversed. The method can be applied to square and rectangular boards of all sizes. It is illustrated sufficiently by De Moivre's solution which is given on the following page, where the numbers indicate the order in which the cells are occupied successively. I place by its side a somewhat similar re-entrant solution, due to Euler, for a board of 36 cells. If a chess-board is used it is convenient to place a counter on each cell as the knight leaves it.

The next serious attempt to deal with the subject was made by Euler[†] in 1759: it was due to a suggestion made by L. Bertrand of Geneva, who subsequently (in 1778) issued an account of it. This method is applicable to boards of any shape and size, but in general the solutions

---

    by C.F. de Jaenisch, 3 vols., St Petersburg, 1862–3; and General Parmentier, *Association Française pour l'avancement des Sciences*, 1891, 1892, 1894.

[*] I do not know where they were published originally; they were quoted by Ozanam and Montucla, see Ozanam, 1803 edition, vol. I, p. 178; 1840 edition, p. 80.

[†] *Mémoires de Berlin* for 1759, Berlin, 1766, pp. 310–337; or *Commentationes Arithmeticae Collectae*, St Petersburg, 1849, vol. I, pp. 337–355.

UNICURSAL PROBLEMS.

34	49	22	11	36	39	24	1
21	10	35	50	23	12	37	40
48	33	62	57	38	25	2	13
9	20	51	54	63	60	41	26
32	47	58	61	56	53	14	3
19	8	55	52	59	64	27	42
46	31	6	17	44	29	4	15
7	18	45	30	5	16	43	28

De Moivre's Solution.

30	21	6	15	28	19
7	16	29	20	5	14
22	31	8	35	18	27
9	36	17	26	13	4
32	23	2	11	34	25
1	10	33	24	3	12

Euler's Thirty-six Cell Solution.

to which it leads are not symmetrical and their mutual connexion is not apparent.

Euler commenced by moving the knight at random over the board until it has no move open to it. With care this will leave only a few cells not traversed: denote them by $a, b, \ldots$. His method consists in establishing certain rules by which these vacant cells can be interpolated into various parts of the circuit, and also by which the circuit can be made re-entrant.

The following example, mentioned by Legendre as one of exceptional difficulty, illustrates the method. Suppose that we have formed

55	58	29	40	27	44	19	22
60	39	56	43	30	21	26	45
57	54	59	28	41	18	23	20
38	51	42	31	8	25	46	17
53	32	37	a	47	16	9	24
50	3	52	33	36	7	12	15
1	34	5	48	b	14	c	10
4	49	2	35	6	11	d	13

Figure i.

22	25	50	39	52	35	60	57
27	40	23	36	49	58	53	34
24	21	26	51	38	61	56	59
41	28	37	48	3	54	33	62
20	47	42	13	32	63	4	55
29	16	19	46	43	2	7	10
18	45	14	31	12	9	64	5
15	30	17	44	1	6	11	8

Figure ii.

Example of Euler's Method.

the route given in figure i; namely, 1, 2, 3, ..., 59, 60; and that there are four cells left untraversed, namely, a, b, c, d.

We begin by making the path 1 to 60 re-entrant. The cell 1 commands a cell p, where p is 32, 52, or 2. The cell 60 commands a cell q, where q is 29, 59, or 51. Then, if any of these values of p and q differ by unity, we can make the route re-entrant. This is the case here if $p = 52$, $q = 51$. Thus the cells $1, 2, 3, \ldots, 51$; $60, 59, \ldots, 52$ form a re-entrant route of 60 moves. Hence, if we replace the numbers $60, 59, \ldots, 52$ by $52, 53, \ldots, 60$, the steps will be numbered consecutively. I recommend the reader who wishes to follow the subsequent details of Euler's argument to construct this square on a piece of paper before proceeding further.

Next, we add the cells a, b, d to this route. In the new diagram of 60 cells formed as above the cell a commands the cells there numbered 51, 53, 41, 25, 7, 5, and 3. It is indifferent which of these we select: suppose we take 51. Then we must make 51 the last cell of the route of 60 cells, so that we can continue with a, b, d. Hence, if the reader will add 9 to every number on the diagram he has constructed, and then replace $61, 62, \ldots, 69$ by $1, 2, \ldots, 9$, he will have a route which starts from the cell occupied originally by 60, the 60th move is on to the cell occupied originally by 51, and the 61st, 62nd, 63rd moves will be on the cells a, b, d respectively.

It remains to introduce the cell c. Since c commands the cell now numbered 25, and 63 commands the cell now numbered 24, this can be effected in the same way as the first route was made re-entrant. In fact the cells numbered $1, 2, \ldots, 24$; $63, 62, \ldots, 25, c$ form a knight's path. Hence we must replace $63, 62, \ldots, 25$ by the numbers $25, 26, \ldots, 63$, and then we can fill up c with 64. We have now a route which covers the whole board.

Lastly, it remains to make this route re-entrant. First, we must get the cells 1 and 64 near one another. This can be effected thus. Take one of the cells commanded by 1, such as 28, then 28 commands 1 and 27. Hence the cells $64, 63, \ldots, 28$; $1, 2, \ldots, 27$ form a route; and this will be represented in the diagram if we replace the cells numbered $1, 2, \ldots, 27$ by $27, 26, \ldots, 1$.

The cell now occupied by 1 commands the cells 26, 38, 54, 12, 14, 16, 28; and the cell occupied by 64 commands the cells 13, 43, 63, 55. The cells 13 and 14 are consecutive, and therefore the cells 64, 63,

..., 14; 1, 2, ..., 13 form a route. Hence we must replace the numbers 1, 2, ..., 13 by 13, 12, ..., 1, and we obtain a re-entrant route covering the whole board, which is represented in the second of the diagrams given on page 160. Euler showed how seven other re-entrant routes can be deduced from any given re-entrant route.

It is not difficult to apply the method so as to form a route which begins on one given cell and ends on any other given cell.

Euler next investigated how his method could be modified so as to allow of the imposition of additional restrictions.

An interesting example of this kind is where the first 32 moves are confined to one half of the board. One solution of this is delineated below. The order of the first 32 moves can be determined by Euler's

58	43	60	37	52	41	62	35
49	46	57	42	61	36	53	40
44	59	48	51	38	55	34	63
47	50	45	56	33	64	39	54
22	7	32	1	24	13	18	15
31	2	23	6	19	16	27	12
8	21	4	29	10	25	14	17
3	30	9	20	5	28	11	26

Euler's Half-board Solution.

50	45	62	41	60	39	54	35
63	42	51	48	53	36	57	38
46	49	44	61	40	59	34	55
43	64	47	52	33	56	37	58
26	5	24	1	20	15	32	11
23	2	27	8	29	12	17	14
6	25	4	21	16	19	10	31
3	22	7	28	9	30	13	18

Roget's Half-board Solution.

method. It is obvious that, if to the number of each such move we add 32, we shall have a corresponding set of moves from 33 to 64 which would cover the other half of the board; but in general the cell numbered 33 will not be a knight's move from that numbered 32, nor will 64 be a knight's move from 1.

Euler however proceeded to show how the first 32 moves might be determined so that, if the half of the board containing the corresponding moves from 33 to 64 was twisted through two right angles, the two routes would become united and re-entrant. If x and y are the numbers of a cell reckoned from two consecutive sides of the board, we may call the cell whose distances are respectively x and y from the opposite sides a complementary cell. Thus the cells (x, y) and $(9 - x, 9 - y)$ are complementary, where x and y denote respectively the column and row occupied by the cell. Then in Euler's solution the numbers in complementary cells differ by 32: for instance, the cell $(3,7)$ is complementary to the cell $(6,2)$, the one is occupied by 57, the other by 25.

Roget's method, which is described later, can be also applied to give half-board solutions. The result is indicated on the facing page. The close of Euler's memoir is devoted to showing how the method could be applied to crosses and other rectangular figures. I may note in particular his elegant re-entrant symmetrical solution for a square of 100 cells.

The next attempt of any special interest is due to Vandermonde[*], who reduced the problem to arithmetic. His idea was to cover the board by two or more independent routes taken at random, and then to connect the routes. He defined the position of a cell by a fraction x/y, whose numerator x is the number of the cell from one side of the board, and whose denominator y is its number from the adjacent side of the board; this is equivalent to saying that x and y are the co-ordinates of a cell. In a series of fractions denoting a knight's path, the differences between the numerators of two consecutive fractions can be only one or two, while the corresponding difference between their denominators must be two or one respectively. Also x and y cannot be less than 1 or greater than 8. The notation is convenient, but Vandermonde applied it merely to obtain a particular solution of the problem for a board of 64 cells: the method by which he effected this is analogous to that established by Euler, but it is applicable only to squares of an even order. The route that he arrives at is defined in his notation by the following fractions.

$$\frac{5}{5}, \frac{4}{3}, \frac{2}{4}, \frac{4}{5}, \frac{5}{3}, \frac{7}{4}, \frac{8}{2}, \frac{6}{1}, \frac{7}{3}, \frac{8}{1}, \frac{6}{2}, \frac{8}{3}, \frac{7}{1}, \frac{5}{2}, \frac{6}{4}, \frac{8}{5}, \frac{7}{7}, \frac{5}{8}, \frac{6}{6}, \frac{5}{4}, \frac{4}{6}, \frac{2}{5}, \frac{1}{7}, \frac{3}{8}, \frac{2}{6},$$

$$\frac{1}{8}, \frac{3}{7}, \frac{1}{6}, \frac{2}{8}, \frac{4}{7}, \frac{3}{5}, \frac{1}{4}, \frac{2}{2}, \frac{4}{1}, \frac{3}{3}, \frac{1}{2}, \frac{3}{1}, \frac{2}{3}, \frac{1}{1}, \frac{3}{2}, \frac{1}{4}, \frac{2}{2}, \frac{4}{3}, \frac{3}{1}, \frac{1}{2}, \frac{2}{4}, \frac{4}{5}, \frac{3}{7}, \frac{1}{8}, \frac{2}{6}, \frac{4}{4}, \frac{3}{6}, \frac{4}{5}, \frac{5}{6},$$

$$\frac{7}{5}, \frac{8}{7}, \frac{6}{8}, \frac{7}{6}, \frac{8}{8}, \frac{6}{7}, \frac{8}{6}, \frac{7}{8}, \frac{5}{7}, \frac{6}{5}, \frac{8}{4}, \frac{7}{2}, \frac{5}{1}, \frac{6}{3}.$$

The path is re-entrant but unsymmetrical. Had he transferred the first three fractions to the end of this series he would have obtained two symmetrical circuits of thirty-two moves joined unsymmetrically, and might have been enabled to advance further in the problem. Vandermonde also considered the case of a route in a cube.

---

[*] *L'Histoire de l'Académie des Sciences* for 1771, Paris, 1774, pp. 566-574.

In 1773 Collini[*] proposed the exclusive use of symmetrical routes arranged without reference to the initial cell, but connected in such a manner as to permit of our starting from it. This is the foundation of the modern manner of attacking the problem. The method was re-invented in 1825 by Pratt[†], and in 1840 by Roget, and has been subsequently employed by various writers. Neither Collini nor Pratt showed skill in using this method. The rule given by Roget is described later.

One of the most ingenious of the solutions of the knight's path is that given in 1823 by Warnsdorff[‡]. His rule is that the knight must be moved always to one of the cells from which it will command the fewest squares not already traversed. The solution is not symmetrical and not re-entrant; moreover it is difficult to trace practically. The rule has not been proved to be true, but no exception to it is known: apparently it applies also to all rectangular boards which can be covered completely by a knight. It is somewhat curious that in most cases a single false step, except in the last three or four moves, will not affect the result.

Warnsdorff added that when, by the rule, two or more cells are open to the knight, it may be moved to either or any of them indifferently. This is not so, and with great ingenuity two or three cases of failure have been constructed, but it would require exceptionally bad luck to happen accidentally on such a route.

The above methods have been applied to boards of various shapes, especially to boards in the form of rectangles, crosses, and circles[§].

All the more recent investigations impose additional restrictions: such as to require that the route shall be re-entrant, or more generally that it shall begin and terminate on given cells.

The best complete solution with which I am acquainted—and one which I believe is not generally known—is due to Roget[‖]. It divides the whole route into four circuits, which can be combined so as to enable us to begin on any cell and terminate on any other cell of a

---

[*] *Solution du Problème du Cavalier au Jeu des échecs*, Mannheim, 1773.
[†] *Studies of Chess*, sixth edition, London, 1825.
[‡] *Des Rösselsprunges einfachste und allgemeinste Lösung*, Schmalkalden, 1823: see Jaenisch, vol. II, pp. 56–61, 273–289.
[§] See *ex. gr.* T. Ciccolini's work *Del Cavallo degli Scacchi*, Paris, 1836.
[‖] *Philosophical Magazine*, April, 1840, series 3, vol. XVI, pp. 305–309; see also the *Quarterly Journal of Mathematics* for 1877, vol. XIV, pp. 354–359. Some solutions, founded on Roget's method, are given in the *Leisure Hour*, Sept. 13, 1873, pp. 587–590; see also *Ibid.*, Dec. 20, 1873, pp. 813–815.

different colour. Hence, if we like to select this last cell at a knight's move from the initial cell, we obtain a re-entrant route. On the other hand, the rule is applicable only to square boards containing $(4n)^2$ cells: for example, it could not be used on the board of the French *jeu des dames*, which contains 100 cells.

Roget began by dividing the board of 64 cells into four quarters. Each quarter contains 16 cells, and these 16 cells can be arranged in 4 groups, each group consisting of 4 cells which form a closed knight's path. All the cells in each such path are denoted by the same letter l, e, a, or p, as the case may be. The path of 4 cells indicated by the consonants l and the path indicated by the consonants p are diamond-shaped: the paths indicated respectively by the vowels e and a are square-shaped, as may be seen by looking at one of the four quarters in figure i below.

l	e	a	p	l	e	a	p
a	p	l	e	a	p	l	e
e	l	p	a	e	l	p	a
p	a	e	l	p	a	e	l
l	e	a	p	l	e	a	p
a	p	l	e	a	p	l	e
e	l	p	a	e	l	p	a
p	a	e	l	p	a	e	l

34	51	32	15	38	53	18	3
31	14	35	52	17	2	39	54
50	33	16	29	56	37	4	19
13	30	49	36	1	20	55	40
48	63	28	9	44	57	22	5
27	12	45	64	21	8	41	58
62	47	10	25	60	43	6	23
11	26	61	46	7	24	59	42

Roget's Solution (i).   *Roget's Solution* (ii).

Now all the 16 cells on a complete chess-board which are marked with the same letter can be combined into one circuit, and wherever the circuit begins we can make it end on any other cell in the circuit, provided it is of a different colour to the initial cell. If it is indifferent on what cell the circuit terminates we may make the circuit re-entrant, and in this case we can make the direction of motion round each group (of 4 cells) the same. For example, all the cells marked p can be arranged in the circuit indicated by the successive numbers 1 to 16 in figure ii above. Similarly all the cells marked a can be combined into the circuit indicated by the numbers 17 to 23; all the l cells into the circuit 33 to 48; and all the e cells into the circuit 49 to 64. Each of the circuits indicated above is symmetrical and re-entrant. The consonant and the vowel circuits are said to be of opposite kinds.

The general problem will be solved if we can combine the four circuits into a route which will start from any given cell, and terminate on the 64th move on any other given cell of a different colour. To effect this Roget gave the two following rules.

First. If the initial cell and the final cell are denoted the one by a consonant and the other by a vowel, take alternately circuits indicated by consonants and vowels, beginning with the circuit of 16 cells indicated by the letter of the initial cell and concluding with the circuit indicated by the letter of the final cell.

Second. If the initial cell and the final cell are denoted both by consonants or both by vowels, first select a cell, Y, in the same circuit as the final cell, Z, and one move from it, next select a cell, X, belonging to one of the opposite circuits and one move from Y. This is always possible. Then, leaving out the cells Z and Y, it always will be possible, by the rule already given, to travel from the initial cell to the cell X in 62 moves, and thence to move to the final cell on the 64th move.

In both cases however it must be noticed that the cells in each of the first three circuits will have to be taken in such an order that the circuit does not terminate on a corner, and it may be desirable also that it should not terminate on any of the border cells. This will necessitate some caution. As far as is consistent with these restrictions it is convenient to make these circuits re-entrant, and to take them and every group in them in the same direction of rotation.

As an example, suppose that we are to begin on the cell numbered 1 in figure ii on the previous page, which is one of those in a p circuit, and to terminate on the cell numbered 64, which is one of those in an e circuit. This falls under the first rule: hence first we take the 16 cells marked p, next the 16 cells marked a, then the 16 cells marked l, and lastly the 16 cells marked e. One way of effecting this is shown in the diagram. Since the cell 64 is a knight's move from the initial cell the route is re-entrant. Also each of the four circuits in the diagram is symmetrical, re-entrant, and taken in the same direction, and the only point where there is any apparent breach in the uniformity of the movement is in the passage from the cell numbered 32 to that numbered 33.

A rule for re-entrant routes, similar to that of Roget, has been given by various subsequent writers, especially by De Polignac[*] and by Laquière[†], who have stated it at much greater length. Neither of these authors seems to have been aware of Roget's theorems. De Polignac, like Roget, illustrates the rule by assigning letters to the various squares in the way explained above, and asserts that a similar rule is applicable to all even squares.

Roget's method can be also applied to two half-boards, as indicated in the figure given above on page 162.

Another way of dividing the board into closed circuits which can be connected was given in 1843 by Moon[‡]. He divided the board into a

a	b	c	d	a	b	c	d
c	d	a	b	c	d	a	b
b	a	A	B	C	D	d	c
d	c	C	D	A	B	b	a
a	b	B	A	D	C	c	d
c	d	D	C	B	A	a	b
b	a	d	c	b	a	d	c
d	c	b	a	d	c	b	a

63	22	15	40	1	42	59	18
14	39	64	21	60	17	2	43
37	62	23	16	41	4	19	58
24	13	38	61	20	57	44	3
11	36	25	52	29	46	5	56
26	51	12	33	8	55	30	45
35	10	49	28	53	32	47	6
50	27	34	9	48	7	54	31

Moon's Solution.   *Jaenisch's Solution.*

central square containing 4^2 cells and a surrounding annulus (see figure on this page). The annulus may be divided into four closed circuits, each containing 12 cells: these are marked respectively with the letters a, b, c, d. The central square may be divided similarly into four closed circuits, each containing 4 cells, denoted by the letters A, B, C, D. We can connect these routes as follows. If we are moving outwards from the central square to the annulus we can go from a cell A either to b or to c or to d (but not to a) and similarly for the other letters. If we are moving inwards from the annulus to the central square we must go from a to D, or d to A, or b to C, or c to B, as the case may be. Thus if the initial cell is on a, we might take either of the cycles

---

[*] *Comptes Rendus*, April, 1861; and *Bulletin de la Société Mathématique de France*, 1881, vol. IX, pp. 17–24.

[†] *Bulletin de la Société Mathématique de France*, 1880, vol. VIII, pp. 82–102, 132–158.

[‡] *Cambridge Mathematical Journal*, 1843, vol. III, pp. 233–236.

a D b C d A c B, or *a D c B d A b C*. By following these rules we always can connect the routes into one path, but in general it will not be re-entrant. It is convenient to take the cells in each circuit in one and the same direction, but a circuit in the outer annulus must not end in a corner cell, and to avoid this we may have to alter the direction in which a circuit is taken.

Moon's rule can be modified to cover the case of any doubly even square board, and the path can be made to begin and end on any two given squares, but I do not propose to go further into details.

The method which Jaenisch gives as the most fundamental is not very different from that of Roget. It leads to eight forms, similar to that in the diagram printed on the previous page, in which the sum of the numbers in every column and every row is 260; but although symmetrical it is not in my opinion so easy to reproduce as that given by Roget.

It is as yet impossible to say how many solutions of the problem exist. Legendre[*] mentioned the question, but Minding[†] was the earliest writer to attempt to answer it. More recent investigations have shown that on the one hand the number of possible routes is less[‡] than the number of combinations of 168 things taken 63 at a time, and on the other hand is greater than 31,054144—since this latter number is the number of re-entrant paths of a particular type[§].

There are many similar problems in which it is required to determine routes by which a piece moving according to certain laws (*ex. gr.* a chess-piece such as a king, knight, &c.) can travel from a given cell over a board so as to occupy successively all the cells, or certain specified cells, once and only once, and terminate its route in a given cell.

Euler's method can be applied to find routes of this kind: for instance, he applied it to find a re-entrant route by which a piece that moved two cells forward like a castle and then one cell like a bishop would occupy in succession all the black cells on the board. As another instance, a castle, placed on a chess-board of n^2 cells, can always be moved in such a manner that it shall move successively on to every cell once and only once; moreover, starting on any cell, its path can be made to terminate, if n be even, on any other cell of a different colour,

---

[*] *Théorie des Nombres*, Paris, 2nd edition, 1830, vol. II, p. 165.
[†] *Cambridge and Dublin Mathematical Journal*, 1852, vol. VII, pp. 147–156; and *Crelle's Journal*, 1853, vol. XLIV, pp. 73–82.
[‡] Jaenisch, vol. II, p. 268.
[§] *Bulletin de la Société Mathématique de France*, 1881, vol. IX, pp. 1–17.

and, if n be odd, on any other cell of the same colour[*]. But it will suffice to have discussed the classical problem of the determination of a knight's path on an ordinary chess-board, and I need not enter on the discussion of other similar problems.

---

[*] *L'Intermédiaire des mathématiciens*, Paris, July, 1901, p. 153.

PART II.

Miscellaneous Essays and Problems.

"No man of science should think it a waste of time to learn something of the history of his own subject; nor is the investigation of laborious methods now fallen into disuse, or of errors once commonly accepted the least valuable of mental disciplines."

"The most worthless book of a bygone day is a record worthy of preservation. Like a telescopic star, its obscurity may render it unavailable for most purposes; but it serves, in hands which know how to use it, to determine the places of more important bodies."

(DE MORGAN.)

CHAPTER VII.

THE MATHEMATICAL TRIPOS.

The Mathematical Tripos has played so prominent a part in the history of education at Cambridge and of mathematics in England, that a sketch of its development* may be interesting to general readers.

So far as mathematics is concerned the history of the University before Newton may be summed up very briefly. The University was founded towards the end of the twelfth century. Throughout the middle ages the studies were organised on lines similar to those at Paris and Oxford. To qualify for a degree it was necessary to perform various exercises, and especially to keep a number of *acts* or to oppose acts kept by other students. An act consisted in effect of a debate in Latin, thrown, at any rate in later times, into syllogistic form. It was commenced by one student, the *respondent*, stating some proposition, often propounded in the form of a thesis, which was attacked by one or more *opponents*, the discussion being controlled by a graduate. The teaching was largely in the hands of young graduates—every master of arts being compelled to reside and teach for at least one year—though no doubt Colleges and private hostels supplemented this instruction in the case of their own students.

---

* The following pages are mostly summarised from my *History of the Study of Mathematics at Cambridge*, Cambridge 1889. The subject is also treated in Whewell's *Liberal Education*, Cambridge, three parts, 1845, 1850, 1853; Wordsworth's *Scholae Academicae*, Cambridge, 1877; my own *Origin and History of the Mathematical Tripos*, Cambridge, 1880; and Dr Glaisher's Presidential Address to the London Mathematical Society, *Transactions*, vol XVIII, 1886, pp. 4–38.

The Reformation in England was mainly the work of Cambridge divines, and in the University the Renaissance was warmly welcomed. In spite of the disorder and confusion of the Tudor period, new studies and a system of professional instruction were introduced. Probably the science (as distinct from the art) of mathematics, save so far as involved in the quadrivium, was still an exotic study, but it was not wholly neglected. Tonstall, subsequently the most eminent English arithmetician of his time, migrated, perhaps about 1495, from Balliol College, Oxford, to King's Hall, Cambridge, and in 1530 the University appointed a mathematical lecturer in the person of Paynell of Pembroke Hall. Most of the subsequent English mathematicians of the Tudor period seem to have been educated at Cambridge; of these I may mention Record, who migrated, probably about 1535, from Oxford, Dee, Digges, Blundeville, Buckley, Billingsley, Hill, Bedwell, Hood, Richard and John Harvey, Edward Wright, Briggs, and Oughtred. The Elizabethan statutes restricted liberty of thought and action in many ways, but, in spite of the civil and religious disturbances of the early half of the 17th century the mathematical school continued to grow. Horrox, Seth Ward, Foster, Rooke, Gilbert Clerke, Pell, Wallis, Barrow, Dacres, and Morland may be cited as prominent Cambridge mathematicians of the time.

Newton's mathematical career dates from 1665; his reputation, abilities, and influence attracted general attention to the subject. He created a school of mathematics and mathematical physics, among the earliest members of which I note the names of Laughton, Samuel Clarke, Craig, Flamsteed, Whiston, Saunderson, Jurin, Taylor, Cotes, and Robert Smith. Since then Cambridge has been regarded, as in a special sense, the home of English mathematicians, and from 1706 onwards we have fairly complete accounts of the course of reading and work of mathematical students there.

Until less than a century ago the form of the method of qualifying for a degree remained substantially unaltered, but the subject-matter of the discussions varied from time to time with the prevalent studies of the place.

After the Renaissance some of the statutable exercises were "huddled," that is, were reduced to a mere form. To huddle an act, the proctor generally asked some question such as *Quid est nomen* to which the answer usually expected was *Nescio*. In these exercises considerable license was allowed, particularly if there were any play on the words involved. For example, T. Brasse, of Trinity, was accosted with the question, *Quid est aes?* to which he answered, *Nescio nisi finis examinationis.* It should be added that retorts such as these were only allowed in the pretence exercises, and a candidate who in the actual examination was asked to give a definition of happiness and replied an exemption from Payne—that being the name of the moderator then presiding—was plucked for want of discrimination in time and place. In earlier years even the farce of huddling seems to have been unnecessary, for it was said in 1675 that it was not uncommon for the proctors to take "cautions for the performance of the statutable exercises, and accept the forfeit of the money so deposited in lieu of their performance."

In medieval times acts had been usually kept on some scholastic question or on a proposition taken from the *Sentences*. About the end of the fifteenth century religious questions, such as the interpretation of Biblical texts, began to be introduced, some fifty or sixty years later the favourite subjects were drawn either from dogmatic theology or from philosophy. In the seventeenth century the questions were usually philosophical, but in the eighteenth century, under the influence of the Newtonian school, a large proportion of them were mathematical.

Further details about these exercises and specimens of acts kept in the 18th century are given in my *History of Mathematics at Cambridge*. Here I will only say that they provided an admirable training in the art of presenting an argument, and in dialectical skill in attack and defence. The mental strain in a contested act was severe. De Morgan, describing his act kept in 1826, wrote[*],

> I was badgered for two hours with arguments given and answered in Latin,—or what we call Latin—against Newton's first section, Lagrange's derived functions, and Locke on innate principles. And though I took off everything, and was pronounced by the moderator to have disputed *magno honore*, I never had such a strain of thought in my life. For the inferior opponents were made as sharp as their betters by their tutors, who kept lists of queer objections drawn from all quarters.

Had the language of the discussions been changed to English, as was repeatedly urged from 1774 onwards, these exercises might have been retained with advantage, but the barbarous Latin and the syllogistic form in which they were carried on prejudiced their retention.

About 1830 a custom grew up for the respondent and opponents to meet previously and arrange their arguments together. The discussions

---

[*] *Budget of Paradoxes*, by A. De Morgan, London, 1872, p. 305.

then became an elaborate farce, and were a mere public performance of what had been already rehearsed. Accordingly the moderators of 1839 took the responsibility of abandoning them. This action was singularly high-handed, since a report of May 30, 1838, had recommended that they should be continued, and there was no reason why they should not have been reformed and retained as a useful feature in the scheme of study.

On the result of the acts a list of those qualified to receive degrees was drawn up. This list was not arranged strictly in order of merit, because the proctors could insert names anywhere in it, but by the beginning of the 18th century this power had become restricted to the right reserved to the vice-chancellor, the senior regent, and each proctor to place in the list one candidate anywhere he liked—a right which continued to exist till 1828, though it was not exercised after 1797. Subject to the granting of these honorary degrees, this final list was arranged in order of merit into wranglers and senior optimes, junior optimes, and poll-men. The bachelors on receiving their degrees took seniority according to their order on this list. The title *wrangler* is derived from these contentious discussions; the title *optime* from the customary compliment given by the moderator to a successful disputant, *Domine...*, *optime disputasti*, or even *optime quidem disputasti*, and the title of *poll-man* from the description of this class as οἱ πολλοί.

The final exercises for the B.A. degree were never huddled, and until 1839 were carried out strictly. University officials were responsible for approving the subject-matter of these acts. Stupid men offered some irrefutable truism, but the ambitious student courted reputation by affirming some paradox. Probably all honour men kept acts, but poll-men were deemed to comply with the regulations by keeping opponencies. The proctors were responsible for presiding at these acts, or seeing that competent graduates did so. In and after 1649 two examiners were specially appointed for this purpose. In 1680[*] these examiners were appointed by the Senate with the title of moderator, and with the joint stipend of four shillings for everyone graduating as B.A. during their year of office. In 1688 the joint stipend of the moderators was fixed at £40 a year. The moderators, like the proctors, were nominated by the Colleges in rotation.

From the earliest times the proctors had the power of questioning

---

[*] See Grace of October 25, 1680.

a candidate at the end of a disputation, and probably all candidates for a degree attended the public schools on certain days to give an opportunity to the proctors, or any master that liked, to examine them[*], though the opportunity was not always used. Different candidates attended on different days. Probably such examinations were conducted in Latin. But soon after 1710[†] the moderators or proctors began the custom of summoning on one day in January all candidates whom they proposed to question. The examination was held in public, and from it the Senate-House Examination arose. The examination at this time did not last more than one day, and was, there can be no doubt, partly on philosophy and partly on mathematics. It is believed that it was always conducted in English, and it is likely that its rapid development was largely due to this.

This introduction of a regular oral examination seems to have been largely due to the fact that when, in 1710, George I gave the Ely library to the University, it was decided to assign for its reception the old Senate-House—now the Catalogue Room in the Library—and to build a new room for the meetings of the Senate. Pending the building of the new Senate-House the books were stored in the Schools. As the Schools were thus rendered unavailable for keeping acts, considerable difficulty was found in arranging for all the candidates to keep the full number of statutable exercises, and thus obtaining opportunities to compare them one with another: hence the introduction of a supplementary oral examination. The advantages of this examination as providing a ready means of testing the knowledge and abilities of the candidates were so patent that it was retained when the necessity for some system of the kind had passed away, and finally it became systematized into an organized test to which all questionists were subjected.

In 1731 the University raised the joint stipend of the moderators to £60 "in consideration of their additional trouble in the Lent Term." This would seem to indicate that the Senate-House Examination had then taken formal shape, and perhaps that a definite scheme for its conduct had become customary.

---

[*]   *Ex. gr.* see De la Pryme's account of his graduation in 1694, *Surtees Society*, vol. LIV, 1870, p. 32.

[†]   W. Reneu, in his letters of 1708–1710 describing the course for the B.A. degree, makes no mention of the Senate-House examination, and I think it is a reasonable inference that it had not then been established.

As long as the order of the list of those approved for degrees was settled on the result of impressions derived from acts kept by the different candidates at different times and on different subjects, it was impossible to arrange the men in strict order of merit, nor was much importance attached to the order. But, with the introduction of an examination of all the candidates on one day, much closer attention was paid to securing a strict order of merit, and more confidence was felt in the published order. It seems to have been consequent on this that in and after 1747 the final lists were freely circulated, and it was further arranged that the names of the honorary optimes should be indicated. In the lists given in the Calendars issued subsequent to 1799 these names are struck out. It is only in exceptional cases that we are acquainted with the true order for the earlier tripos lists, but in a few cases the relative positions of the candidates are known; for example, in 1680 Bentley came out as third though he was put down as sixth in the list of wranglers.

Of the detailed history of the examination until the middle of the eighteenth century we know nothing. From 1750 onwards, however, we have more definite accounts of it. At this time, it would seem that all the men from each College were taken together as a class, and questions passed down by the proctors or moderators till they were answered: but the examination remained entirely oral, and technically was regarded as subsidiary to the discussions which had been previously held in the schools. As each class contained men of very different abilities a custom grew up by which every candidate was liable to be taken aside to be questioned by any M.A. who wished to do so, and this was regarded as an important part of the examination. The subjects were mathematics and philosophy. The examination now continued for two days and a half. At the conclusion of the second day the moderators received the reports of those masters of arts who had voluntarily taken part in the examination, and provisionally settled the final list; while the last half-day was used in revising and re-arranging the order of merit.

Richard Cumberland has left an account of the tests to which he was subjected when he took his B.A. degree in 1751. Clearly the disputations still played an important part, and it is difficult to say what weight was attached to the subsequent Senate-House examination; his reference to it is only of a general character. After saying that he kept two acts and two opponencies he continues[*]:

> The last time I was called upon to keep an act in the schools I sent in three questions to the Moderator, which he withstood as being all mathematical, and required me to conform to the usage of proposing one metaphysical question in the place of that, which I should think fit to withdraw. This was ground I never liked to take, and I appealed against his requisition: the act was accordingly put by till the matter of right should be ascertained by the statutes of the university, and in the result of that enquiry it was given for me, and my question stood.... I yielded now to advice, and paid attention to my health, till we were cited to the senate house to be examined for our Bachelor's degree. It was hardly ever my lot during that examination to enjoy any respite. I seemed an object singled out as every man's mark, and was kept perpetually at the table under the process of question and answer.

It was found possible by means of the new examination to differentiate the better men more accurately than before; and accordingly, in 1753, the first class was subdivided into two, called respectively wranglers and senior optimes, a division which is still maintained.

The semi-official examination by M.A.s was regarded as the more important part of the test, and the most eminent residents in the University took part in it. Thus John Fenn, of Caius, 5th wrangler in 1761, writes[†]:

> On the following Monday, Tuesday, and Wednesday, we sat in the Senate-house for public examination; during this time I was officially examined by the Proctors and Moderators, and had the honor of being taken out for examination by Mr Abbot, the celebrated mathematical tutor of St John's College, by the eminent professor of mathematics Mr Waring, of Magdalene, and by Mr Jebb of Peterhouse, a man thoroughly versed in the academical studies.

This irregular examination by any master who chose to take part in it constantly gave rise to accusations of partiality.

In 1763 the traditional rules for the conduct of the examination took more definite shape. Henceforth the examiners used the disputations only as a means of classifying the men roughly. On the result of their "acts," and probably partly also of their general reputation, the candidates were divided into eight classes, each arranged in alphabetical order. The subsequent position of the men in the class was determined solely by the Senate-House examination. The first two classes

---

[*] *Memoirs of Richard Cumberland*, London, 1806, pp. 78, 79.
[†] Quoted by C. Wordsworth, *Scholae Academicae*, Cambridge, 1877, pp. 30–31.

comprised all who were expected to be wranglers, the next four classes included the other candidates for honours, and the last two classes consisted of poll-men only. Practically anyone placed in either of the first two classes was allowed, if he wished, to take an aegrotat senior optime, and thus escape all further examination: this was called gulphing it. All the men from one College were no longer taken together, but each class was examined separately and *vivâ voce*; and hence, since all the students comprised in each class were of about equal attainments, it was possible to make the examination more effective. Richard Watson, of Trinity, claimed that this change was made by him when acting as moderator in 1763. He says[*]:

> There was more room for partiality... then [*i.e.* in 1759] than there is now; and I attribute the change, in a great degree, to an alteration which I introduced the first year I was moderator [*i.e.* in 1763], and which has been persevered in ever since. At the time of taking their Bachelor of Arts' degree, the young men are examined in classes, and the classes are now formed according to the abilities shown by individuals in the schools. By this arrangement, persons of nearly equal merits are examined in the presence of each other, and flagrant acts of partiality cannot take place. Before I made this alteration, they were examined in classes, but the classes consisted of members of the same College, and the best and worst were often examined together.

It is probable that before the examination in the Senate-House began a candidate, if manifestly placed in too low a class, was allowed the privilege of challenging the class to which he was assigned. Perhaps this began as a matter of favour, and was only granted in exceptional cases, but a few years later it became a right which every candidate could exercise; and I think that it is partly to its development that the ultimate predominance of the tripos over the other exercises for the degree is due.

In the same year, 1763, it was decided that the relative position of the senior and second wranglers, namely, Paley, of Christ's, and Frere, of Caius, was to be decided by the Senate-House examination and not by the disputations. Henceforward distinction in the Senate-House examination was regarded as the most important honour open to undergraduates.

In 1768 Dr Smith, of Trinity College, founded prizes for mathematics and natural philosophy open to two commencing bachelors. The

---

[*] *Anecdotes of the Life of Richard Watson by Himself*, London, 1817, pp. 18, 19.

examination followed immediately after the Senate-House examination, and the distinction, being much coveted, tended to emphasize the mathematical side of the normal University education of the best men. Since 1883 the prizes have been awarded on the result of dissertations[*].

Until now the Senate-House examination had been oral, but about this time, *circ.* 1770, it began to be the custom to dictate some or all of the questions and to require answers to be written. Only one question was dictated at a time, and a fresh one was not given out until some student had solved that previously read: a custom which by causing perpetual interruptions to take down new questions must have proved very harassing. We are perhaps apt to think that an examination conducted by written papers is so natural that the custom is of long continuance, but I know no record of any in Europe earlier than the eighteenth century. Until 1830 the questions for the Smith's Prize were dictated.

The following description of the Senate-House examination as it existed in 1772 is given by Jebb[†].

> The moderators, some days before the arrival of the time prescribed by the vice-chancellor, meet for the purpose of forming the students into divisions of six, eight, or ten, according to their performance in the schools, with a view to the ensuing examination.
>
> Upon the first of the appointed days, at eight o'clock in the morning, the students enter the senate-house, preceded by a master of arts from each college, who... is called the "father" of the college...
>
> After the proctors have called over the names, each of the moderators sends for a division of the students: they sit with him round a table, with pens, ink, and paper, before them: he enters upon his task of examination, and does not dismiss the set till the hour is expired. This examination has now for some years been held in the english language.
>
> The examination is varied according to the abilities of the students. The moderator generally begins with proposing some questions from the six books of Euclid, plain trigonometry, and the first rules of algebra. If any person fails in an answer, the question goes to the next. From the elements of mathematics, a transition is made to the four branches of philosophy, viz. mechanics, hydrostatics, apparent astronomy, and optics, as explained in the works of Maclaurin, Cotes, Helsham, Hamilton, Rutherforth, Keill, Long, Ferguson, and Smith. If the moderator finds the set of questionists, under examination, capable of answering

---

[*] See Grace of October 25, 1885; and the *Cambridge University Reporter*, October 23 and 30, 1883.
[†] *The Works of J. Jebb*, London, 1787, vol. ii, pp. 290–297.

him, he proceeds to the eleventh and twelfth books of Euclid, conic sections, spherical trigonometry, the higher parts of algebra, and sir Isaac Newton's Principia; more particularly those sections, which treat of the motion of bodies in eccentric and revolving orbits; the mutual action of spheres, composed of particles attracting each other according to various laws; the theory of pulses, propagated through elastic mediums; and the stupendous fabric of the world. Having closed the philosophical examination, he sometimes asks a few questions in Locke's Essay on the human understanding, Butler's Analogy, or Clarke's Attributes. But as the highest academical distinctions are invariably given to the best proficients in mathematics and natural philosophy, a very superficial knowledge in morality and metaphysics will suffice.

When the division under examination is one of the highest classes, problems are also proposed, with which the student retires to a distant part of the senate-house; and returns, with his solution upon paper, to the moderator, who, at his leisure, compares it with the solutions of other students, to whom the same problems have been proposed.

The extraction of roots, the arithmetic of surds, the invention of divisers, the resolution of quadratic, cubic, and biquadratic equations; together with the doctrine of fluxions, and its application to the solution of questions "de maximis et minimis," to the finding of areas, to the rectification of curves, the investigation of the centers of gravity and oscillation, and to the circumstances of bodies, agitated, according to various laws, by centripetal forces, as unfolded, and exemplified, in the fluxional treatises of Lyons, Saunderson, Simpson, Emerson, Maclaurin, and Newton, generally form the subject matter of these problems.

When the clock strikes nine, the questionists are dismissed to breakfast: they return at half past nine, and stay till eleven; they go in again at half past one, and stay till three; and, lastly, they return at half-past three, and stay till five.

The hours of attendance are the same upon the subsequent day.

On the third day they are finally dismissed at eleven.

During the hours of attendance, every division is twice examined in form, once by each of the moderators, who are engaged for the whole time in this employment.

As the questionists are examined in divisions of only six or eight at a time, but a small portion of the whole number is engaged, at any particular hour, with the moderators; and, therefore, if there were no further examination, much time would remain unemployed.

But the moderator's inquiry into the merits of the candidates forms the least material part of the examination.

The "fathers" of the respective colleges, zealous for the credit of the societies, of which they are the guardians, are incessantly employed in examining those students, who appear most likely to contest the palm of glory with their sons.

This part of the process is as follows:

> The father of a college takes a student of a different college aside, and, sometimes for an hour and an half together, strictly examines him in every part of mathematics and philosophy, which he professes to have read.
>
> After he hath, from this examination, formed an accurate idea of the student's abilities and acquired knowledge, he makes a report of his absolute or comparative merit to the moderators, and to every other father who shall ask him the question.
>
> Besides the fathers, all masters of arts, and doctors, of whatever faculty they be, have the liberty of examining whom they please; and they also report the event of each trial, to every person who shall make the inquiry.
>
> The moderators and fathers meet at breakfast, and at dinner. From the variety of reports, taken in connection with their own examination, the former are enabled, about the close of the second day, so far to settle the comparative merits of the candidates, as to agree upon the names of four-and-twenty, who to them appear most deserving of being distinguished by marks of academical approbation.
>
> These four-and-twenty [wranglers and senior optimes] are recommended to the proctors for their private examination; and, if approved by them, and no reason appears against such placing of them from any subsequent inquiry, their names are set down in two divisions, according to that order, in which they deserve to stand; are afterwards printed; and read over upon a solemn day, in the presence of the vice-chancellor, and of the assembled university.
>
> The names of the twelve [junior optimes], who, in the course of the examination, appear next in desert, are also printed, and are read over, in the presence of the vice-chancellor, and of the assembled university, upon a day subsequent to the former...
>
> The students, who appear to have merited neither praise nor censure, [the poll-men], pass unnoticed: while those, who have taken no pains to prepare themselves for the examination, and have appeared with discredit in the schools, are distinguished by particular tokens of disgrace.

Jebb's statement about the number of wranglers and senior optimes is only approximate.

It may be added that it was now frankly recognized that the examination was competitive[*]. Also that though it was open to any member of the Senate to take part in it, yet the determination of the relative merit of the students was entirely in the hands of the moderators[†]. Although the examination did not occupy more than three days it must have been a severe physical trial to anyone who was delicate. It was held in winter and in the Senate-House. That building was then noted

---

[*] "Emulation, which is the principle upon which the plan is constructed." *The Works of J. Jebb*, London, 1787, vol. iii, p. 261.

[†] *The Works of J. Jebb*, London, 1787, vol. iii, p. 272.

for its draughts, and was not warmed in any way: and according to tradition, on one occasion the candidates on entering in the morning found the ink in the pots on their desks frozen.

The University was not altogether satisfied[*] with the scheme in force, and in 1779[†] the scheme of examination was amended in various respects. In particular the examination was extended to four days, a third day being given up entirely to natural religion, moral philosophy, and Locke. It was further announced[‡] that a candidate would not receive credit for advanced subjects unless he had satisfied the examiners in Euclid and elementary Natural Philosophy.

A system of brackets or "classes quam minimae" was now introduced. Under this system the examiners issued on the morning of the fourth day a provisional list of men who had obtained honours, with the names of those of about equal merit bracketed, and that day was devoted to arranging the names in each bracket in order of merit: the examiners being given explicit authority to invite the assistance of others in this work. Whether at this time a candidate could request to be re-examined with the view of being moved from one bracket to another is uncertain, but later this also was allowed.

Under the scheme of 1779 also the number of examiners was increased to four, the moderators of one year becoming, as a matter of course, the examiners of the next. Thus of the four examiners in each year, two had taken part in the examination of the previous year, and the continuity of the system of examination was maintained. The names of the moderators appear on the tripos lists, but the names of the examiners were not printed on the lists till some years later.

The right of any M.A. to take part in the examination was not affected, though henceforth it was exercised more sparingly, and I believe was not insisted on after 1785. But it became a regular custom for the moderators to invite particular M.A.s to examine and compare specified candidates. Milner, of Queens', was constantly asked to assist in this way.

It was not long before it became an established custom that a candidate, who was dissatisfied with the class in which he had been placed as the result of his disputations, might challenge it before the

---

[*] See Graces of July 5, 1773, and of February 17, 1774.
[†] See Graces of March 19, 20, 1779.
[‡] Notice issued by the Vice-Chancellor, dated May 19, 1779.

examination began. This power seems to have been used but rarely; it was, however, a recognition of the fact that a place in the tripos list was to be determined by the Senate-House examination alone, and the examiners soon acquired the habit of settling the preliminary classes without exclusive reference to the previous disputations.

The earliest papers actually set in the Senate-House, and now extant, are two problem papers set in 1785 and 1786 by W. Hodson, of Trinity, then a proctor. The autograph copies from which he gave out the questions were luckily preserved, and are in the library[*] of Trinity College. They must be almost the last problem papers which were dictated, instead of being printed and given as a whole to the candidates.

The problem paper in 1786 was as follows:

1. To determine the velocity with which a Body must be thrown, in a direction parallel to the Horizon, so as to become a secondary planet to the Earth; as also to describe a parabola, and never return.

2. To demonstrate, supposing the force to vary as $\frac{1}{D^2}$, how far a body must fall both within and without the Circle to acquire the Velocity with which a body revolves in a Circle.

3. Suppose a body to be turned (*sic*) upwards with the Velocity with which it revolves in an Ellipse, how high will it ascend? The same is asked supposing it to move in a parabola.

4. Suppose a force varying first as $\frac{1}{D^3}$, secondly in a greater ratio than $\frac{1}{D^2}$ but less than $\frac{1}{D^3}$, and thirdly in a less ratio than $\frac{1}{D^2}$, in each of these Cases to determine whether at all, and where the body parting from the higher Apsid will come to the lower.

5. To determine in what situation of the moon's Apsid they go most forwards, and in what situation of her Nodes the Nodes go most backwards, and why?

6. In the cubic equation $x^3 + qx + r = 0$ which wants the second term; supposing $x = a + b$ and $3ab = -q$, to determine the value of x.

7. To find the fluxion of $x^r \times (y^n + z^m)^{1/q}$.

8. To find the fluent of $\frac{a\dot{x}}{a+x}$.

9. To find the fluxion of the mth power of the Logarithm of x.

10. Of right-angled Triangles containing a given Area to find that whereof the sum of the two legs $AB + BC$ shall be the least possible. [This and the two following questions are illustrated by diagrams. The angle at B is the right angle.]

11. To find the Surface of the Cone ABC. [The cone is a right one on a circular base.]

12. To rectify the arc DB of the semicircle DBV.

---

[*] The *Challis Manuscripts*, III, 61.

In cases of equality in the Senate-House examination the acts were still taken into account in settling the tripos order: and in 1786 when the second, third, and fourth wranglers came out equal in the examination a memorandum was published that the second place was given to that candidate who *dialectis magis est versatus*, and the third place to that one who *in scholis sophistarum melius disputavit*.

There seem to have been considerable intervals in the examination by the moderators, and the examinations by the extraneous examiners took place in these intervals. Those candidates who at any time were not being examined occupied themselves with amusements, provided they were not too boisterous and obvious: probably dice and cards played a large part in them. Gunning in an amusing account of his examination in 1788 talks of games with a teetotum[*] in which he took part on the Wednesday (when Locke and Paley formed the subjects of examination), but "which was carried on with great spirit... by considerable numbers during the whole of the examination."

About this time, 1790, the custom of printing the problem papers was introduced, but until 1828 the other papers continued to be dictated. Since 1827 all the papers have been printed.

I insert here the following letter[†] from William Gooch, of Caius, in which he describes his examination in the Senate-House in 1791. It must be remembered that it is the letter of an undergraduate addressed to his father and mother, and was not intended either for preservation or publication: a fact which certainly does not detract from its value.

Monday $\frac{1}{4}$ aft. 12.

We have been examin'd this Morning in pure Mathematics & I've hitherto kept just about even with Peacock which is much more than I expected. We are going at 1 o'clock to be examin'd till 3 in Philosophy.

From 1 till 7 I did more than Peacock; But who did most at Moderator's Rooms this Evening from 7 till 9, I don't know yet;—but I did above three times as much as the Senr Wrangler last year, yet I'm afraid not so much as Peacock.

Between One & three o'Clock I wrote up 9 sheets of Scribbling Paper so you may suppose I was pretty fully employ'd.

Tuesday Night.

I've been shamefully us'd by Lax to-day;—Tho' his anxiety for Peacock must (of course) be very great, I never suspected that his Partially (*sic*) w^d get the better of his Justice. I had entertain'd too high an opinion of

---

[*] H. Gunning, *Reminiscences*, second edition, London, 1855, vol. i, p. 82.
[†] C. Wordsworth, *Scholae Academicae*, Cambridge, 1877, pp. 322–23.

him to suppose it.—he gave Peacock a long private Examination & then came to me (I hop'd) on the same subject, but 'twas only to *Bully* me as much as he could,—whatever I said (tho' right) he tried to convert into Nonsense by seeming to misunderstand me. However I don't entirely dispair of being first, tho' you see Lax seems determin'd that I shall not.—I had no Idea (before I went into the Senate-House) of being able to contend at all with Peacock.

Wednesday evening.

Peacock & I are still in perfect Equilibrio & the Examiners themselves can give no guess yet who is likely to be first;—a New Examiner (Wood of St. John's, who is reckon'd the first Mathematician in the University, for Waring doesn't reside) was call'd solely to examine Peacock & me only.—but by this new Plan nothing is yet determin'd.—So Wood is to examine us again to-morrow morning.

Thursday evening.

Peacock is declar'd first & I second,—Smith of this Coll. is either 8th or 9th & Lucas is either 10th or 11th.—Poor Quiz Carver is one of the οἱ πολλοί;—I'm perfectly *satisfied* that the Senior Wranglership is Peacock's due, but *certainly* not so very indisputably as Lax pleases to represent it—I understand that *he* asserts 'twas 5 to 4 in Peacock's favor. Now Peacock & I have explain'd to each other how we went on, & can *prove indisputably* that it wasn't 20 to 19 in his favor;—I *cannot* therefore be displeas'd for being plac'd second, tho' I'm provov'd (*sic*) with Lax for his false report (so much beneath the Character of a Gentleman.)—

N.B. it is my very *particular Request* that you don't mention Lax's behaviour to me to any one.

Such was the form ultimately taken by the Senate-House examination, a form which it substantially retained without alteration for nearly half-a-century. It soon became the sole test by which candidates were judged. The University was not obliged to grant a degree to anyone who performed the statutable exercises, and it was open to the University to refuse to pass a supplicat for the B.A. degree unless the candidate had presented himself for the Senate-House examination. In 1790 James Blackburn, of Trinity, a questionist of exceptional abilities, was informed that in spite of his good disputations he would not be allowed a degree unless he also satisfied the examiners in the tripos. He accordingly solved one "very hard problem," though in consequence of a dispute with the authorities he refused to attempt any more.[*]

It will be recollected that the examination was now compulsory on all candidates pursuing the normal course for the B.A. degree. In 1791

---

[*] Gunning, *Reminiscences*, second edition, London, 1855, vol. i, p. 182.

the University laid down rules[*] for its conduct, so far as it concerned poll-men, decreeing that those who passed were to be classified in four divisions or classes, the names in each class to be arranged alphabetically, but not to be printed on the official tripos lists. The classes in the final lists must be distinguished from the eight preliminary classes issued before the commencement of the examination. The men in the first six preliminary classes were expected to take honours; those in the seventh and eighth preliminary classes were *primâ facie* poll-men.

In 1799 the moderators announced[†] that for the future they would require every candidate to show a competent knowledge of the first book of Euclid, arithmetic, vulgar and decimal fractions, simple and quadratic equations, and Locke and Paley. Paley's works seem to be held in esteem by modern divines, and his *Evidences*, though not his *Philosophy*, still remains (1905) one of the subjects of the Previous Examination, but his contemporaries thought less highly of his writings, or at any rate of his Philosophy. Thus Best is quoted by Wordsworth[‡] as saying of Paley's *Philosophy*, "The tutors of Cambridge no doubt neutralize by their judicious remarks, when they read it to their pupils, all that is pernicious in its principles": so also Richard Watson, Bishop of Llandaff, in his anecdotal autobiography[§], says, in describing the Senate-House examination in which Paley was senior wrangler, that Paley was afterwards known to the world by many excellent productions, "though there are some... principles in his philosophy which I by no means approve."

In 1800 the moderators extended to all men in the first four preliminary classes the privilege of being allowed to attempt the problem papers: hitherto this privilege had been confined to candidates placed in the first two classes. Until 1828 the problem papers were set in the evenings, and in the rooms of the moderator.

The *University Calendars* date from 1796, and from 1802 to 1882 inclusive contain the printed tripos papers of the previous January. The papers from 1801 to 1820 and from 1838 to 1849 inclusive were also published in separate volumes, which are to be found in most public libraries. No problems were ever set to the men in the seventh and

---

[*] See Grace of April 8, 1791.
[†] Communicated by the moderators to fathers of Colleges on January 18, 1799, and agreed to by the latter.
[‡] C. Wordsworth, *Scholae Academicae*, Cambridge, 1877, p. 123.
[§] *Anecdotes of the Life of R. Watson*, London, 1817, p. 19.

eighth preliminary classes, which contained the poll-men. None of the bookwork papers of this time are now extant, but it is believed that they contained but few riders. Many of the so-called problems were really pieces of bookwork or easy riders: it must however be remembered that the text-books then in circulation were inferior and incomplete as compared with modern ones.

The *Calendar* of 1802 contains a diffuse account of the examination. It commences as follows:

> On the Monday morning, a little before eight o'clock, the students, generally about a hundred, enter the Senate-House, preceded by a master of arts, who on this occasion is styled the father of the College to which he belongs. On two pillars at the entrance of the Senate-House are hung the classes and a paper denoting the hours of examination of those who are thought most competent to contend for honours. Immediately after the University clock has struck eight, the names are called over, and the absentees, being marked, are subject to certain fines. The classes to be examined are called out, and proceed to their appointed tables, where they find pens, ink, and paper provided in great abundance. In this manner, with the utmost order and regularity, two-thirds of the young men are set to work within less than five minutes after the clock has struck eight. There are three chief tables, at which six examiners preside. At the first, the senior moderator of the present year and the junior moderator of the preceding year. At the second, the junior moderator of the present and the senior moderator of the preceding year. At the third, two moderators of the year previous to the two last, or two examiners appointed by the Senate. The two first tables are chiefly allotted to the six first classes; the third, or largest, to the οἱ πολλοί.
>
> The young men hear the propositions or questions delivered by the examiners; they instantly apply themselves; demonstrate, prove, work out and write down, fairly and legibly (otherwise their labour is of little avail) the answers required. All is silence; nothing heard save the voice of the examiners; or the gentle request of some one, who may wish a repetition of the enunciation. It requires every person to use the utmost dispatch; for as soon as ever the examiners perceive anyone to have finished his paper and subscribed his name to it another question is immediately given...
>
> The examiners are not seated, but keep moving round the tables, both to judge how matters proceed and to deliver their questions at proper intervals. The examination, which embraces arithmetic, algebra, fluxions, the doctrine of infinitesimals and increments, geometry, trigonometry, mechanics, hydrostatics, optics, and astronomy, in all their various gradations, is varied according to circumstances: no one can anticipate a question, for in the course of five minutes he may be dragged from Euclid to Newton, from the humble arithmetic of Bonnycastle to the abstruse analytics of Waring. While this examination is proceeding at the three

> tables between the hours of eight and nine, printed problems are delivered to each person of the first and second classes; these he takes with him to any window he pleases, where there are pens, ink, and paper prepared for his operations.

The examination began at eight. At nine o'clock the papers had to be given up, and half-an-hour was allowed for breakfast. At half-past nine the candidates came back, and were examined in the way described above till eleven, when the Senate-House was again cleared. An interval of two hours then took place. At one o'clock all returned to be again examined. At three the Senate-House was cleared for half-an-hour, and, on the return of the candidates, the examination was continued till five. At seven in the evening the first four classes went to the senior moderator's rooms to solve problems. They were finally dismissed for the day at nine, after eight hours of examination. The work of Tuesday was similar to that of Monday: Wednesday was partly devoted to logic and moral philosophy. At eight o'clock on Thursday morning a first list was published with all candidates of about equal merits bracketed. Until nine o'clock a candidate had the right to challenge anyone above him to an examination to see which was the better. At nine a second list came out, and a candidate's right of challenge was then confined to the bracket immediately above his own. If he proved himself the equal of the man so challenged his name was transferred to the upper bracket. To challenge and then to fail to substantiate the claim to removal to a higher bracket was considered rather ridiculous. Revised lists were published at 11 a.m., 3 p.m., and 5 p.m., according to the results of the examination during that day. At five the whole examination ended. The proctors, moderators, and examiners then retired to a room under the Public Library to prepare the list of honours, which was sometimes settled without much difficulty in a few hours, but sometimes not before 2 a.m. or 3 a.m. the next morning. The name of the senior wrangler was generally announced at midnight, and the rest of the list the next morning. In 1802 there were eighty-six candidates for honours, and they were divided into fifteen brackets, the first and second brackets containing each one name only, and the third bracket four names.

It is clear from the above account that the competition fostered by the examination had developed so much as to threaten to impair its usefulness as guiding the studies of the men. On the other hand, there can be no doubt that the carefully devised arrangements for obtaining an accurate order of merit stimulated the best men to throw all their energies into the work for the examination. It is easy to point out the usual double-edged result of a strict order of merit. The problem before the University was to retain its advantages while checking any abuses to which it might lead.

It was the privilege of the moderators to entertain the proctors and some of the leading resident mathematicians the night before the issue of the final list, and to communicate that list in confidence to their guests. This pleasant custom survived till 1884. I revived the practice in 1890 when acting as senior moderator, but it seems to have now ceased.

In 1806 Sir Frederick Pollock was senior wrangler, and in 1869 in answer to an appeal from De Morgan for an account of the mathematical study of men at the beginning of the century he wrote a letter[*] which is sufficiently interesting to bear reproduction:

> I shall write in answer to your inquiry, *all* about my books, my studies, and my degree, and leave you to settle all about the proprieties which my letter may give rise to, as to egotism, modesty, &c. The only books I read the first year were Wood's *Algebra* (as far as quadratic equations), Bonnycastle's ditto, and *Euclid* (Simpson's). In the second year I read Wood (beyond quadratic equations), and Wood and Vince, for what they called the *branches*. In the third year I read the *Jesuit's* Newton and Vince's *Fluxions*; these were all the *books*, but there were certain MSS. floating about which I copied—which belonged to Dealtry, second wrangler in Kempthorne's year. I have no doubt that I had read less and seen fewer books than any senior wrangler of about my time, or any period since; but what I knew I knew thoroughly, and it was completely at my fingers' ends. I consider that I was the last *geometrical* and *fluxional* senior wrangler; I was not up to the *differential* calculus, and never acquired it. I went up to college with a knowledge of Euclid and algebra to quadratic equations, nothing more; and I never read any second year's lore during my first year, nor any third year's lore during my second; my *forte* was, that what I *did* know I *could produce at any moment with* PERFECT *accuracy*. I could repeat the first book of Euclid word by word and letter by letter. During my first year I was not a '*reading*' man (so called); I had no expectation of honours or a fellowship, and I attended all the lectures on all subjects—Harwood's anatomical, Woollaston's chemical, and Farish's mechanical lectures—but the examination at the end of the first year revealed to me my powers. I was not only in the first class, but it was generally understood I was *first* in the first class; neither I nor any one for me expected I should get in at all. Now, as I had taken no pains to prepare (taking, however, marvellous pains while the examination was going

---

[*] *Memoir of A. de Morgan*, London, 1882, pp. 387–392.

on), I knew better than any one else the value of my *examination qualities* (great rapidity and perfect accuracy); and I said to myself, 'If you're not an ass, you'll be senior wrangler;' and *I took to 'reading' accordingly*. A curious circumstance occurred when the Brackets came out in the Senate-house declaring the result of the examination: I saw at the top the name of Walter *bracketed alone* (as he was); in the bracket below were *Fiott, Hustler, Jephson*. I looked down and could not find my own name till I got to Bolland, when my pride took fire, and I said, 'I must have beaten *that man*, so I will look up again;' and on looking up carefully I found the nail had been passed through my name, and I was at the top bracketed *alone*, even above Walter. You may judge what my feelings were at this discovery; it is the only instance of two such brackets, and it made my fortune—that is, made me independent, and gave me an immense college reputation. It was said I was more than half of the examination before any one else. The two moderators were Hornbuckle, of St John's, and Brown (Saint Brown), of Trinity. The Johnian congratulated me. I said perhaps I might be challenged; he said, 'Well, if you are you're quite safe—you may sit down and do nothing, and no one would get up to you in a whole day.'...

Latterly the Cambridge examinations seem to turn upon very different matters from what prevailed in my time. I think a Cambridge education has for its object to make good members of society—not to extend science and make profound mathematicians. The tripos questions in the Senate-house ought not to go beyond certain limits, and geometry ought to be cultivated and encouraged much more than it is.

To this De Morgan replied:

Your letter suggests much, because it gives possibility of answer. The *branches* of algebra of course mainly refer to the second part of Wood, now called the theory of equations. Waring was his guide. Turner—whom you must remember as head of Pembroke, senior wrangler of 1767—told a young man in the hearing of my informant to be sure and attend to quadratic equations. 'It was a quadratic,' said he, 'made me senior wrangler.' It seems to me that the Cambridge *revivers* were Waring, Paley, Vince, Milner.

You had Dealtry's MSS. He afterwards published a very good book on fluxions. He merged his mathematical fame in that of a Claphamite Christian. It is something to know that the tutor's MS. was in vogue in 1800–1806.

Now—how did you get your conic sections? How much of Newton did you read? From Newton direct, or from tutor's manuscript?

Surely Fiott was our old friend Dr Lee. I missed being a pupil of Hustler by a few weeks. He retired just before I went up in February 1823. The echo of Hornbuckle's answer to you about the challenge has lighted on Whewell, who, it is said, wanted to challenge Jacob, and was answered that he could not beat [him] if he were to write the whole day and the other wrote nothing. I do not believe that Whewell would have listened to any such dissuasion.

I doubt your being the last fluxional senior wrangler. So far as I know, Gipps, Langdale, Alderson, Dicey, Neale, may contest this point with you.

The answer of Sir Frederick Pollock to these questions is dated August 7, 1869, and is as follows.

> You have put together as *revivers* five very different men. Woodhouse was better than Waring, who could not prove Wilson's (Judge of C.P.) guess about the property of prime numbers; but Woodhouse (I think) did prove it, and a beautiful proof it is. Vince was a bungler, and I think utterly insensible of mathematical beauty.
>
> Now for your questions. I did not get my conic sections from Vince. I copied a MS. of Dealtry. I fell in love with the cone and its sections, and everything about it. I have never forsaken my favourite pursuit; I delighted in such problems as two spheres touching each other and also the inside of a hollow cone, &c. As to Newton, I read a good deal (men *now* read nothing), but I read much of the notes. I detected a blunder which nobody seemed to be aware of. Tavel, tutor of Trinity, was not; and he argued very favourably of me in consequence. The application of the Principia I got from MSS. The blunder was this: in calculating the resistance of a globe at the end of a cylinder oscillating in a resisting medium they had forgotten to notice that there is a difference between the resistance to a globe and a circle of the same diameter.
>
> The story of Whewell and Jacob cannot be true. Whewell was a very, *very* considerable man, I think not a *great* man. I have no doubt Jacob beat him in accuracy, but the supposed answer *cannot* be true; it is a mere echo of what actually passed between me and Hornbuckle on the day the Tripos came out—for the truth of which I vouch. I think the examiners are taking too *practical* a turn; it is a waste of time to calculate *actually* a longitude by the help of logarithmic tables and lunar observations. It would be a fault not to know *how*, but a greater to be handy at it.

A few minor changes in the Senate-House examinations were made in 1808[*]. A fifth day was added to the examination. Of the five days thus given up to it three were devoted to mathematics, one to logic, philosophy, and religion, and one to the arrangement of the brackets. Apart from the evening paper the examination on each of the first three days lasted six hours. Of these eighteen hours, eleven were assigned to book-work and seven to problems. The problem papers were set from 6 to 10 in the evening.

---

[*] See Graces, December 15, 1808.

A letter from Whewell dated January 19, 1816, describes his examination in the Senate-House*.

> Jacob. Whewell. Such is the order in which we are fixed after a week's examination... I had before been given to understand that a great deal depended upon being able to write the greatest possible quantity in the smallest time, but of the rapidity which was actually necessary I had formed the most distant idea. I am upon no occasion a quick writer, and upon subjects where I could not go on without sometimes thinking a little I soon found myself considerably behind. I was therefore surprised, and even astonished, to find myself bracketed off, as it is called, in the second place; that is, on the day when a new division of the classes is made for the purpose of having a closer examination of the respective merits of men who come pretty near to each other, I was not classed with anybody, but placed alone in the second bracket. The man who is at the head of the list is of Caius College, and was always expected to be very high, though I do not know that anybody expected to see him so decidedly superior as to be bracketed off by himself.

The tendency to cultivate mechanical rapidity was a grave evil, and lasted long after Whewell's time. According to rumour the highest honours in 1845 were obtained, to the general regret of the University, by assiduous practice in writing†.

The devotion of the Cambridge school to geometrical and fluxional methods has led to its isolation from contemporary continental mathematicians. Early in the nineteenth century the evil consequence of this began to be recognized; and it was felt to be little less than a scandal that the researches of Lagrange, Laplace, and Legendre were unknown to many Cambridge mathematicians save by repute. An attempt to explain the notation and methods of the calculus as used on the Continent was made by R. Woodhouse, who stands out as the apostle of the new movement. It is doubtful if he could have brought analytical methods into vogue by himself; but his views were enthusiastically adopted by three students, Peacock, Babbage, and Herschel, who succeeded in carrying out the reforms he had suggested. They created an Analytical Society which Babbage explained was formed to advocate "the principles of pure *d*-ism as opposed to the *dot*-age of the University." The character of the instruction in mathematics at the University has at all times largely depended on the text-books then in use, and the impor-

---

* S. Douglas, *Life of W. Whewell*, London, 1881, p. 20.

† For a contemporary account of this see C.A. Bristed, *Five Years in an English University*, New York, 1852, pp. 233–239.

tance of good books of this class was emphasized by a traditional rule that questions should not be set on a new subject in the tripos unless it had been discussed in some treatise suitable and available for Cambridge students[*]. Hence the importance attached to the publication of the work on analytical trigonometry by Woodhouse in 1809, and of the works on the differential calculus issued by members of the Analytical Society in 1816 and 1820.

In 1817 Peacock, who was moderator, introduced the symbols for differentiation into the papers set in the Senate-House examination. But his colleague continued to use the fluxional notation. Peacock himself wrote on March 17 of 1817 (*i.e.* shortly after the examination) on the subject as follows[†]:

> I assure you... that I shall never cease to exert myself to the utmost in the cause of reform, and that I will never decline any office which may increase my power to effect it. I am nearly certain of being nominated to the office of Moderator in the year 1818–19, and as I am an examiner in virtue of my office, for the next year I shall pursue a course even more decided than hitherto, since I shall feel that men have been prepared for the change, and will then be enabled to have acquired a better system by the publication of improved elementary books. I have considerable influence as a lecturer, and I will not neglect it. It is by silent perseverance only that we can hope to reduce the many-headed monster of prejudice, and make the University answer her character as the loving mother of good learning and science.

In 1818 all candidates for honours, that is, all men in the first six preliminary classes, were allowed to attempt the problems: this change was made by the moderators.

In 1819 G. Peacock, who was again moderator, induced his colleague to adopt the new notation. It was employed in the next year by Whewell, and in the following year by Peacock again. Henceforth the calculus in its modern language and analytical methods were freely used, new subjects were introduced, and for many years the examination provided a mathematical training fairly abreast of the times.

By this time the disputations had ceased to have any immediate effect on a man's place in the tripos. Thus Whewell[‡], writing about his duties as moderator in 1820, said:

---

[*] See *ex. gr.*, the Grace of November 14, 1827, referred to below.
[†] *Proceedings of the Royal Society*, London, 1859, vol. ix, pp. 538–9.
[‡] *Whewell's Writings and Correspondence*, ed. Todhunter, London, 1876, vol. ii, p. 36.

> You would get very exaggerated ideas of the importance attached to it [an Act] if you were to trust Cumberland; I believe it was formerly more thought of than it is now. It does not, at least immediately, produce any effect on a man's place in the tripos, and is therefore considerably less attended to than used to be the case, and in most years is not very interesting after the five or six best men: so that I look for a considerable exercise of, or rather demand for, patience on my part. The other part of my duty in the Senate House consists in manufacturing wranglers, senior optimes, etc. and is, while it lasts, very laborious.

Of the examination itself in this year he wrote as follows[*]:

> The examination in the Senate House begins to-morrow, and is rather close work while it lasts. We are employed from seven in the morning till five in the evening in giving out questions and receiving written answers to them; and when that is over, we have to read over all the papers which we have received in the course of the day, to determine who have done best, which is a business that in numerous years has often kept the examiners up the half of every night; but this year is not particularly numerous. In addition to all this, the examination is conducted in a building which happens to be a very beautiful one, with a marble floor and a highly ornamented ceiling; and as it is on the model of a Grecian temple, and as temples had no chimneys, and as a stove or a fire of any kind might disfigure the building, we are obliged to take the weather as it happens to be, and when it is cold we have the full benefit of it—which is likely to be the case this year. However, it is only a few days, and we have done with it.

A sketch of the examination in the previous year from the point of view of an examinee was given by J.M.F. Wright[†], but there is nothing of special interest in it.

Sir George Airy[‡] gave the following sketch of his recollections of the reading and studies of undergraduates of his time and of the tripos of 1823, in which he had been senior wrangler:

> At length arrived the Monday morning on which the examination for the B.A. degree was to begin.... We were all marched in a body to the Senate-House and placed in the hands of the Moderators. How the "candidates for honours" were separated from the οἱ πολλοί I do not know, I presume that the Acts and the Opponencies had something to do with it. The honour candidates were divided into six groups: and of these Nos. 1 and 2 (united), Nos. 3 and 4 (united), and Nos. 5 and 6 (united), received the questions of one Moderator. No. 1, Nos. 2 and 3 (united),

---

[*] S. Douglas, *Life of Whewell*, London, 1881, p. 56.
[†] *Alma Mater*, London, 1827, vol. ii, pp. 58–98.
[‡] See *Nature*, vol. 35, Feb. 24, 1887, pp. 397–399.

Nos. 4 and 5 (united), and No. 6, received those of the other Moderator. The Moderators were reversed on alternate days. There were no printed question-papers: each examiner had his bound manuscript of questions, and he read out his first question; each of the examinees who thought himself able proceeded to write out his answer, and then orally called out "Done." The Moderator, as soon as he thought proper, proceeded with another question. I think there was only one course of questions on each day (terminating before 3 o'clock, for the Hall dinner). The examination continued to Friday mid-day. On Saturday morning, about 8 o'clock, the list of honours (manuscript) was nailed on the door of the Senate-House.

It must be remembered that for students pursuing the normal course the Senate-House examination still provided the only avenue to a degree. That examination involved a knowledge of the elements of moral philosophy and theology, an acquaintance with the rules of formal logic, and the power of reading and writing scholastic Latin, but mathematics was the predominant subject, and this led to a certain one-sidedness in education. The evil of this was generally recognized, and in 1822 various reforms were introduced in the University curriculum; in particular the Previous Examination was established for students in their second year, the subjects being prescribed Greek and Latin works, a Gospel, and Paley's *Evidences*. Set classical books were introduced in the final examination of poll-men; and another honour or tripos examination was established for classical students. These alterations came into effect in 1824; and henceforth the Senate-House examination, so far as it related to mathematical students, was known as the Mathematical Tripos.

In 1827 the scheme of examination in the Mathematical Tripos was revised. By regulations* which came into operation in January, 1828, another day was added, so that the examination extended over four days, exclusive of the day of arranging the brackets; the number of hours of examination was twenty-three, of which seven were assigned to problems. On the first two days all the candidates had the same questions proposed to them, inclusive of the evening problems, and the examination on those days excluded the higher and more difficult parts of mathematics, in order, in the words of the report, "that the candidates for honours may not be induced to pursue the more abstruse and profound mathematics, to the neglect of more elementary knowledge." Accordingly, only such questions as could be solved without

---

* See the Grace, November 14, 1827.

the aid of the differential calculus were set on the first day, and those set on the second day involved only its elementary applications. The classes were reduced to four, determined as before by the exercises in the schools. The regulations of 1827 definitely prescribed that all the papers should be printed. They are also noticeable as being the last which gave the examiners power to ask *vivâ voce* questions, though such questions were restricted to "propositions contained in the mathematical works commonly in use in the University, or examples and explanations of such propositions." It was further recommended that no paper should contain more questions than well-prepared students could be expected to answer within the time allowed for it, but that if any candidate, before the end of the time, had answered all the questions in the paper, the examiners might propose additional questions *vivâ voce*. The power of granting honorary optime degrees now ceased; it had already fallen into abeyance. Henceforth the examination was conducted under definite rules, and I no longer concern myself with the traditions of the examination.

In the same year as these changes became effective the examination for the poll degree was separated from the tripos with different sets of papers and a different schedule of subjects[*]. It was, however, still nominally considered as forming part of the Senate-House examination, and until 1858 those who obtained a poll degree were arranged in four classes, described as fourth, fifth, sixth, and seventh, as if in continuation of the junior optimes or third class of the tripos. The year 1828 therefore shews us the Senate-House examination dividing into two distinct parts; one known as the mathematical tripos, the other as the poll examination. In 1851[1] the classical tripos was made independent of the mathematical tripos, and thus provided a separate avenue to a degree. Historically, the examination usually known as "the General" represents the old Senate-House examination for the poll-men, but gradually it has been moved to an earlier period in the normal course taken by the men. In 1852 another set of examinations, at first called "the professor's examinations," and now somewhat modified and known as "the Specials," was instituted for all poll-men to take before they could qualify for a degree. In 1858 the fiction that the poll-examinations were part of

---

[*] See Grace, May 21, 1828, confirming a Report of March 27, 1828.

---

1. '1850' corrected to '1851' as per errata sheet

the Senate-House examination was abandoned, and subsequently they have been treated as providing an independent method of obtaining the degree: thus now the mathematical tripos is the sole representative of the old Senate-House examination. Since 1858 numerous other ways of obtaining the degree have been established, and it is now possible to get it by shewing proficiency in very special, or even technical subjects.

Further changes in the mathematical tripos were introduced in 1833[*]. The duration of the examination, before the issue of the brackets, was extended to five days, and the number of hours of examination on each day was fixed at five and a-half. Seven and a-half hours were assigned to problems. The examination on the first day was confined to subjects that did not require the differential calculus, and only the simplest applications of the calculus were permitted on the second and third days. During the first four days of the examination the same papers were set to all the candidates alike, but on the fifth day the examination was conducted according to classes. No reference was made to *vivâ voce* questions, and the preliminary classification of the brackets only survived in a permission to re-examine candidates if it were found necessary. This permissive rule remained in force till 1848, but I believe that in fact it was never used. In December, 1834, a few unimportant details were amended.

Mr Earnshaw, the senior moderator in 1836, informed me that he believed that the tripos of that year was the earliest one in which all the papers were marked, and that in previous years the examiners had partly relied on their impression of the answers given.

New regulations came into force[†] in 1839. The examination now lasted for six days, and continued as before for five hours and a-half each day. Eight and a-half hours were assigned to problems. Throughout the whole examination the same papers were set to all candidates, and no reference was made to any preliminary classes. It was no doubt in accordance with the spirit of these changes that the acts in the schools should be abolished, but they were discontinued by the moderators of 1839 without the authority of the Senate. The examination was for the future confined[‡] to mathematics.

---

[*] See the Grace of April 6, 1832.

[†] See Grace of May 30, 1838.

[‡] Under a badly-worded grace passed on May 11, 1842, on the recommendation of a syndicate on theological studies, candidates for mathematical honours were, after 1846, required to attend the poll examination on Paley's Moral Philosophy,

In the same year in which the new scheme came into force a proposal to again reopen the subject was rejected (March 6).

The difficulty of bringing professorial lectures into relation with the needs of students has more than once been before the University. The desirability of it was emphasized by a Syndicate in February, 1843, which recommended conferences at stated intervals between the mathematical professors and examiners. This report foreshadowed the creation of a Mathematical Board, but it was rejected by the Senate on March 31.

A few years later the scheme of the examination was again reconstructed by regulations* which came into effect in 1848. The duration of the examination was extended to eight days. The examination lasted in all forty-four and a-half hours, twelve of which were devoted to problems. The first three days were assigned to specified elementary subjects; in the papers set on these days riders were to be set as well as bookwork, but the methods of analytical geometry and the calculus were excluded. After the first three days there was a short interval, at the end of which the examiners issued a list of those who had so acquitted themselves as to deserve mathematical honours. Only those whose names were contained in this list were admitted to the last five days of the examination, which was devoted to the higher parts of mathematics. After the conclusion of the examination the examiners, taking into account the whole eight days, brought out the list arranged in order of merit. No provision was made for any rearrangement of this list corresponding to the examination of the brackets. The arrangements of 1848 remained in force till 1873.

In the same year as these regulations came into force, a Board of Mathematical Studies (consisting of the mathematical professors, and the moderators and examiners for the current year and the two preceding years) was constituted† by the Senate. From that time forward their minutes supply a permanent record of the changes gradually introduced into the tripos. I do not allude to subsequent changes which only concern unimportant details of the examination.

In May, 1849, the Board issued a report in which, after giving a

---

the New Testament and Ecclesiastical History. This had not been the intention of the Senate, and on March 14, 1855, a grace was passed making this clear.

* See Grace of May 13, 1846, confirming a report of March 23, 1846.

† See Grace of October 31, 1848.

review of the past and existing state of the mathematical studies in the University, they recommended that the mathematical theories of electricity, magnetism, and heat should not be admitted as subjects of examination. In the following year they issued a second report, in which they recommended the omission of elliptical integrals, Laplace's coefficients, capillary attraction, and the figure of the earth considered as heterogeneous, as well as a definite limitation of the questions in lunar and planetary theory. In making these recommendations the Board were only giving expression to what had become the practice in the examination.

I may, in passing, mention a curious attempt which was made in 1853 and[1] 1854 to assist candidates in judging of the relative difficulty of the questions asked. This was effected by giving to the candidates, at the same time as the examination paper, a slip of paper on which the marks assigned for the bookwork and rider for each question were printed. I mention the fact merely because these things are rapidly forgotten and not because it is of any intrinsic value. I possess a complete set of slips which came to me from Dr Todhunter.

In 1856 there was an amusing difference of opinion between the Vice-Chancellor and the moderators. The Vice-Chancellor issued a notice to say that for the convenience of the University he had directed the tripos lists to be published at 8.0 a.m. as well as at 9.0 a.m., but when the University arrived at 8.0 the moderators said that they should not read the list until 9.0.

Considerable changes in the scheme of examination were introduced in 1873. On December 5, 1865, the Board had recommended the addition of Laplace's coefficients and the figure of the earth considered as heterogeneous as subjects of the examination; the report does not seem to have been brought before the Senate, but attention was called to the fact that certain departments of mathematics and mathematical physics found no place in the tripos schedules, and were neglected by most students. Accordingly a syndicate was appointed on June 6, 1867, to consider the matter, and a scheme drawn up by them was approved in 1868[*] and came into effect in 1873. The new scheme of examination was framed on the same lines as that of 1848. The subjects in the first

---

[*] See Grace of June 2, 1868. It was carried by a majority of only five in a house of 75.

---

1. '1853 and' inserted as per errata sheet

three days were left unchanged, but an extra day was added, devoted to the elements of mathematical physics. The essence of the modification was the greatly extended range of subjects introduced into the schedule of subjects for the last five days, and their arrangement in divisions, the marks awarded to the five divisions being approximately those awarded to the three days in proportion to 2, 1, 1, 1, 2/3 to 1 respectively. Under the new regulations the number of examiners was increased from four to five.

The assignment of marks to groups of subjects was made under the impression that the best candidates would concentrate their abilities on a selection of subjects from the various divisions. But it was found that, unless the questions were made extremely difficult, more marks could be obtained by reading superficially all the subjects in the five divisions than by attaining real proficiency in a few of the higher ones: while the wide range of subjects rendered it practically impossible to thoroughly cover all the ground in the time allowed. The failure was so pronounced that in 1877 another syndicate was appointed to consider the mathematical studies and examinations of the University. They presented an elaborate scheme, but on May 13, 1878, some of the most important parts of it were rejected and their subsequent proposals, accepted on November 21, 1878 (by 62 to 49), represented a compromise which pleased few members of the Senate[*].

Under the new scheme which came into force in 1882 the tripos was divided into two portions: the first portion was taken at the end of the third year of residence, the range of subjects being practically the same as in the regulations of 1848, and the result brought out in the customary order of merit. The second portion was held in the following January, and was open only to those who had been wranglers in the preceding June. This portion was confined to higher mathematics and appealed chiefly to specialists. The result was brought out in three classes, each arranged in alphabetical order. The moderators and examiners conducted the whole examination without any extraneous aid.

In the next year or two further amendments were made[†], moving the second part to the June of the fourth year, throwing it open to

---

[*] See Graces of May 17, 1877; May 29, 1878; and November 21, 1878: and the *Cambridge University Reporter*, April 2, May 14, June 4, October 29, November 12, and November 26, 1878.

[†] See the Graces of December 13, 1883; June 12, 1884; February 10, 1885; October 29, 1885; and June 1, 1886.

all men who had graduated in the tripos of the previous June, and transferring the conduct of the examination in Part 2 to four examiners nominated by the Board: this put it largely under the control of the professors. The range of subjects of Part 2 was also greatly extended, and candidates were encouraged to select only a few of them. It was further arranged that Part 1 might be taken at the end of a man's second year of residence, though in that case it would not qualify for a degree. A student who availed himself of this leave could take Part 2 at the end either of his third or of his fourth year as he pleased. The tripos is still (1905) carried on under the scheme of 1886.

The general effect of these changes was to destroy the homogeneity of the tripos. Objections to the new scheme were soon raised. Especially, it was said—whether rightly or wrongly—that Part 1 contained too many technical subjects to serve as a general educational training for any save mathematicians; that the distinction of a high place in the historic list produced on its results tended to prevent the best men taking it in their second year, though by this time they had read sufficiently to be able to do so; and that Part 2 was so constructed as to appeal only to professional mathematicians, and that thus the higher branches of mathematics were neglected by all save a few specialists.

Whatever value be attached to these opinions, the number of students studying mathematics fell rapidly under the scheme of 1886. In 1899 the Board proposed[*] further changes. These seemed to some members of the Senate to be likely to still further decrease the number of men who took up the subject as one of general education. At any rate the two main proposals were rejected (February 15, 1900) by votes of 151 to 130 and 161 to 129.

The curious origin of the term tripos has been repeatedly told, and an account of it may fitly close this chapter. Formerly there were three principal occasions on which questionists were admitted to the title or degree of bachelor. The first of these was the comitia priora, held on Ash-Wednesday, for the best men in the year. The next was the comitia posteriora, which was held a few weeks later, and at which any student who had distinguished himself in the quadragesimal exercises subsequent to Ash-Wednesday had his seniority reserved to him. Lastly, there was the comitia minora, for students who had in no special way distinguished themselves. In the fifteenth century an important part

---

[*] See Reports dated November 7, 1899, and January 20, 1900.

in the ceremony on each of these occasions was taken by a certain "ould bachilour," who sat upon a three-legged stool or tripos before the proctors and tested the abilities of the would-be graduates by arguing some question with the "eldest son," who was selected from them as their representative. To assist the latter in what was often an unequal contest his "father," that is, the officer of his college who was to present him for his degree, was allowed to come to his assistance.

Originally the ceremony was a serious one, and had a certain religious character. It took place in Great St Mary's Church, and marked the admission of the student to a position with new responsibilities, while the season of Lent was chosen with a view to bring this into prominence. The Puritan party objected to the observance of such ecclesiastical ceremonies, and in the course of the sixteenth century they introduced much license and buffoonery into the proceedings. The part played by the questionist became purely formal. A serious debate still sometimes took place between the father of the senior questionist and a regent master who represented the University; but the discussion was prefaced by a speech by the bachelor, who came to be called Mr Tripos just as we speak of a judge as the bench, or of a rower as an oar. Ultimately public opinion permitted Mr Tripos to say pretty much what he pleased, so long as it was not dull and was scandalous. The speeches he delivered or the verses he recited were generally preserved by the Registrary, and were known as the tripos verses: originally they referred to the subjects of the disputations then propounded. The earliest copies now extant are those for 1575.

The University officials, to whom the personal criticisms in which the tripos indulged were by no means pleasing, repeatedly exhorted him to remember "while exercising his privilege of humour, to be modest withal." In 1740, says Mr Mullinger[*], "the authorities after condemning the excessive license of the tripos announced that the comitia at Lent would in future be conducted in the Senate-House; and all members of the University, of whatever order or degree, were forbidden to assail or mock the disputants with scurrilous jokes or unseemly witticisms. About the year 1747–8, the moderators initiated the practice of printing the honour lists on the back of the sheets containing the tripos verses, and after the year 1755 this became the invariable prac-

---

[*] J.B. Mullinger, *The University of Cambridge*, Cambridge, vol. i, 1873, pp. 175, 176.

tice. By virtue of this purely arbitrary connection these lists themselves became known as the tripos; and eventually the examination itself, of which they represented the results, also became known by the same designation."

The tripos ceased to deliver his speech about 1750, but the issue of tripos verses continued for nearly 150 years longer. During the latter part of this time they consisted of four sets of verses, usually in Latin, but occasionally in Greek, in which current topics in the University were treated lightly or seriously as the writer thought fit. They were written for the proctors and moderators by undergraduates or commencing bachelors, who were supposed each to receive a pair of white kid gloves in recognition of their labours. Thus gradually the word tripos changed its meaning "from a thing of wood to a man, from a man to a speech, from a speech to sets of verses, from verses to a sheet of coarse foolscap paper, from a paper to a list of names, and from a list of names to a system of examination*."

In 1895 the proctors and moderators, without consulting the Senate, sent in no verses, and thus, in spite of widespread regret, an interesting custom of many centuries standing was destroyed. No doubt it may be argued that the custom had never been embodied in statute or ordinance, and thus was not obligatory. Also it may be said that its continuance was not of material benefit to anybody. I do not think that such arguments are conclusive, and personally I regret the disappearance of historic ties unless it can be shown that they cause inconvenience, which of course in this case could not be asserted.

---

* Wordsworth, *Scholae Academicae*, Cambridge, 1877, p. 21.

CHAPTER VIII.

THREE GEOMETRICAL PROBLEMS.

AMONG the more interesting geometrical problems of antiquity are three questions which attracted the special attention of the early Greek mathematicians. Our knowledge of geometry is derived from Greek sources, and thus these questions have attained a classical position in the history of the subject. The three questions to which I refer are (i) the duplication of a cube, that is, the determination of the side of a cube whose volume is double that of a given cube; (ii) the trisection of an angle; and (iii) the squaring of a circle, that is, the determination of a square whose area is equal to that of a given circle—each problem to be solved by a geometrical construction involving the use of straight lines and circles only, that is, by Euclidean geometry.

With the restriction last mentioned all three problems are insoluble[*]. To duplicate a cube the length of whose side is a, we have to find a line of length x, such that $x^3 = 2a^3$. Again, to trisect a given angle, we may proceed to find the sine of the angle, say a, then, if x is the sine of an angle equal to one-third of the given angle, we have $4x^3 = 3x - a$. Thus the first and second problems, when considered analytically, require the solution of a cubic equation; and since a construction by means of circles (whose equations are of the form $x^2 + y^2 + ax + by + c = 0$) and straight lines (whose equations are of the form $\alpha x + \beta y + \gamma = 0$) cannot be equivalent to the solution of a cubic equation, it is inferred that the problems are insoluble if in our

---

[*] F. Klein, *Vorträge über ausgewählte Fragen der Elementargeometrie*, Leipzig, 1895.

constructions we are restricted to the use of circles and right lines. If the use of the conic sections is permitted, both of these questions can be solved in many ways. The third problem is different in character, but under the same restrictions it also is insoluble.

I propose to give some of the constructions which have been proposed for solving the first two of these problems. To save space, I shall not draw the necessary diagrams, and in most cases I shall not add the proofs: the latter present but little difficulty. I shall conclude with some historical notes on approximate solutions of the quadrature of the circle.

The Duplication of the Cube[*].

The problem of the duplication of the cube was known in ancient times as the Delian problem, in consequence of a legend that the Delians had consulted Plato on the subject. In one form of the story, which is related by Philoponus[†], it is asserted that the Athenians in 430 B.C., when suffering from the plague of eruptive typhoid fever, consulted the oracle at Delos as to how they could stop it. Apollo replied that they must double the size of his altar which was in the form of a cube. To the unlearned suppliants nothing seemed more easy, and a new altar was constructed either having each of its edges double that of the old one (from which it followed that the volume was increased eight-fold) or by placing a similar cubic altar next to the old one. Whereupon, according to the legend, the indignant god made the pestilence worse than before, and informed a fresh deputation that it was useless to trifle with him, as his new altar must be a cube and have a volume exactly double that of his old one. Suspecting a mystery the Athenians applied to Plato, who referred them to the geometricians. The insertion of Plato's name is an obvious anachronism. Eratosthenes[‡] relates a somewhat similar story, but with Minos as the propounder of the problem.

---

[*]   See *Historia Problematis de Cubi Duplicatione* by N.T. Reimer, Göttingen, 1798; and *Historia Problematis Cubi Duplicandi* by C.H. Biering, Copenhagen, 1844: also *Das Delische Problem*, by A. Sturm, Linz, 1895–7. Some notes on the subject are given in my *History of Mathematics*.
[†]   *Philoponus ad Aristotelis Analytica Posteriora*, bk. I, chap. vii.
[‡]   *Archimedis Opera cum Eutocii Commentariis*, ed. Torelli, Oxford, 1792, p. 144; ed. Heiberg, Leipzig, 1880–1, vol. III, pp. 104–107.

In an Arab work, the Greek legend was distorted into the following extraordinarily impossible piece of history, which I cite as a curiosity of its kind. "Now in the days of Plato," says the writer, "a plague broke out among the children of Israel. Then came a voice from heaven to one of their prophets, saying, 'Let the size of the cubic altar be doubled, and the plague will cease'; so the people made another altar like unto the former, and laid the same by its side. Nevertheless the pestilence continued to increase. And again the voice spake unto the prophet, saying, 'They have made a second altar like unto the former, and laid it by its side, but that does not produce the duplication of the cube.' Then applied they to Plato, the Grecian sage, who spake to them, saying, 'Ye have been neglectful of the science of geometry, and therefore hath God chastised you, since geometry is the most sublime of all the sciences.' Now, the duplication of a cube depends on a rare problem in geometry, namely...". And then follows the solution of Apollonius, which is given later.

If a is the length of the side of the given cube and x that of the required cube, we have $x^3 = 2a^3$, that is, $x : a = \sqrt[3]{2} : 1$. It is probable that the Greeks were aware that the latter ratio is incommensurable, in other words, that no two integers can be found whose ratio is the same as that of $\sqrt[3]{2} : 1$, but it did not therefore follow that they could not find the ratio by geometry: in fact, the side and diagonal of a square are instances of lines whose numerical measures are incommensurable.

I proceed now to give some of the geometrical constructions which have been proposed for the duplication of the cube[*]. With one exception, I confine myself to those which can be effected by the aid of the conic sections.

Hippocrates[†] (circ. 420 B.C.) was perhaps the earliest mathematician who made any progress towards solving the problem. He did not give a geometrical construction, but he reduced the question to that of finding two means between one straight line (a), and another twice as long $(2a)$. If these means are x and y, we have $a : x = x : y = y : 2a$, from which it follows that $x^3 = 2a^3$. It is in this form that the problem

---

[*] On the application to this problem of the traditional Greek methods of analysis by Hero and Philo (leading to the solution by the use of Apollonius's circle), by Nicomedes (leading to the solution by the use of the conchoid), and by Pappus (leading to the solution by the use of the cissoid), see *Geometrical Analysis* by J. Leslie, Edinburgh, second edition, 1811, pp. 247–250, 453.

[†] Proclus, ed. Friedlein, pp. 212, 213.

is always presented now. Formerly any process of solution by finding these means was called a mesolabum.

One of the first solutions of the problem was that given by Archytas[*] in or about the year 400 B.C. His construction is equivalent to the following. On the diameter OA of the base of a right circular cylinder describe a semicircle whose plane is perpendicular to the base of the cylinder. Let the plane containing this semicircle rotate round the generator through O, then the surface traced out by the semicircle will cut the cylinder in a tortuous curve. This curve will itself be cut by a right cone, whose axis is OA and semi-vertical angle is (say) $60°$, in a point P, such that the projection of OP on the base of the cylinder will be to the radius of the cylinder in the ratio of the side of the required cube to that of the given cube. Of course the proof given by Archytas is geometrical; and it is interesting to note that in it he shows himself familiar with the results of the propositions Euc. III, 18, III, 35, and XI, 19. To show analytically that the construction is correct, take OA as the axis of x, and the generator of the cylinder drawn through O as axis of z, then with the usual notation, in polar coordinates, if a is the radius of the cylinder, we have for the equation of the surface described by the semicircle $r = 2a\sin\theta$; for that of the cylinder $r\sin\theta = 2a\cos\varphi$; and for that of the cone $\sin\theta\cos\varphi = \frac{1}{2}$. These three surfaces cut in a point such that $\sin^3\theta = \frac{1}{2}$, and therefore $(r\sin\theta)^3 = 2a^3$. Hence the volume of the cube whose side is $r\sin\theta$ is twice that of the cube whose side is a.

The construction attributed to Plato[†] (circ. 360 B.C.) depends on the theorem that, if CAB and DAB are two right-angled triangles, having one side, AB, common, their other sides, AD and BC, parallel, and their hypothenuses, AC and BD, at right angles, then if these hypothenuses cut in P, we have $PC : PB = PB : PA = PA : PD$. Hence, if such a figure can be constructed having $PD = 2PC$, the problem will be solved. It is easy to make an instrument by which the figure can be drawn.

The next writer whose name is connected with the problem is Menaechmus[‡], who in or about 340 B.C. gave two solutions of it.

In the first of these he pointed out that two parabolas having a common vertex, axes at right angles, and such that the latus rectum

---

[*]   *Archimedis Opera*, ed. Torelli, p. 143; ed. Heiberg, vol. III, pp. 98–103.
[†]   Ibid., ed. Torelli, p. 135; ed. Heiberg, vol. III, pp. 66–71.
[‡]   Ibid., ed. Torelli, pp. 141–143; ed. Heiberg, vol. III, pp. 92–99.

of the one is double that of the other will intersect in another point whose abscissa (or ordinate) will give a solution. If we use analysis this is obvious; for, if the equations of the parabolas are $y^2 = 2ax$ and $x^2 = ay$, they intersect in a point whose abscissa is given by $x^3 = 2a^3$. It is probable that this method was suggested by the form in which Hippocrates had cast the problem: namely, to find x and y so that $a : x = x : y = y : 2a$, whence we have $x^2 = ay$ and $y^2 = 2ax$.

The second solution given by Menaechmus was as follows. Describe a parabola of latus rectum l. Next describe a rectangular hyperbola, the length of whose real axis is $4l$, and having for its asymptotes the tangent at the vertex of the parabola and the axis of the parabola. Then the ordinate and the abscissa of the point of intersection of these curves are the mean proportionals between l and $2l$. This is at once obvious by analysis. The curves are $x^2 = ly$ and $xy = 2l^2$. These cut in a point determined by $x^3 = 2l^3$ and $y^3 = 4l^3$. Hence $l : x = x : y = y : 2l$.

The solution of Apollonius[*], which was given about 220 B.C., was as follows. The problem is to find two mean proportionals between two given lines. Construct a rectangle $OADB$, of which the adjacent sides OA and OB are respectively equal to the two given lines. Bisect AB in C. With C as centre describe a circle cutting OA produced in a and cutting OB produced in b, so that aDb shall be a straight line. If this circle can be so described, it will follow that $OA : Bb = Bb : Aa = Aa : OB$, that is, Bb and Aa are the two mean proportionals between OA and OB. It is impossible to construct the circle by Euclidean geometry, but Apollonius gave a mechanical way of describing it.

The only other construction of antiquity to which I will refer is that given by Diocles and Sporus[†]. It is as follows. Take two sides of a rectangle OA, OB, equal to the two lines between which the means are sought. Suppose OA to be the greater. With centre O and radius OA describe a circle. Let OB produced cut the circumference in C and let AO produced cut it in D. Find a point E on BC so that if DE cuts AB produced in F and cuts the circumference in G, then $FE = EG$. If E can be found, then OE is the first of the means between OA and OB. Diocles invented the cissoid in order to determine E, but it can be found equally conveniently by the aid of conics.

---

[*] *Archimedis Opera*, ed. Torelli, p. 137; ed. Heiberg, vol. III, pp. 76–79. The solution is given in my *History of Mathematics*, London, 1901, p. 84.
[†] *Ibid.*, ed. Torelli, pp. 138, 139, 141; ed. Heiberg, vol. III, pp. 78–84, 90–93.

In more modern times several other solutions have been suggested. I may allude in passing to three given by Huygens[*], but I will enunciate only those proposed respectively by Vieta, Descartes, Gregory of St Vincent, and Newton.

Vieta's construction is as follows[†]. Describe a circle, centre O, whose radius is equal to half the length of the larger of the two given lines. In it draw a chord AB equal to the smaller of the two given lines. Produce AB to E so that $BE = AB$. Through A draw a line AF parallel to OE. Through O draw a line $DOCFG$, cutting the circumference in D and C, cutting AF in F, and cutting BA produced in G, so that $GF = OA$. If this line can be drawn then $AB : GC = GC : GA = GA : CD$.

Descartes pointed out[‡] that the curves $x^2 = ay$ and $x^2 + y^2 = ay + bx$ cut in a point (x, y) such that $a : x = x : y = y : b$. Of course this is equivalent to the first solution given by Menaechmus, but Descartes preferred to use a circle rather than a second conic.

Gregory's construction was given in the form of the following theorem[§]. The hyperbola drawn through the point of intersection of two sides of a rectangle so as to have the two other sides for its asymptotes meets the circle circumscribing the rectangle in a point whose distances from the asymptotes are the mean proportionals between two adjacent sides of the rectangle. This is the geometrical expression of the proposition that the curves $xy = ab$ and $x^2 + y^2 = ay + bx$ cut in a point (x, y) such that $a : x = x : y = y : b$.

One of the constructions proposed by Newton is as follows[‖]. Let OA be the greater of two given lines. Bisect OA in B. With centre O and radius OB describe a circle. Take a point C on the circumference so that BC is equal to the other of the two given lines. From O draw ODE cutting AC produced in D, and BC produced in E, so that the intercept $DE = OB$. Then $BC : OD = OD : CE = CE : OA$. Hence OD and CE are two mean proportionals between any two lines BC and OA.

---

[*] *Opera Varia*, Leyden, 1724, pp. 393–396.
[†] *Opera Mathematica*, ed. Schooten, Leyden, 1646, prop, v, pp. 242–243.
[‡] *Geometria*, bk. III, ed. Schooten, Amsterdam, 1659, p. 91.
[§] Gregory of St Vincent, *Opus Geometricum Quadraturae Circuli*, Antwerp, 1647, bk. VI, prop. 138, p. 602.
[‖] *Arithmetica Universalis*, Ralphson's (second) edition, 1728, p. 242; see also pp. 243, 245.

The Trisection of an Angle*.

The trisection of an angle is the second of these classical problems, but tradition has not enshrined its origin in romance. The following two constructions are among the oldest and best known of those which have been suggested; they are quoted by Pappus[†], but I do not know to whom they were due originally.

The first of them is as follows. Let AOB be the given angle. From any point P in OB draw PM perpendicular to OA. Through P draw PR parallel to OA. On MP take a point Q so that if OQ is produced to cut PR in R then $QR = 2 \cdot OP$. If this construction can be made, then $AOR = \frac{1}{3}AOB$. The solution depends on determining the position of R. This was effected by a construction which may be expressed analytically thus. Let the given angle be $\tan^{-1}(b/a)$. Construct the hyperbola $xy = ab$, and the circle $(x-a)^2 + (y-b)^2 = 4(a^2+b^2)$. Of the points where they cut, let x be the abscissa which is greatest, then $PR = x - a$, and $\tan^{-1}(b/x) = \frac{1}{3}\tan^{-1}(b/a)$.

The second construction is as follows. Let AOB be the given angle. Take $OB = OA$, and with centre O and radius OA describe a circle. Produce AO indefinitely and take a point C on it external to the circle so that if CB cuts the circumference in D then CD shall be equal to OA. Draw OE parallel to CDB. Then, if this construction can be made, $AOE = \frac{1}{3}AOB$. The ancients determined the position of the point C by the aid of the conchoid: it could be also found by the use of the conic sections.

I proceed to give a few other solutions; confining myself to those effected by the aid of conics.

Among the other constructions given by Pappus[‡] I may quote the following. Describe a hyperbola whose eccentricity is two. Let its centre be C and its vertices A and A'. Produce CA' to S so that $A'S = CA'$. On AS describe a segment of a circle to contain the given angle. Let

---

* On the bibliography of the subject see the supplements to *L'Intermédiaire des Mathématiciens*, Paris, May and June, 1904.

† Pappus, *Mathematicae Collectiones*, bk. IV, props. 32, 33 (ed. Commandino, Bonn, 1670, pp. 97–99). On the application to this problem of the traditional Greek methods of analysis see *Geometrical Analysis*, by J. Leslie, Edinburgh, second edition, 1811, pp. 245–247.

‡ Pappus, bk. IV, prop. 34, pp. 99–104.

the orthogonal bisector of AS cut this segment in O. With centre O and radius OA or OS describe a circle. Let this circle cut the branch of the hyperbola through A' in P. Then $SOP = \frac{1}{3}SOA$.

In modern times one of the earliest of the solutions by a direct use of conics was suggested by Descartes, who effected it by the intersection of a circle and a parabola. His construction[*] is equivalent to finding the points of intersection other than the origin, of the parabola $y^2 = \frac{1}{4}x$ and the circle $x^2 + y^2 - \frac{13}{4}x + 4ay = 0$. The ordinates of these points are given by the equation $4y^3 = 3y - a$. The smaller positive root is the sine of one-third of the angle whose sine is a. The demonstration is ingenious.

One of the solutions proposed by Newton is practically equivalent to the third one which is quoted above from Pappus. It is as follows[†]. Let A be the vertex of one branch of a hyperbola whose eccentricity is two, and let S be the focus of the other branch. On AS describe the segment of a circle containing an angle equal to the supplement of the given angle. Let this circle cut the S branch of the hyperbola in P. Then PAS will be equal to one-third of the given angle.

The following elegant solution is due to Clairaut[‡]. Let AOB be the given angle. Take $OA = OB$, and with centre O and radius OA describe a circle. Join AB, and trisect it in H, K, so that $AH = HK = KB$. Bisect the angle AOB by OC cutting AB in L. Then $AH = 2 \cdot HL$. With focus A, vertex H, and directrix OC, describe a hyperbola. Let the branch of this hyperbola which passes through H cut the circle in P. Draw PM perpendicular to OC and produce it to cut the circle in Q. Then by the focus and directrix property we have $AP : PM = AH : HL = 2 : 1$, $\therefore AP = 2 \cdot PM = PQ$. Hence, by symmetry, $AP = PQ = QR$. $\therefore AOP = POQ = QOR$.

I may conclude by giving the solution which Chasles[§] regards as the most fundamental. It is equivalent to the following proposition. If OA and OB are the bounding radii of a circular arc AB, then a rectangular hyperbola having OA for a diameter and passing through the point of intersection of OB with the tangent to the circle at A will pass through

---

[*] *Geometria*, bk. III, ed. Schooten, Amsterdam, 1659, p. 91.
[†] *Arithmetica Universalis*, problem XLII, Ralphson's (second) edition, London, 1728, p. 148; see also pp. 243–245.
[‡] I believe that this was first given by Clairaut, but I have mislaid my reference. The construction occurs as an example in the *Geometry of Conics*, by C. Taylor, Cambridge, 1881, No. 308, p. 126.
[§] *Traité des sections coniques*, Paris, 1865, art. 37, p. 36.

one of the two points of trisection of the arc.

The Quadrature of the Circle[*].

The object of the third of the classical problems was the determination of a side of a square whose area should be equal to that of a given circle.

The investigation, previous to the last two hundred years, of this question was fruitful in discoveries of allied theorems, but in more recent times it has been abandoned by those who are able to realize what is required. The history of this subject has been treated by competent writers in such detail that I shall content myself with a very brief allusion to it.

Archimedes showed[†] (what possibly was known before) that the problem is equivalent to finding the area of a right-angled triangle whose sides are equal respectively to the perimeter of the circle and the radius of the circle. Half the ratio of these lines is a number, usually denoted by π.

That this number is incommensurable had been long suspected, and has been now demonstrated. The earliest analytical proof of it was given by Lambert[‡] in 1761; in 1803 Legendre[§] extended the proof to show that π^2 was also incommensurable; and recently Lindemann[||] has

---

[*] See Montucla's *Histoire des Recherches sur la Quadrature du Cercle*, edited by P.L. Lacroix, Paris, 1831; also various articles by A. De Morgan, and especially his *Budget of Paradoxes*, London, 1872. A popular sketch of the subject has been compiled by H. Schubert, *Die Quadratur des Zirkels*, Hamburg, 1889; and since the publication of the earlier editions of these *Recreations* Prof. F. Rudio of Zurich has given an analysis of the arguments of Archimedes, Huygens, Lambert, and Legendre on the subject, with an introduction on the history of the problem, Leipzig, 1892.

[†] *Archimedis Opera*, Κύκλου μέτρησις, prop. I, ed. Torelli, pp. 203–205; ed. Heiberg, vol. I, pp. 258–261, vol. III, pp. 269–277.

[‡] *Mémoires de l'Académie de Berlin* for 1761, Berlin, 1768, pp. 265–322.

[§] Legendre's *Geometry*, Brewster's translation, Edinburgh, 1824, pp. 239–245.

[||] Ueber die Zahl π, *Mathematische Annalen*, Leipzig, 1882, vol. XX, pp. 213–225. The proof leads to the conclusion that, if x is a root of a rational integral algebraical equation, then e^x cannot be rational: hence, if πi was the root of such an equation, $e^{\pi i}$ could not be rational; but $e^{\pi i}$ is equal to -1, and therefore is rational; hence πi cannot be the root of such an algebraical equation, and therefore neither can π.

shown that π cannot be the root of a rational algebraical equation.

An earlier attempt by James Gregory to give a geometrical demonstration of this is worthy of notice. Gregory proved[*] that the ratio of the area of any arbitrary sector to that of the inscribed or circumscribed polygons is not expressible by a finite number of algebraical terms. Hence he inferred that the quadrature was impossible. This was accepted by Montucla, but it is not conclusive, for it is conceivable that some particular sector might be squared, and this particular sector might be the whole circle.

In connection with Gregory's proposition above cited, I may add that Newton[†] proved that in any closed oval an arbitrary sector bounded by the curve and two radii cannot be expressed in terms of the co-ordinates of the extremities of the arc by a finite number of algebraical terms. The argument is condensed and difficult to follow: the same reasoning would show that a closed oval curve cannot be represented by an algebraical equation in polar co-ordinates. From this proposition no conclusion as to the quadrature of the circle is to be drawn, nor did Newton draw any. In the earlier editions of this work I expressed an opinion that the result presupposed a particular definition of the word oval, but on more careful reflection I think that the conclusion is valid without restriction.

With the aid of the quadratrix, or the conchoid, or the cissoid, the quadrature of the circle is easy, but the construction of those curves assumes a knowledge of the value of π, and thus the question is begged.

I need hardly add that, if π represented merely the ratio of the circumference of a circle to its diameter, the determination of its numerical value would have but slight interest. It is however a mere accident that π is defined usually in that way, and it really represents a certain number which would enter into analysis from whatever side the subject was approached.

I recollect a distinguished professor explaining how different would be the ordinary life of a race of beings born, as easily they might be, so that the fundamental processes of arithmetic, algebra and geometry were different to those which seem to us so evident, but, he added, it is impossible to conceive of a universe in which e and π should not exist.

---

[*] *Vera Circuli et Hyperbolae Quadratura*, Padua, 1668: this is reprinted in Huygens's *Opera Varia*, Leyden, 1724, pp. 405–462.

[†] *Principia*, bk. I, section VI, lemma XXVIII.

I have quoted elsewhere an anecdote, which perhaps will bear repetition, that illustrates how little the usual definition of π suggests its properties. De Morgan was explaining to an actuary what was the chance that a certain proportion of some group of people would at the end of a given time be alive; and quoted the actuarial formula, involving π, which, in answer to a question, he explained stood for the ratio of the circumference of a circle to its diameter. His acquaintance, who had so far listened to the explanation with interest, interrupted him and exclaimed, "My dear friend, that must be a delusion, what can a circle have to do with the number of people alive at the end of a given time?" In reality the fact that the ratio of the length of the circumference of a circle to its diameter is the number denoted by π does not afford the best analytical definition of π, and is only one of its properties.

The use of a single symbol to denote this number $3.14159\ldots$ seems to have been introduced about the beginning of the eighteenth century. W. Jones[*] in 1706 represented it by π; a few years later[†] John Bernoulli denoted it by c; Euler in 1734 used p, and in 1736 used c; Chr. Goldback in 1742 used π; and after the publication of Euler's *Analysis* the symbol π was generally employed.

The numerical value of π can be determined by either of two methods with as close an approximation to the truth as is desired.

The first of these methods is geometrical. It consists in calculating the perimeters of polygons inscribed in and circumscribed about a circle, and assuming that the circumference of the circle is intermediate between these perimeters[‡]. The approximation would be closer if the areas and not the perimeters were employed. The second and modern method rests on the determination of converging infinite series for π.

We may say that the π-calculators who used the first method regarded π as equivalent to a geometrical ratio, but those who adopted the modern method treated it as the symbol for a certain number which enters into numerous branches of mathematical analysis.

It may be interesting if I add here a list of some of the approximations to the value of π given by various writers[§]. This will indicate

---

[*] *Synopsis Palmariorum Matheseos*, London, 1706, pp. 243, 263 *et seq.*

[†] See notes by G. Eneström in the *Bibliotheca Mathematica*, Stockholm, 1889, vol. III, p. 28; *Ibid.*, 1890, vol. IV, p. 22.

[‡] The history of this method has been written by K.E.I. Selander, *Historik öfver Ludolphska Talet*, Upsala, 1868.

[§] For the methods used in classical times and the results obtained, see the notices incidentally those who have studied the subject to the best advantage.

The ancient Egyptians[*] took 256/81 as the value of π, this is equal to 3.1605...; but the rougher approximation of 3 was used by the Babylonians[†] and by the Jews[‡]. It is not unlikely that these numbers were obtained empirically.

We come next to a long roll of Greek mathematicians who attacked the problem. Whether the researches of the members of the Ionian School, the Pythagoreans, Anaxagoras, Hippias, Antipho, and Bryso led to numerical approximations for the value of π is doubtful, and their investigations need not detain us. The quadrature of certain lunes by Hippocrates of Chios is ingenious and correct, but a value of π cannot be thence deduced; and it seems likely that the later members of the Athenian School concentrated their efforts on other questions.

It is probable that Euclid[§], the illustrious founder of the Alexandrian School, was aware that π was greater than 3 and less than 4, but he did not state the result explicitly.

The mathematical treatment of the subject began with Archimedes, who proved that π is less than $3\frac{1}{7}$ and greater than $3\frac{10}{71}$, that is, it lies between 3.1428... and 3.1408.... He established[‖] this by inscribing in a circle and circumscribing about it regular polygons of 96 sides, then determining by geometry the perimeters of these polygons, and finally assuming that the circumference of the circle was intermediate between these perimeters: this leads to the result

---

of their authors in M. Cantor's *Geschichte der Mathematik*, Leipzig, vol. I, 1880. For medieval and modern approximations, see the article by A. De Morgan on the Quadrature of the Circle in vol. XIX of the *Penny Cyclopaedia*, London, 1841; with the additions given by B. de Haan in the *Verhandelingen* of Amsterdam, 1858, vol. IV, p. 22: the conclusions were tabulated, corrected, and extended by Dr J.W.L. Glaisher in the *Messenger of Mathematics*, Cambridge, 1873, vol. II, pp. 119–128; and *Ibid.*, 1874, vol. III, pp. 27–46.

[*] *Ein mathematisches Handbuch der alten Aegypter* (*i.e.* the Rhind papyrus), by A. Eisenlohr, Leipzig, 1877, arts. 100–109, 117, 124.

[†] Oppert, *Journal Asiatique*, August, 1872, and October, 1874.

[‡] 1 Kings, ch. 7, ver. 23; 2 Chronicles, ch. 4, ver. 2.

[§] These results can be deduced from Euc. IV, 15, and IV, 8: see also book XII, prop. 16.

[‖] *Archimedis Opera*, Κύκλου μέτρησις, prop. III, ed. Torelli, Oxford, 1792, pp. 205–216; ed. Heiberg, Leipzig, 1880, vol. I, pp. 263-271.

$6336/2017\frac{1}{4} < \pi < 14688/4673\frac{1}{2}$,[1] from which he deduced the limits given above. This method is equivalent to using the proposition $\sin\theta < \theta < \tan\theta$, where $\theta = \pi/96$: the values of $\sin\theta$ and $\tan\theta$ were deduced by Archimedes from those of $\sin\frac{1}{3}\pi$ and $\tan\frac{1}{3}\pi$ by repeated bisections of the angle. With a polygon of n sides this process gives a value of π correct to at least the integral part of $(2\log n - 1.19)$ places of decimals. The result given by Archimedes is correct to 2 places of decimals. His analysis leads to the conclusion that the perimeters of these polygons for a circle whose diameter is 4970 feet would lie between 15610 feet and 15620 feet—actually it is about 15613 feet 9 inches.

Apollonius discussed these results, but his criticisms have been lost.

Hero of Alexandria gave[*] the value 3, but he quoted[†] the result 22/7: possibly the former number was intended only for rough approximations.

The only other Greek approximation that I need mention is that given by Ptolemy[‡], who asserted that $\pi = 3°8'30''$. This is equivalent to taking $\pi = 3 + \frac{8}{60} + \frac{30}{3600} = 3\frac{17}{120} = 3.141\dot{6}$.

The Roman surveyors seem to have used 3, or sometimes 4, for rough calculations. For closer approximations they often employed $3\frac{1}{8}$ instead of $3\frac{1}{7}$, since the fractions then introduced are more convenient in duodecimal arithmetic. On the other hand Gerbert[§] recommended the use of 22/7.

Before coming to the medieval and modern European mathematicians it may be convenient to note the results arrived at in India and the East.

Baudhayana[||] took 49/16 as the value of π.

Arya-Bhata[¶], circ. 530, gave 62832/20000, which is equal to 3.1416. He showed that, if a is the side of a regular polygon of n sides inscribed in a circle of unit diameter, and if b is the side of a regular inscribed polygon of $2n$ sides, then $b^2 = \frac{1}{2} - \frac{1}{2}(1-a^2)^{\frac{1}{2}}$. From the side of an

---

[*] *Mensurae*, ed. Hultsch, Berlin, 1864, p. 188.
[†] *Geometria*, ed. Hultsch, Berlin, 1864, pp. 115, 136.
[‡] *Almagest*, bk. VI, chap. 7; ed. Halma, vol. I, p. 421.
[§] *Œuvres de Gerbert*, ed. Olleris, Clermont, 1867, p. 453.
[||] The *Sulvasutras* by G. Thibaut, *Asiatic Society of Bengal*, 1875, arts. 26–28.
[¶] *Leçons de calcul d'Aryabhata*, by L. Rodet in the *Journal Asiatique*, 1879, series 7, vol. XIII, pp. 10, 21.

---

1. Inserted 14688/

inscribed hexagon, he found successively the sides of polygons of 12, 24, 48, 96, 192, and 384 sides. The perimeter of the last is given as equal to $\sqrt{9.8694}$, from which his result was obtained by approximation.

Brahmagupta*, circ. 650, gave $\sqrt{10}$, which is equal to 3.1622.... He is said to have obtained this value by inscribing in a circle of unit diameter regular polygons of 12, 24, 48, and 96 sides, and calculating successively their perimeters, which he found to be $\sqrt{9.65}$, $\sqrt{9.81}$, $\sqrt{9.86}$, $\sqrt{9.87}$ respectively; and to have assumed that as the number of sides is increased indefinitely the perimeter would approximate to $\sqrt{10}$.

Bhaskara, circ. 1150, gave two approximations. One[†]—possibly copied from Arya-Bhata, but said to have been calculated afresh by Archimedes's method from the perimeters of regular polygons of 384 sides—is 3927/1250, which is equal to 3.1416: the other[‡] is $\frac{754}{240}$, which is equal to $3.141\dot{6}$, but it is uncertain whether this was not given only as an approximate value.

Among the Arabs the values 22/7, $\sqrt{10}$, and 62832/20000 were given by Alkarisimi[§], circ. 830; and no doubt were derived from Indian sources. He described the first as an approximate value, the second as used by geometricians, and the third as used by astronomers.

In Chinese works the values 3, $\frac{22}{7}$, $\frac{157}{50}$ are said to occur: probably the last two results were copied from the Arabs.

Returning to European mathematicians, we have the following successive approximations to the value of π: many of those prior to the eighteenth century having been calculated originally with the view of demonstrating the incorrectness of some alleged quadrature.

Leonardo of Pisa[‖], in the thirteenth century, gave for π the value $1440/458\frac{1}{3}$ which is equal to 3.1418.... In the fifteenth century, Purbach[¶] gave or quoted the value 62832/20000, which is equal to $3.141\dot{6}$; Cusa believed that the accurate value was $\frac{3}{4}(\sqrt{3}+\sqrt{6})$ which is equal to 3.1423...; and, in 1464, Regiomontanus[**] is said to have given a

---

* *Algebra... from Brahmegupta and Bhascara*, trans. by H.T. Colebrooke, London, 1817, chap. XII, art. 40, p. 308.
† *Ibid.*, p. 87.
‡ *Ibid.*, p. 95.
§ *The Algebra of Mohammed ben Musa*, ed. by F. Rosen, London, 1831, pp. 71–72.
‖ Boncompagni's *Scritti di Leonardo*, vol. II (*Practica Geometriae*), Rome, 1862, p. 90.
¶ Appendix to the *De Triangulis* of Regiomontanus, Basle, 1541, p. 131.
** In his correspondence with Cardinal Cusa, *De Quadratura Circuli*, Nuremberg,

value equal to 3.14243.

Vieta[*], in 1579, showed that π was greater than $31415926535/10^{10}$, and less than $31415926537/10^{10}$. This was deduced from the perimeters of the inscribed and circumscribed polygons of 6×2^{16} sides, obtained by repeated use of the formula $2\sin^2 \frac{1}{2}\theta = 1 - \cos\theta$. He also gave[†] a result equivalent to the formula

$$\frac{2}{\pi} = \frac{\sqrt{2}}{2} \frac{\sqrt{(2+\sqrt{2})}}{2} \frac{\sqrt{\{2+\sqrt{(2+\sqrt{2})}\}}}{2} \cdots .$$

The father of Adrian Metius[‡], in 1585, gave $355/113$, which is equal to $3.14159292\ldots$, and is correct to 6 places of decimals. This was a curious and lucky guess, for all that he proved was that π was intermediate between $377/120$ and $333/106$, whereon he jumped to the conclusion that he should obtain the true fractional value by taking the mean of the numerators and the mean of the denominators of these fractions.

In 1593 Adrian Romanus[§] calculated the perimeter of the inscribed regular polygon of $1073,741824$ (*i.e.* 2^{30}) sides, from which he determined the value of π correct to 15 places of decimals.

L. van Ceulen devoted no inconsiderable part of his life to the subject. In 1596[||] he gave the result to 20 places of decimals: this was calculated by finding the perimeters of the inscribed and circumscribed regular polygons of 60×2^{33} sides, obtained by the repeated use of a theorem of his discovery equivalent to the formula $1-\cos A = 2\sin^2 \frac{1}{2}A$. I possess a finely executed engraving of him of this date, with the result printed round a circle which is below his portrait. He died in 1610, and by his directions the result to 35 places of decimals (which was as far as he had then calculated it) was engraved on his tombstone[¶] in St Peter's

---

1533, wherein he proved that Cusa's result was wrong. I cannot quote the exact reference, but the figures are given by competent writers and I have no doubt are correct.

[*] *Canon Mathematicus seu ad Triangula*, Paris, 1579, pp. 56, 66: probably this work was printed for private circulation only, it is very rare.

[†] *Vietae Opera*, ed. Schooten, Leyden, 1646, p. 400.

[‡] *Arithmeticae libri duo et Geometriae*, by A. Metius, Leyden, 1626, pp. 88, 89. [Probably issued originally in 1611.]

[§] *Ideae Mathematicae*, Antwerp, 1593: a rare work, which I have never been able to consult.

[||] *Vanden Circkel*, Delf, 1596, fol. 14, p. 1; or *De Circulo*, Leyden, 1619, p. 3.

[¶] The inscription is quoted by Prof. de Haan in the *Messenger of Mathematics*, 1874, vol. III, p. 25.

Church, Leyden. His posthumous arithmetic[*] contains the result to 32 places; this was obtained by calculating the perimeter of a polygon, the number of whose sides is 2^{62}, *i.e.* $4,611686,018427,387904$. Van Ceulen also compiled a table of the perimeters of various regular polygons.

Willebrord Snell[†], in 1621, obtained from a polygon of 2^{30} sides an approximation to 34 places of decimals. This is less than the numbers given by van Ceulen, but Snell's method was so superior that he obtained his 34 places by the use of a polygon from which van Ceulen had obtained only 14 (or perhaps 16) places. Similarly, Snell obtained from a hexagon an approximation as correct as that for which Archimedes had required a polygon of 96 sides, while from a polygon of 96 sides he determined the value of π correct to seven decimal places instead of the two places obtained by Archimedes. The reason is that Archimedes, having calculated the lengths of the sides of inscribed and circumscribed regular polygons of n sides, assumed that the length of $1/n$th of the perimeter of the circle was intermediate between them; whereas Snell constructed from the sides of these polygons two other lines which gave closer limits for the corresponding arc. His method depends on the theorem $3\sin\theta/(2+\cos\theta) < \theta < (2\sin\frac{1}{3}\theta + \tan\frac{1}{3}\theta)$, by the aid of which a polygon of n sides gives a value of π correct to at least the integral part of $(4\log n - .2305)$ places of decimals, which is more than twice the number given by the older rule. Snell's proof of his theorem is incorrect, though the result is true.

Snell also added a table[‡] of the perimeters of all regular inscribed and circumscribed polygons, the number of whose sides is 10×2^n where n is not greater than 19 and not less than 3. Most of these were quoted from van Ceulen, but some were recalculated. This list has proved useful in refuting circle-squarers. A similar list was given by James Gregory[§].

In 1630 Grienberger[‖], by the aid of Snell's theorem, carried the

---

[*] *De Arithmetische en Geometrische Fondamenten*, Leyden, 1615, p. 163; or p. 144 of the Latin translation by W. Snell, published at Leyden in 1615 under the title *Fundamenta Arithmetica et Geometrica*. This was reissued, together with a Latin translation of the *Vanden Circkel*, in 1619, under the title *De Circulo*; in which see pp. 3, 29–32, 92.

[†] *Cyclometricus*, Leyden, 1621, p. 55.

[‡] It is quoted by Montucla, ed. 1831, p. 70.

[§] *Vera Circuli et Hyperbolae Quadratura*, prop. 29, quoted by Huygens, *Opera Varia*, Leyden, 1724, p. 447.

[‖] *Elementa Trigonometrica*, Rome, 1630, end of preface.

approximation to 39 places of decimals. He was the last mathematician who adopted the classical method of finding the perimeters of inscribed and circumscribed polygons. Closer approximations serve no useful purpose. Proofs of the theorems used by Snell and other calculators in applying this method were given by Huygens in a work[*] which may be taken as closing the history of this method.

In 1656 Wallis[†] proved that
$$\frac{\pi}{2} = \frac{2 \cdot 2 \cdot 4 \cdot 4 \cdot 6 \cdot 6 \cdots}{1 \cdot 3 \cdot 3 \cdot 5 \cdot 5 \cdot 7 \cdot 7 \cdots},$$
and quoted a proposition given a few years earlier by Viscount Brouncker to the effect that
$$\frac{\pi}{4} = 1 + \frac{1^2}{2+} \frac{3^2}{2+} \frac{5^2}{2+} \cdots,$$
but neither of these theorems was used to any large extent for calculation.

Subsequent calculators have relied on converging infinite series, a method that was hardly practicable prior to the invention of the calculus, though Descartes[‡] had indicated a geometrical process which was equivalent to the use of such a series. The employment of infinite series was proposed by James Gregory[§], who established the theorem that $\theta = \tan\theta - \frac{1}{3}\tan^3\theta + \frac{1}{5}\tan^5\theta - \cdots$, the result being true only if θ lies between $-\frac{1}{4}\pi$ and $\frac{1}{4}\pi$.

The first mathematician to make use of Gregory's series for obtaining an approximation to the value of π was Abraham Sharp[‖], who,

---

[*] *De Circula Magnitudine Inventa*, 1654; *Opera Varia*, pp. 351–387. The proofs are given in G. Pirie's *Geometrical Methods of Approximating the Value of* π, London, 1877, pp. 21–23.

[†] *Arithmetica Infinitorum*, Oxford, 1656, prop. 191. An analysis of the investigation by Wallis was given by Cayley, *Quarterly Journal of mathematics*, 1889, vol. XXIII, pp. 165–169.

[‡] See Euler's paper in the *Novi Commentarii Academiae Scientiarum*, St Petersburg, 1763, vol. VIII, pp. 157–168.

[§] See the letter to Collins, dated Feb. 15, 1671, printed in the *Commercium Epistolicum*, London, 1712, p. 25, and in the Macclesfield Collection, *Correspondence of Scientific Men of the Seventeenth Century*, Oxford, 1841, vol. II, p. 216.

[‖] See *Life of A. Sharp* by W. Cudworth, London, 1889, p. 170. Sharp's work is given in one of the preliminary discourses (p. 53 *et seq.*) prefixed to H. Sherwin's *Mathematical Tables*. The tables were issued at London in 1705: probably the discourses were issued at the same time, though the earliest copies I have seen were printed in 1717.

in 1699, on the suggestion of Halley, determined it to 72 places of decimals (71 correct). He obtained this value by putting $\theta = \tfrac{1}{6}\pi$ in Gregory's series.

Machin[*], earlier than 1706, gave the result to 100 places (all correct). He calculated it by the formula

$$\tfrac{1}{4}\pi = 4\tan^{-1}\tfrac{1}{5} - \tan^{-1}\tfrac{1}{239}.$$

De Lagny[†], in 1719, gave the result to 127 places of decimals (112 correct), calculating it by putting $\theta = \tfrac{1}{6}\pi$ in Gregory's series.

Hutton[‡], in 1776, and Euler[§], in 1779, suggested the use of the formulae $\tfrac{1}{4}\pi = \tan^{-1}\tfrac{1}{2} + \tan^{-1}\tfrac{1}{3}$ or $\tfrac{1}{4}\pi = 5\tan^{-1}\tfrac{1}{7} + 2\tan^{-1}\tfrac{3}{79}$, but neither carried the approximation as far as had been done previously.

Vega, in 1789[||], gave the value of π to 143 places of decimals (126 correct); and, in 1794[¶], to 140 places (136 correct).

Towards the end of the last century Baron Zach saw in the Radcliffe Library, Oxford, a manuscript by an unknown author which gives the value of π to 154 places of decimals (152 correct).

In 1841 Rutherford[**] calculated it to 208 places of decimals (152 correct), using the formula $\tfrac{1}{4}\pi = 4\tan^{-1}\tfrac{1}{5} - \tan^{-1}\tfrac{1}{70} + \tan^{-1}\tfrac{1}{99}$.

In 1844 Dase[††] calculated it to 205 places of decimals (200 correct), using the formula $\tfrac{1}{4}\pi = \tan^{-1}\tfrac{1}{2} + \tan^{-1}\tfrac{1}{5} + \tan^{-1}\tfrac{1}{8}$.

In 1847 Clausen[‡‡] carried the approximation to 250 places of decimals (248 correct), calculating it independently by the formulae $\tfrac{1}{4}\pi = 2\tan^{-1}\tfrac{1}{3} + \tan^{-1}\tfrac{1}{7}$ and $\tfrac{1}{4}\pi = 4\tan^{-1}\tfrac{1}{5} - \tan^{-1}\tfrac{1}{239}$.

In 1853 Rutherford[§§] carried his former approximation to 440 places of decimals (all correct), and William Shanks prolonged the approximation to 530 places. In the same year Shanks published an approximation

---

[*] W. Jones's *Synopsis Palmariorum*, London, 1706, p. 243; and Maseres, *Scriptores Logarithmici*, London, 1796, vol. III, pp. vii–ix, 155–164.
[†] *Histoire de l'Académie* for 1719, Paris, 1721, p. 144.
[‡] *Philosophical Transactions*, 1776, vol. LXVI, pp. 476–492.
[§] *Nova Acta Academiae Scientiarum Petropolitanae* for 1793, St Petersburg, 1798, vol. XI, pp. 133–149: the memoir was read in 1779.
[||] *Nova Acta Academiae Scientiarum Petropolitanae* for 1790, St Petersburg, 1795, vol. IX, p. 41.
[¶] *Thesaurus Logarithmorum* (*logarithmisch-trigonometrischer Tafeln*), Leipzig, 1794, p. 633.
[**] *Philosophical Transactions*, 1841, p. 283.
[††] *Crelle's Journal*, 1844, vol. XXVII, p. 198.
[‡‡] Schumacher, *Astronomische Nachrichten*, vol. XXV, col. 207.
[§§] *Proceedings of the Royal Society*, Jan. 20, 1853, vol. VI, pp. 273-275.

to 607 places[*]: and in 1873 he carried the approximation to 707 places of decimals[†]. These were calculated from Machin's formula.

In 1853 Richter, presumably in ignorance of what had been done in England, found the value of π to 333 places[‡] of decimals (330 correct); in 1854 he carried the approximation to 400 places[§]; and in 1855 carried it to 500 places[‖].

Of the series and formulae by which these approximations have been calculated, those used by Machin and Dase are perhaps the easiest to employ. Other series which converge rapidly are the following,

$$\frac{\pi}{6} = \frac{1}{2} + \frac{1}{2} \cdot \frac{1}{3 \cdot 2^3} + \frac{1 \cdot 3}{2 \cdot 4} \cdot \frac{1}{5 \cdot 2^5} + \cdots,$$

and

$$\frac{\pi}{4} = 2 + 22\tan^{-1}\frac{1}{28} + \tan^{-1}\frac{1}{443} - 5\tan^{-1}\frac{1}{1393} - 10\tan^{-1}\frac{1}{11018},$$

the latter of these is due to Mr Escott[¶].

As to those writers who believe that they have squared the circle their number is legion and, in most cases, their ignorance profound, but their attempts are not worth discussing here. "Only prove to me that it is impossible," said one of them, "and I will set about it immediately"; and doubtless the statement that the problem is insoluble has attracted much attention to it.

Among the geometrical ways of approximating to the truth the following is one of the simplest. Inscribe in the given circle a square, and to three times the diameter of the circle add a fifth of a side of the square, the result will differ from the circumference of the circle by less than one-seventeen-thousandth part of it.

An approximate value of π has been obtained experimentally by the theory of probability. On a plane a number of equidistant parallel straight lines, distance apart a, are ruled; and a stick of length l, which

---

[*] *Contributions to Mathematics*, W. Shanks, London, 1853, pp. 86, 87.
[†] *Proceedings of the Royal Society*, 1872–3, vol. XXI, p. 318; 1873–4, vol. XXII, p. 45.
[‡] *Grunert's Archiv*, vol. XXI, p. 119.
[§] *Ibid.*, vol. XXIII, p. 476: the approximation given in vol. XXII, p. 473, is correct only to 330 places.
[‖] *Ibid.*, vol. XXV, p. 472; and *Elbingen Anzeigen*, No. 85.
[¶] *L'Intermédiaire des mathématiciens*, Paris, Dec. 1896, vol. III, p. 276.

is less than a, is dropped on to the plane. The probability that it will fall so as to lie across one of the lines is $2l/\pi a$. If the experiment is repeated many hundreds of times, the ratio of the number of favourable cases to the whole number of experiments will be very nearly equal to this fraction: hence the value of π can be found. In 1855 Mr A. Smith[*] of Aberdeen made 3204 trials, and deduced $\pi = 3.1553$. A pupil of Prof. De Morgan[*], from 600 trials, deduced $\pi = 3.137$. In 1864 Captain Fox[†] made 1120 trials with some additional precautions, and obtained as the mean value $\pi = 3.1419$.

Other similar methods of approximating to the value of π have been indicated. For instance, it is known that if two numbers are written down at random, the probability that they will be prime to each other is $6/\pi^2$. Thus, in one case[‡] where each of 50 students wrote down 5 pairs of numbers at random, 154 of the pairs were found to consist of numbers prime to each other. This gives $6/\pi^2 = 154/250$, from which we get $\pi = 3.12$.

---

[*] A. De Morgan, *Budget of Paradoxes*, London, 1872, pp. 171, 172 [quoted from an article by De Morgan published in 1861].
[†] *Messenger of Mathematics*, Cambridge, 1873, vol. II, pp. 113, 114.
[‡] Note on π by R. Chartres. *Philosophical Magazine*, London, series 6, vol. XXXIX, March, 1904, p. 315.

CHAPTER IX.

MERSENNE'S NUMBERS.

ONE of the unsolved riddles of higher arithmetic, to which I have alluded in Chapter I, is the discovery of the method by which Mersenne or his contemporaries determined values of p which make a number of the form $2^p - 1$ a prime. It is convenient to describe such primes as *Mersenne's Numbers*. In this chapter, for shortness, I use N to denote a number of the form $2^p - 1$. In a memoir in the *Messenger of Mathematics* in 1891 I gave a brief sketch of the history of the problem. I here repeat the facts in somewhat more detail, and add a sketch of methods used in attacking the problem.

Mersenne's enunciation of the results associated with his name is in the preface to his *Cogitata**. The passage is as follows:

> "Vbi fuerit operae pretium aduertere XXVIII numeros a Petro Bungo pro perfectis exhibitos, capite XXVIII, libri de Numeris, non esse omnes Perfectos, quippe 20 sunt imperfecti, adeovt [adeunt?] solos octo perfectos habeat...qui sunt è regione tabulae Bungi, 1, 2, 3, 4, 8, 10, 12, et 29: quique soli perfecti sunt, vt qui Bungum habuerint, errori medicinam faciant.
>
> Porrò numeri perfecti adeo rari sunt, vt vndecim dumtaxat potuerint hactenus inueniri: hoc est, alii tres a Bongianis differentes: neque enim vllus est alius perfectus ab illis octo, nisi superes exponentem numerum 62, progressionis duplae ab 1 incipientis. Nonus enim perfectus est potestas exponentis 68 minus 1. Decimus, potestas exponentis 128, minus 1. Vndecimus denique, potestas 258, minus 1, hoc est potestas 257, vnitate decurtata, multiplicata per potestatem 256.

---

* *Cogitata Physico-Mathematica*, Paris, 1644, praefatio generalis, article 19.

> Qui vndecim alios repererit, nouerit se analysim omnem, quae fuerit hactenus, superasse: memineritque interea nullum esse perfectum à 17000 potestate ad 32000; & nullum potestatum interuallum tantum assignari posse, quin detur illud absque perfectis. Verbi gratia, si fuerit exponens 1050000, nullus erit numerus progressionis duplae vsque ad 2090000, qui perfectis numeris seruiat, hoc est qui minor vnitate, primus existat.
>
> Vnde clarum est quàm rari sint perfecti numeri, & quàm merito viris perfectis comparentur; esseque vnam ex maximis totius Matheseos difficultatibus, praescriptam numerorum perfectorum multitudinum exhibere; quemadmodum & agnoscere num dati numeri 15, aut 20 caracteribus constantes, sint primi necne, cùm nequidem saeculum integrum huic examini, quocumque modo hactenus cognito, sufficiat.

It is evident that, if p is not a prime, then N is composite, and two or more of its factors can be written down by inspection. Hence we may confine ourselves to prime values of p. Mersenne, in effect, asserted that the only values of p, not greater than 257, which make N a prime, are 1, 2, 3, 5, 7, 13, 17, 19, 31, 67, 127, 257: I assume that the number 67 is a misprint for 61. With this correction we have no reason to doubt the truth of the statement, but it has not been definitely established.

There are 56 primes not greater than 257. The determination of the prime or composite character of N for the 9 cases when p is less than 20 presents no difficulty: in only one of them is N composite. For 2 of the remaining 47 cases (namely, when $p = 23$ and 37) the decomposition of N had been given by Fermat. For 9 of them (namely, when $p = 29, 43, 73, 83, 131, 179, 191, 239, 251$) the factors of N were given by Euler. He also proved that N was prime when $p = 31$. Plana gave the factors of N when $p = 41$. Landry and Le Lasseur discovered the factors in 10 cases (namely, when $p = 47, 53, 59, 79, 97, 113, 151, 211, 223,$ and 233), but their analysis has not been published. Seelhoff showed that N was prime when $p = 61$, Cunningham gave the factors when $p = 197$, and Cole the factors when $p = 67$. Statements have been made that the composite character of N when $p = 89$, and its prime character when $p = 127$ have been proved, but the proofs have not been published or verified.

Thus there are 21 values of p for which Mersenne's statement still awaits verification. These are 71, 89, 101, 103, 107, 109, 127, 137, 139, 149, 157, 163, 167, 173, 181, 193, 199, 227, 229, 241, 257. For these values N is (according to Mersenne) prime when $p = 127$, and 257, and is composite for the other values, but as explained above it is probable that the character of N is known when $p = 89$ and 127.

To put the matter in another way. According to Mersenne's statement (corrected by the substitution of 61 for 67), 44 of the 56 primes less than 258 make N composite and the remaining 12 primes make N prime. In 25 out of the 44 cases in which N is said to be composite we know its factors, and in 19 cases the statement is still unverified. In 10 out of the 12 cases in which he said that N was prime his statement has been verified, and in 2 cases it is still unverified.

From the wording of the last clause in the above quotation it has been conjectured that the result had been communicated to Mersenne, and that he published it without being aware of how it was proved. In itself this seems probable. He was a good mathematician, but not an exceptional genius. It would be strange if he established a proposition which has baffled Euler, Lagrange, Legendre, Gauss, Jacobi, and other mathematicians of the first rank; but if the proposition is due to Fermat, with whom Mersenne was in constant correspondence, the case is altered, and not only is the absence of a demonstration explained, but we cannot be sure that we have attacked the problem on the best lines.

The known results as to the prime or composite character of N, and in the latter case its smallest factor, are given in the table on the facing page. The cases that remain as yet unverified are marked with an asterisk.

Before describing the methods used for attacking the problem it will be convenient to state in more detail when and by whom these results were established.

The factors (if any) of such values of N as are less than a million can be verified easily: they have been known for a long time, and I need not allude to them in detail.

The factors of N when $p = 11, 23$, and 37 had been indicated by Fermat*, some four years prior to the publication of Mersenne's work, in a letter dated October 18, 1640. The passage is as follows:

> En la progression double, si d'un nombre quarré, généralement parlant, vous ôtez 2 ou 8 ou 32 &c., les nombres premiers moindres de l'unite qu'un multiple du quaternaire, qui mesureront le reste, feront l'effet requis. Comme de 25, qui est un quarré, ôtez 2; le reste 23 mesurera la 11e puissance -1; ôtez 2 de 49, le reste 47 mesurera la 23e puissance -1. Ôtez 2 de 225, le reste 223 mesurera la 37e puissance -1, &c.

---

* *Oeuvres de Fermat*, Paris, vol. II, 1894, p. 210; or *Opera Mathematica*, Toulouse, 1679, p. 164; or Brassinne's *Précis*, Paris, 1853, p. 144.

MERSENNE'S NUMBERS.

p	Value of $N = 2^p - 1$		
1	1	prime	
2	3	prime	
3	7	prime	
5	31	prime	
7	127	prime	
11	$2047 = 23 \times 89$	composite	
13	8191	prime	
17	131071	prime	
19	524287	prime	
23	$8388607 = 47 \times 178481$	composite	Fermat
29	$536870911 = 233 \times 1103 \times 2089$	composite	Euler
31	2147483647	prime	Euler
37	$137438953471 = 223 \times 616318177$	composite	Fermat
41	$2199023255551 = 13367 \times 164511353$	composite	Plana
43	$8796093022207 = 431 \times 9719 \times 2099863$	composite	Euler
47	$2351 \times 4513 \times 13264529$	composite	Landry
53	$6361 \times 69431 \times 20394401$	composite	Landry
59	$179951 \times 3203431780337$	composite	Landry
61	2305843009213693951	prime	Seelhoff
67	$\equiv 0\ (193707721)$	composite	Cole
71	2361183241434822606847	$*$	
73	$\equiv 0\ (439)$	composite	Euler
79	$\equiv 0\ (2687)$	composite	Le Lasseur
83	$\equiv 0\ (167)$	composite	Euler
89	618970019642690137449562111	$*$	
97	$\equiv 0\ (11447)$	composite	Le Lasseur
101	2535301200456458802993406410751	$*$	
103	10141204801825835211973625643007	$*$	
107	162259276829213363391578010288127	$*$	
109	649037107316853453566312041152511	$*$	
113	$\equiv 0\ (3391)$	composite	Le Lasseur
127	170141183460469231731687303715884105727	$*$	
131	$\equiv 0\ (263)$	composite	Euler
137	....	$*$	
139	....	$*$	
149	....	$*$	
151	$\equiv 0\ (18121)$	composite	Le Lasseur
157	....	$*$	
163	....	$*$	
167	....	$*$	

p	Value of $N = 2^p - 1$		
173	....	*	
179	$\equiv 0\ (359)$	composite	Euler
181	....	*	
191	$\equiv 0\ (383)$	composite	Euler
193	....	*	
197	$\equiv 0\ (7487)$	composite	Cunningham
199	....	*	
211	$\equiv 0\ (15193)$	composite	Le Lasseur
223	$\equiv 0\ (18287)$	composite	Le Lasseur
227	....	*	
229	....	*	
233	$\equiv 0\ (1399)$	composite	Le Lasseur
239	$\equiv 0\ (479)$	composite	Euler
241	....	*	
251	$\equiv 0\ (503)$	composite	Euler
257	....	*	

The factors of N when $p = 29$, 43, and 73 were given by Euler[*] in 1732. The fact that N is composite for the values $p = 83$, 131, 179, 191, 239, and 251 follows from a proposition enunciated, at the same time, by Euler to the effect that, if $4n + 3$ and $8n + 7$ are primes, then $2^{4n+3} - 1 \equiv 0 \pmod{8n + 7}$. This was proved by Lagrange[†] in his classical memoir of 1775. The proposition also covers the cases of $p = 11$ and $p = 23$. This is the only general theorem on the subject which is yet established.

The fact that N is prime when $p = 31$ was proved by Euler[‡] in 1771. Fermat had asserted, in the letter mentioned above, that the only possible prime factors of $2^p \pm 1$, when p was prime, were of the form $np + 1$, where n is an integer. This was proved by Euler[§] in 1748, who added that, since $2^p \pm 1$ is odd, every factor of it must be odd, and therefore if p is odd n must be even. But if p is a given number we can define n much more closely, and Euler showed that the prime

---

[*] *Commentarii Academiae Scientiarum Petropolitanae*, 1738, vol. VI, p. 103; or *Commentationes Arithmeticae Collectae*, vol. I, p. 2.

[†] *Nouveaux Mémoires de l'Académie des Sciences de Berlin*, 1775, pp. 323–356.

[‡] *Histoire de l'Académie des Sciences* for 1772, Berlin, 1774, p. 36. See also Legendre, *Théorie des Nombres*, third edition, Paris, 1830, vol. I, pp. 222–229.

[§] *Novi Commentarii Academiae Scientiarum Petropolitanae*, vol. I, p. 20; or *Commentationes Arithmeticae Collectae*, St Petersburg, 1849, vol. I, pp. 55, 56.

factors (if any) of $2^{31} - 1$ were necessarily primes of the form $248n + 1$ or $248n + 63$; also they must be less than $\sqrt{2^{31} - 1}$, that is, than 46339. Hence it is necessary to try only forty divisors to see if $2^{31} - 1$ is a prime or composite.

The factors of N when $p = 41$ were given by Plana[*] in 1859. He showed that the prime factors (if any) are primes of the form $328n+1$ or $328n+247$, and lie between 1231 and $\sqrt{2^{41} - 1}$, that is, 1048573. Hence it is necessary to try only 513 divisors to see if $2^{41} - 1$ is composite: the seventeenth of these divisors gives the required factors. This is the same method of attacking the problem which was used by Euler in 1771, but it would be very laborious to employ it for values of p greater than 41. Plana[†] added the forms of the prime divisors of N, if p is not greater than 101.

That N is prime when $p = 127$ seems to have been verified by Lucas[‡] in 1876 and 1877. The demonstration has not been published.

The discovery of the factors of N for the values $p = 47, 53$, and 59 is due apparently to the late F. Landry, who established theorems on the factors (if any) of numbers of certain forms. Instead of publishing his results he issued a challenge to all mathematicians to solve the general problem. This is contained in a rare pamphlet published at Paris in 1867, of which I possess a copy, in which the factors of certain numbers are given, and on page 8 of which it is implied that he had obtained the factors of $2^p - 1$ when $p = 47, 53$, and 59. He seems to have communicated his results to Lucas, who quoted them in the memoir cited below[§].

The factors of N when $p = 79$ and 113 were given first by Le Lasseur, and were quoted by Lucas in the same memoir[§].

A factor of N when $p = 233$ was discovered later by Le Lasseur, and was quoted by Lucas in 1882[∥].

---

[*] G.A.A. Plana, *Memorie della Reale Accademia delle Scienze di Torino*, Series 2, vol. XX, 1863, p. 130.

[†] *Ibid.*, p. 137.

[‡] *Sur la Théorie des Nombres Premiers*, Turin, 1876, p. 11; and *Recherches sur les Ouvrages de Léonard de Pise*, Rome, 1877, p. 26, quoted by Lieut.-Colonel A.J.C. Cunningham, *Proceedings of the London Mathematical Society*, Nov. 14, 1895, vol. XXVII, p. 54.

[§] *American Journal of Mathematics*, 1878, vol. I, p. 236.

[∥] *Récréations*, 1882–3, vol. I, p. 241.

The factors of N when $p = 97, 151, 211$, and 223 were determined subsequently by Le Lasseur, and were quoted by Lucas[*] in 1883.

That N is prime when $p = 61$ had been conjectured by Landry and in 1886 a demonstration was offered by Seelhoff[†]. His demonstration is open to criticism, but the fact has been verified by others[‡], and may be accepted as proved.

That N is composite when $p = 89$ seems to have been verified by Lucas[§] in 1891, but the demonstration has not been published, nor have the actual factors been discovered.

That 7487 is a factor of N when $p = 197$ was shown by A.J.C. Cunningham in 1895[||].

That N is not prime when $p = 67$ seems to have been verified by Lucas[¶] in 1876 and 1877. The composite nature of[1] $N = 2^p - 1$ when $p = 67$ was confirmed by E. Fauquembergue[**], and was also implied by Lucas[††] in 1891. The factors were given by F.N. Cole[‡‡] in 1903.

Bickmore in the memoir[§§] cited below showed that 1433 is another factor of N if $p = 179$; and that 1913 and 5737 are other factors of N if $p = 239$.

I turn next to consider the methods by which these results can be obtained. It is impossible to believe that the statement made by

---

[*] *Récréations*, vol. II, p. 230.
[†] P.H.H. Seelhoff, *Zeitschrift für Mathematik und Physik*, 1886, vol. XXXI, p. 178.
[‡] See Weber-Wellstein, *Encyclopaedie der Elementar-Mathematik*, p. 48; and F.N. Cole, *Bulletin of the American Mathematical Society*, December, 1903, p. 136.
[§] *Théorie des Nombres*, Paris, 1891, p. 376.
[||] *Proceedings of the London Mathematical Society*, March 14, 1895, vol. XXVI, p. 261.
[¶] *Sur la Théorie des Nombres Premiers*, Turin, 1876, p. 11, quoted by Lieut-Colonel A.J.C. Cunningham, *Proceedings of the London Mathematical Society*, Nov. 14, 1895, vol. XXVII, p. 54, and *Recherches sur les Ouvrages de Léonard de Pise*, Rome, 1877, p. 26.
[**] *L'Intermédiaire des mathématiciens*, Paris, Sept. 1894, vol. I, p. 148.
[††] *Théorie des Nombres*, Paris, 1891, p. 376.
[‡‡] *On the Factoring of Large Numbers*, *Bulletin of the American Mathematical Society*, December, 1903, pp. 134–137.
[§§] C.E. Bickmore, *Messenger of Mathematics*, Cambridge, 1895, vol. XXV, p. 19.

---

1. Corrected: originally $N\ 2^p = 1$

Mersenne rested on an empirical conjecture, but the riddle as to how it was discovered is still, after nearly 250 years, unsolved.

I cannot offer any solution of the riddle. But it may be interesting to indicate some ways which have been used in attacking the problem. The object is to find a prime divisor q (other than N and 1) of a number N when N is of the form $2^p - 1$ and p is a prime. It can be easily shown that q must be of the form $2pt + 1$. Also q must be of one of the forms $8i \pm 1$: for N is of the form $2A^2 - B^2$, where A is even and B odd, hence[*] any factor of it must be of the form $2a^2 - b^2$, that is, of the form $8i \pm 1$, and 2 must be a quadratic residue of q. The theory of residues is, however, of but little use in finding factors of the cases that still await solution, though the possibility some day of finding a complete series of solutions by properties of residues must not be neglected[†]. Our present knowledge of the means of factorizing N has been summed up in the statement[‡] that a prime factor of the form $2pt + 1$ can be found directly by rules due to Legendre, Gauss, and Jacobi, when $t = 1, 3, 4, 8,$ or 12; and that a factor of the form $2ptt' + 1$ can be found indirectly by a method due to Bickmore when $t = 1, 3, 4, 8,$ or 12, and t' is an odd integer greater than 3. But this only indicates how little has yet been done towards finding a general solution of the problem.

First. There is the simple but crude method of trying all possible prime divisors q which are of the form $2pt + 1$ as well as of one of the forms $8i \pm 1$.

The chief known results for the smaller factors may be summarized by saying that a prime of this form will divide N when $t = 1$, if $p = 11$, 23, 83, 131, 179, 191, 239, or 251; when $t = 3$, if $p = 37, 73,$ or 233; when $t = 5$, if $p = 43$; when $t = 15$, if $p = 113$; when $t = 17$, if $p = 79$; when $t = 19$, if $p = 29$, or 197; when $t = 25$, if $p = 47$; when $t = 41$, if $p = 223$; when $t = 59$, if $p = 97$; when $t = 163$, if $p = 41$; when $t = 1525$, if $p = 59$; when $t = 4$, if $p = 11, 29, 179,$ or 239; when $t = 8$, if $p = 11$; when $t = 12$, if $p = 239$; when $t = 36$, if $p = 29$, or 211; when

---

[*] Legendre, *Théorie des Nombres*, third edition, Paris, 1830, vol. I, § 143. In the case of Mersenne's numbers, $B = b = 1$.

[†] For methods of finding the residue indices of 2 see Bickmore, *Messenger of Mathematics*, May, 1895, vol. XXV, pp. 15–21; see also Lieut-Colonel A.J.C. Cunningham on 2 as a 16-ic residue, *Proceedings of the London Mathematical Society*, 1895–6, vol. XXVII, pp. 85–122.

[‡] *Transactions of the British Association for the Advancement of Science* (Ipswich Meeting), 1895, p. 614.

$t = 60$, if $p = 53$, or 151; and when $t = 1445580$, if $p = 67$.

Of the 25 cases in which we know that Mersenne's statement of the composite character of N is correct all save 3 can be easily verified by trial in this way. For neglecting all values of t not exceeding, say, 60 which make q either composite or not of one of the forms $8i \pm 1$ we have in each case only some 20 or so divisors to try. Of the 3 other cases in which Mersenne's statement of the composite character of N has been verified, one verification ($p = 41$) is due to Plana, who frankly confessed that the result was reached "par un heureux hasard"; a second is due to Landry ($p = 59$), who did not explain how he obtained the factors; and the third is due to Cole ($p = 67$), who established it by the use of quadratic residues of N, involving laborious numerical work.

Of the 10 cases in which we know that Mersenne's statement of the prime character of N is correct all save one may be verified by trial in this way, for the number of possible factors is not large. The exception is the case where p equals 61, which Seelhoff and others have shown to be prime.

Thus practically we may say that simple empirical trials would at once lead us to all the conclusions known except in the case of $p = 41$ due to Plana, of $p = 59$ due to Landry, of $p = 61$ due to Seelhoff, and of $p = 67$ due to Cole. In fact, save for these four results the conclusions of all mathematicians to date could be obtained by anyone by a few hours' arithmetical work.

As p increases the number of factors to be tried increases so fast that, if p is large, it would be practically impossible to apply the test to obtain large factors. This is an important point, for it has been asserted that in the cases still awaiting verification there are no factors less than 50,000. Hence, we may take it as reasonably certain that this cannot have been the method by which the result was originally obtained; nor, as here enunciated, is it likely to give many factors not yet known. Of course it is possible there may be ways by which the number of possible values of t might be further limited, and if we could find them we might thus diminish the number of possible factors to be tried, but it will be observed that the values of N which still await verification are very large, for instance, when $p = 257$, N contains no less than 78 digits.

It is hardly necessary to add that if q is known and is not very large we can determine whether or not it is a factor of N without the labour of performing the division.

For instance, if we want to verify that $q = 13367$ is a factor of N when $p = 41$, we proceed thus. Take the power of 2 nearest to q or to its square-root. We have to the modulus q

$$\begin{aligned} 2^{14} &= 16384 \equiv 3017 \equiv 7 \times 431\,, \\ \therefore\ 2^{28} &\equiv 49(-1377) \equiv -638\,, \\ \therefore\ 2^{27} &\equiv -319\,, \\ \therefore\ 2^{14+27} &\equiv (3017)(-319) \equiv 1\,, \\ \therefore\ 2^{41} &\equiv 1\,. \end{aligned}$$

Second. We can proceed by reducing the problem to the solution of an indeterminate equation.

It is clear that we can obtain a factor of N if we can express it as the difference of the squares, or more generally of the nth powers, of two integers u and v. Further, if we can express a multiple of N, say mN, in this form, we can find a factor of mN and (with certain obvious limitations as to the value of m) this will lead to a factor of N. It may be also added that if m can be found so that N/m is expressible as a continued fraction of a certain form, a certain continuant[*] defined by the form of the continued fraction is a factor of N.

Since N can always be expressed as the difference of two squares, this method seems a natural one by which to attack the problem. If we put

$$N = n^2 + a = (n+b)^2 - (b^2 + 2bn - a),$$

we can make use of the known forms of u and v, and thus obtain an indeterminate equation between two variables x and y of the form

$$x^2 = (2py + H)^2 - 4(K - y)$$

where H and K are numbers which can be easily calculated. Integral values of x and y where $y < K$ will determine values of u and v, and thus give factors of N.

---

[*] See J.G. Birch in the *Messenger of Mathematics*, August, 1902, vol. XXII, pp. 52–55.

We can also attack the problem by indeterminate equations in another way. For the factors must be of the form $2pt+1$ and $8ps+1$, hence

$$(2pt+1)(8ps+1) = N,$$
$$= 2^p - 1,$$
$$= 2(2^{p-1} - 1) + 1,$$
$$\therefore 4s + t + 8pst = (2^{p-1} - 1)/p,$$
$$= (\text{say})\alpha + 8p\beta.$$

Hence $\quad 4s + t = \alpha + 8px$, and $st = \beta - x$,

where $x \not> \beta$ and t is odd. These results again lead to an indeterminate equation.

But, in both cases, unless p is small, the resulting equations are intractable.

Third. A not uncommon method of attacking problems such as this, dealing with the factorization of large numbers, is through the theory of quadratic forms[*]. At best this is a difficult method to use, and we have no reason to think that it would have been employed by a mathematician of the seventeenth century. I here content myself with alluding to it.

Fourth. There is yet another way in which the problem might be attacked. The problem will be solved if we can find an odd prime q so that to it as modulus $2^{p+y} \equiv z$, and $2^y \equiv z$, where y and z may have any values we like to choose. If such values of q, y, and z can be found, we have $2^y(2^p - 1) \equiv 0$. Therefore $2^p = 1$, that is, q is a divisor of N.

For example, to the modulus 23, we have

$$2^8 \equiv 3,$$
$$2^{16} \equiv 3^2.$$

Also $\quad\quad\quad\quad\quad\quad\quad 2^5 \equiv 3^2.$

Therefore $\quad\quad\quad\quad 2^{16} - 2^5 \equiv 0,$
$$\therefore 2^{11} - 1 \equiv 0.$$

Without going further into the matter we may say that the *à priori* determination of the values q, y, and z introduces us to an almost

---

[*] For a sketch of this see G.B. Mathews, *Theory of Numbers*, part 1, Cambridge, 1891, pp. 261-271. See also F.N. Cole's paper, On the Factoring of Large Numbers, *Bulletin of the American Mathematical Society*, December, 1903, pp. 134-137; and *Quadratic Partitions* by A.J.C. Cunningham, London, 1904.

untrodden field. It is just possible (though I should suppose unlikely) that the key to the riddle is to be found on methods of finding q, y, z, to satisfy the above conditions. For instance, if we could say what was the remainder when 2^x was divided by a prime q of the form $2pt + 1$, and if the remainders were the same when $x = u$ and $x = v$, then to the modulus q we should have, $2^u \equiv 2^v$, and therefore $2^{u-v} \equiv 1$.

It should however be noted that Jacobi's *Canon Arithmeticus* and the similar canon drawn up by Cunningham would, if carried far enough, enable us to solve the problem by this method. Cunningham's *Canon* gives the solution of the congruence $2^x \equiv R$ for all prime moduli less than 1000, but it is of no use in determining factors of N larger than 1000. It is however possible that if R or q have certain forms such a canon might be constructed, and thus lead to a solution of the problem.

Fifth. It is noteworthy that the odd values of p specified by Mersenne are primes of one of the four forms $2^q \pm 1$ or $2^q \pm 3$, but it is not the case that all primes of these forms make N a prime, for instance, N is composite if $p = 2^3 + 3 = 11$ or if $p = 2^5 - 3 = 29$.

This fact has suggested to more than one mathematician the possibility that some test as to the prime or composite character of N when p is of one of these forms may be discoverable. Of course this is merely a conjecture. There is however this to say for it, that we know that Fermat[*] had paid attention to numbers of this form.

Sixth. The number N when expressed in the binary scale, consists of 1 repeated p times. This has suggested whether the work connected with the determination of factors of N might not with advantage be expressed in the binary scale. A method based on the use of properties of this scale has been indicated by G. de Longchamps[†], but as there given it would be unlikely to lead to the discovery of large divisors. I am, however, inclined to think that greater advantages would be gained by working in a scale whose radix was $4p$ or may-be $8p$—the resulting numbers being then expressed by a reasonably small number of digits. In fact when expressed in the latter scale in only one out of the 25 cases in which the factors of N are known does the smallest factor contain more than two digits.

---

[*] *Ex. gr.*, see above, page 31.
[†] *Comptes rendus de l'Académie des Sciences*, Paris, Nov. 1877, vol. LXXXV, pp. 950–952.

Seventh. I have reserved to the last the description of the method which seems to me to be the most hopeful.

We know by Fermat's Theorem that if $x+1$ is a prime then $2^x - 1$ is divisible by $x+1$. Hence if $2pt+1$ is a prime we have, to the modulus $2pt+1$

$$2^{2pt} - 1 \equiv 0,$$
$$\therefore (2^p - 1)(1 + 2^p + 2^{2p} + \cdots + 2^{(2t-1)p}) \equiv 0.$$

Hence, a divisor of $2^p - 1$ will be known, if we can find a value of t such that $2pt+1$ is prime and the second factor is prime to it.

This method may be used to establish Euler's theorem of 1732. For if we put $t=1$, and if $2p+1$ is a prime, we have, to the modulus $2p+1$

$$(2^p - 1)(2^p + 1) \equiv 0.$$

Hence $2^p \equiv 1$ if $2^p + 1$ is prime to $2p+1$. This is the case if $p = 4m+3$. Hence $2p+1$ is a factor of N if $p = 11, 23, 83, 131, 179, 191, 239$, and 251, for in these cases $2p+1$ is prime.

The problem of Mersenne's Numbers is a particular case of the determination of the factors of $a^n - 1$. This has been the subject of investigations by many mathematicians: an outline of their conclusions has been given by Bickmore[*]. I ought also to add a reference to the general method suggested by F.W. Lawrence[†] for the factorization of any high number: it is possible that Fermat used some method analogous to this.

Finally, I should add that machines[‡] have been devised for investigating whether a number is prime, but I do not know that any have been constructed suitable for numbers as large as those involved in the numbers in question.

---

[*] *Messenger of Mathematics*, Cambridge, 1895–6, vol. XXV, pp. 1–44; also 1896–7, vol. XXVI, pp. 1–38; see also a note by Mr E.B. Escott in the *Messenger*, 1903–4, vol. XXXIII, p. 49.

[†] *Ibid.*, 1894–5, vol. XXIV, pp. 100–109; *Quarterly Journal of Mathematics*, 1896, vol. XXVIII, pp. 285–311; and *Transactions of the London Mathematical Society*, May 13, 1897, vol. XXVIII, pp. 465–475.

[‡] F.W. Lawrence, *Quarterly Journal of Mathematics*, 1896, already quoted, pp. 310–311; see also C.A. Laisant, *Comptes Rendus Association Français pour l'avancement des sciences*, 1891 (Marseilles), vol. XX, pp. 165–8.

CHAPTER X.

ASTROLOGY.

Astrologers professed to be able to foretell the future, and within certain limits to control it. I propose to give in this chapter a concise account of the rules they used for this purpose[*].

I have not attempted to discuss the astrology of periods earlier than the middle ages, for the technical laws of the ancient astrology are not known with accuracy. At the same time there is no doubt that, as far back as we have any definite historical information, the art was practised in the East; that thence it was transplanted to Egypt, Greece, and Rome; and that the medieval astrology was founded on it. It is probable that the rules did not differ materially from those described in this chapter[†], and it may be added that the more intelligent thinkers of the old world recognised that the art had no valid pretenses to accuracy. I may note also that the history of the development of the art ceases with the general acceptance of the Copernican theory, after which the practice of astrology rapidly became a mere cloak for imposture.

---

[*] I have relied mainly on the *Manual of Astrology* by Raphael—whose real name was R.C. Smith—London, 1828, to which the references to Raphael hereafter given apply; and on Cardan's writings, especially his commentary on Ptolemy's work and his *Geniturarum Exempla*. I am indebted also for various references and gossip to Whewell's *History of the Inductive Sciences*; to various works by Raphael, published in London between 1825 and 1832; and to a pamphlet by M. Uhlemann, entitled *Grundzüge der Astronomie und Astrologie*, Leipzig, 1857.

[†] On the influences attributed to the planets, see *The Dialogue of Bardesan on Fate*, translated by W. Cureton in the *Spicilegium Syriacum*, London, 1855.

All the rules of the medieval astrology—to which I confine myself—are based on the Ptolemaic astronomy, and originate in the *Tetrabiblos*[*] which is said, it may be falsely, to have been written by Ptolemy himself. The art was developed by numerous subsequent writers, especially by Albohazen[†], and Firmicus. The last of these collected the works of most of his predecessors in a volume[‡], which remained a standard authority until the close of the sixteenth century.

I may begin by reminding the reader that though there was a fairly general agreement as to the methods of procedure and interpretation—which alone I attempt to describe—yet there was no such thing as a fixed code of rules or a standard text-book. It is therefore difficult to reduce the rules to any precise and definite form, and almost impossible, within the limits of a chapter, to give detailed references. At the same time the practice of the elements of the art was tolerably well established and uniform, and I feel no doubt that my account, as far as it goes, is substantially correct.

There were two distinct problems with which astrologers concerned themselves. One was the determination in general outline of the life and fortunes of an enquirer: this was known as *natal astrology*, and was effected by the erection of a *scheme of nativity*. The other was the means of answering any specific question about the individual: this was known as *horary astrology*. Both depended on the casting or erecting of a *horoscope*. The person for whom it was erected was known as the *native*.

A horoscope was cast according to the following rules[§]. The space between two concentric and similarly situated squares was divided into twelve spaces, as shown in the annexed diagram. These twelve spaces were known technically as *houses*; they were numbered consecutively 1, 2, ..., 12 (see figure); and were described as the first house, the second house, and so on. The dividing lines were termed *cusps*: the line between the houses 12 and 1 was called the cusp of the first house, the line between the houses 1 and 2 was called the cusp of the second house, and so on, finally the line between the houses 11 and 12 was called the cusp of the twelfth house. Each house had also a name of its

---

[*] There is an English translation by J. Wilson, London [*n.d.*]; and a French translation is given in Halma's edition of Ptolemy's works.
[†] *De judiciis astrorum*, ed. Liechtenstein, Basle, 1571.
[‡] *Astronomicorum*, eight books, Venice, 1499.
[§] Raphael, pp. 91–109.

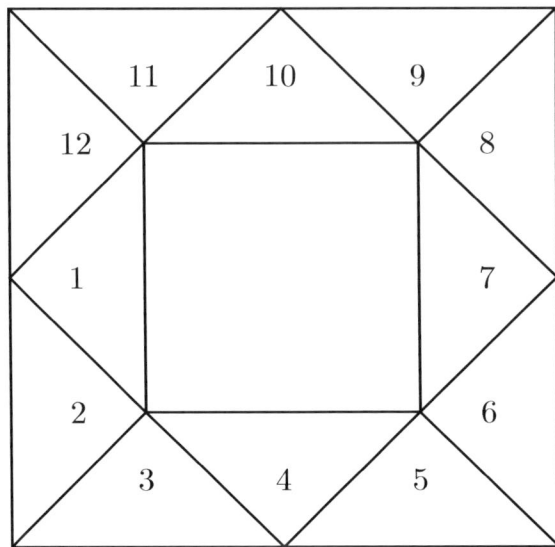

own—thus the first house was called the ascendant house, the eighth house was called the house of death, and so on—but as these names are immaterial for my purpose I shall not define them.

Next, the positions which the various astrological signs and planets had occupied at some definite time and place (for instance, the time and place of birth of the native, if his nativity was being cast) were marked on the celestial sphere. This sphere was divided into twelve equal spaces by great circles drawn through the zenith, the angle between any two consecutive circles being 30°. The first circle was drawn through the East point, and the space between it and the next circle towards the North corresponded to the first house, and sometimes was called the first house. The next space, proceeding from East to North, corresponded to the second house, and so on. Each of the twelve spaces between these circles corresponded to one of the twelve houses, and each of the circles to one of the cusps.

In delineating[*] a horoscope, it was usual to begin by inserting the zodiacal signs. A zodiacal sign extends over 30°, and was marked on the cusp which passed through it: by its side was written a number indicating the distance to which its influence extended in the earlier of the two houses divided by the cusp. Next the position of the planets in these signs were calculated, and each planet was marked in its proper

---

[*] Raphael, pp. 118–131.

house and near the cusp belonging to the zodiacal sign in which the planet was then situated: it was followed by a number indicating its right ascension measured from the beginning of the sign. The name of the native and the date for which the horoscope was cast were inserted usually in the central square. The diagram near the end of this chapter is a facsimile of the horoscope of Edward VI as cast by Cardan and will serve as an illustration of the above remarks.

We are now in a position to explain how a horoscope was *read* or interpreted. Each house was associated with certain definite questions and subjects, and the presence or absence in that house of the various signs and planets gave the answer to these questions or information on these subjects.

These questions cover nearly every point on which information would be likely to be sought. They may be classified roughly as follows. For the answer, so far as it concerns the native, to all questions connected with his life and health, look in house 1; for questions connected with his wealth, refer to house 2; for his kindred and communications to him, refer to 3; for his parents and inheritances, refer to 4; for his children and amusements, refer to 5; for his servants and illnesses, refer to 6; for his marriage and amours, refer to 7; for his death, refer to 8; for his learning, religion and travels, refer to 9; for his trade and reputation, refer to 10; for his friends, refer to 11; and finally for questions connected with his enemies, refer to house 12.

I proceed to describe briefly the influences of the planets, and shall then mention those of the zodiacal signs; I should note however that in practice the signs were in many respects more influential than the planets.

The astrological "planets" were seven in number, and included the Sun and the Moon. They were Saturn or the Great Infortune, Jupiter or the Great Fortune, Mars or the Lesser Infortune, the Sun, Venus or the Lesser Fortune, Mercury, and the Moon: the above order being that of their apparent times of rotation round the earth.

Each of them had a double signification. In the first place it impressed certain characteristics, such as good fortune, feebleness, &c., on the dealings of the native with the subjects connected with the house in which it was located; and in the second place it imported certain objects into the house which would affect the dealings of the native with the subjects of that house.

To describe the exact influence of each planet in each house would involve a long explanation, but the general effect of their presence may be indicated roughly as follows*. The presence of Saturn is malignant: that of Jupiter is propitious: that of Mars is on the whole injurious: that of the Sun indicates respectability and moderate success: that of Venus is rather favourable: that of Mercury implies rapid practical action: and lastly the presence of the Moon merely faintly reflects the influence of the planet nearest her, and suggests rapid changes and fickleness. Besides the planets, the Moon's nodes and some of the more prominent fixed stars† also had certain influences.

These vague terms may be illustrated by taking a few simple cases.

For example, in casting a nativity, the life, health, and general career of the native were determined by the first or ascendant house, whence comes the expression that a man's fortune is in the ascendant. Now the most favourable planet was Jupiter. Therefore, if at the instant of birth Jupiter was in the first house, the native might expect a long, happy, healthy life; and being born under Jupiter he would have a "jovial" disposition. On the other hand, Saturn was the most unlucky of all the planets, and was as potent as malignant. If at the instant of birth he was in the first house, his potency might give the native a long life, but it would be associated with an angry and unhappy temper, a spirit covetous, revengeful, stern, and unloveable, though constant in friendship no less than in hate, which was what astrologers meant by a "saturnine" character. Similarly a native born under Mercury, that is, with Mercury in the first house, would be of a mercurial nature, while anyone born under Mars would have a martial bent.

Moreover it was the prevalent opinion that a jovial person would have his horoscope affected by Jupiter, even if that planet had not been in the ascendant at the time of birth. Thus the horoscope of an adult depended to some extent on his character and previous life. It is hardly necessary to point out how easily this doctrine enabled an astrologer to make the prediction of the heavens agree with facts that were known or probable.

In the same way the other houses are affected. For instance, no astrologer, who believed in the art, would have wished to start on a long journey when Saturn was in the ninth house or house of travels;

---

* Raphael, pp. 70–90; pp. 204–209.
† Raphael, pp. 129–131,

and, if at the instant of birth Saturn was in that house, the native always would incur considerable risk on his journeys.

Moreover every planet was affected to some extent by its aspect (conjunction, opposition, or quadrature) to every other planet according to elaborate rules* which depended on their positions and directions of motion: in particular the angular distance between the Sun and the Moon—sometimes known as the "part of fortune"—was regarded as specially important, and this distance affected the whole horoscope. In general, conjunction was favourable, quadrature unfavourable, and opposition ambiguous.

Each planet not only influenced the subjects in the house in which it was situated, but also imported certain objects into the house. Thus Saturn was associated with grandparents, paupers, beggars, labourers, sextons, and gravediggers. If, for example, he was present in the fourth house, the native might look for a legacy from some such person; if he was present in the twelfth house, the native must be careful of the consequences of the enmity of any such person; and so on.

Similarly Jupiter was associated generally with lawyers, priests, scholars, and clothiers; but, if he was conjoined with a malignant planet, he represented knaves, cheats, and drunkards. Mars indicated soldiers (or, if in a watery sign, sailors on ships of war), masons, doctors, smiths, carpenters, cooks, and tailors; but, if afflicted with Mercury or the Moon, he denoted the presence of thieves. The Sun implied the action of kings, goldsmiths, and coiners; but, if afflicted by a malignant planet, he denoted false pretenders. Venus imported musicians, embroiderers, and purveyors of all luxuries; but, if afflicted, prostitutes and bullies. Mercury imported astrologers, philosophers, mathematicians, statesmen, merchants, travellers, men of intellect, and cultured workmen; but, if afflicted, he signified the presence of pettifoggers, attorneys, thieves, messengers, footmen, and servants. Lastly, the presence of the Moon introduced sailors and those engaged in inferior offices.

I come now to the influence and position of the zodiacal signs. So far as the first house was concerned, the sign of the zodiac which was there present was even more important than the planet or planets, for it was one of the most important indications of the durations of life.

Each sign was connected with certain parts of the body—*ex. gr.* Aries influenced the head, neck and shoulders—and that part of the

---

* Raphael, pp. 132–170.

body was affected according to the house in which the sign was. Further each sign was associated with certain countries and connected the subjects of the house in which the sign was situated with those countries: *ex. gr.* Aries was associated especially with events in England, France, Syria, Verona, Naples, &c.

The sign in the first house determined also the character and appearance of the native*. Thus the character of a native born under Aries (m) was passionate; under Taurus (f) was dull and cruel; under Gemini (m) was active and ingenious; under Cancer (f) was weak and yielding; under Leo (m) was generous, resolute, and ambitious; under Virgo (f) was sordid and mean; under Libra (m) was amorous and pleasant; under Scorpio (f) was cold and reserved; under Sagittarius (m) was generous, active, and jolly; under Capricorn (f) was weak and narrow; under Aquarius (m) was honest and steady; and under Pisces (f) was phlegmatic and effeminate.

Moreover the signs were regarded as alternately masculine and feminine, as indicated above by the letters m or f placed after each sign. A masculine sign is fortunate, and all planets situated in the same house have their good influence rendered thereby more potent and their unfavourable influence mitigated. But all feminine signs are unfortunate, their direct effect is evil, and they tend to nullify all the good influence of any planet which they afflict (*i.e.* with which they are connected), and to increase all its evil influences, while they also import an element of fickleness into the house and often turn good influences into malignant ones. The precise effect of each sign was different on every planet.

I think the above account is sufficient to enable the reader to form a general idea of the manner in which a horoscope was cast and interpreted, and I do not propose to enter into further details. This is the less necessary as the rules—especially as to the relative importance to be assigned to various planets when their influence was conflicting—were so vague that astrologers had little difficulty in finding in the horoscope of a client any fact about his life of which they had information or any trait of character which they suspected him to possess.

That this vagueness was utilized by quacks is notorious, but no doubt many an astrologer in all honesty availed himself of it, whether consciously or unconsciously. It must be remembered also that the rules were laid down at a time when men were unacquainted with any exact

---

* Raphael, pp. 61–69.

science, with the possible exception of mathematics, and further that, if astrology had been reduced to a series of inelastic rules applicable to all horoscopes, the number of failures to predict the future correctly would have rapidly led to a recognition of the folly of the art. As it was, the failures were frequent and conspicuous enough to shake the faith of most thoughtful men. Moreover it was a matter of common remark that astrologers showed no greater foresight in meeting the difficulties of life than their neighbours, while they were neither richer, wiser, nor happier for their supposed knowledge. But though such observations were justified by reason they were often forgotten in times of difficulty and danger. A prediction of the future and the promise of definite advice as to the best course of action, revealed by the heavenly bodies themselves, appealed to the strongest desires of all men, and it was with reluctance that the futility of the advice was gradually recognized.

The objections to the scheme had been stated clearly by several classical writers. Cicero[*] pointed out that not one of the futures foretold for Pompey, Crassus, and Caesar had been verified by their subsequent lives, and added that the planets, being almost infinitely distant, cannot be supposed to affect us. He also alluded to the fact, which was especially pressed by Pliny[†], that the horoscopes of twins are practically identical though their careers are often very different, or as Pliny put it, every hour in every part of the world are born lords and slaves, kings and beggars.

In answer to the latter obvious criticism astrologers replied by quoting the anecdote of Publius Nigidius Figulus, a celebrated Roman astrologer of the time of Julius Caesar. It is said that when an opponent of the art urged as an objection the different fates of persons born in two successive instants, Nigidius bade him make two contiguous marks on a potter's wheel, which was revolving rapidly near them. On stopping the wheel, the two marks were found to be far removed from each other. Nigidius received the name of Figulus, the potter, in remembrance of this story, but his argument, says St Augustine[‡], who gives us the narrative, was as fragile as the ware which the wheel manufactured.

On the other hand Seneca and Tacitus may be cited as being on

---

[*] Cicero, *De Divinatione*, II, 42.
[†] Pliny, *Historia Naturalis*, VII, 49; XXIX, 1.
[‡] St Augustine, *De Civitate Dei*, bk. V, chap. iii; *Opera omnia*, ed. Migne, vol. VII, p. 143.

the whole favourable to the claims of astrology, though both recognized that it was mixed up with knavery and fraud. An instance of successful prediction which is given by the latter of these writers[*] may be used more correctly as an illustration of how the ordinary professors of the art varied their predictions to suit their clients and themselves. The story deals with the first introduction of the astrologer Thrasyllus to the emperor Tiberius. Those who were brought to Tiberius on any important matter were admitted to an interview in an apartment situated on a lofty cliff in the island of Capreae. They reached this place by a narrow path overhanging the sea, accompanied by a single freedman of great bodily strength; and on their return, if the emperor had conceived any doubts of their trustworthiness, a single blow buried the secret and its victim in the ocean. After Thrasyllus had, in this retreat, stated the results of his art as they concerned the emperor, the latter asked the astrologer whether he had calculated how long he himself had to live. The astrologer examined the aspect of the stars, and while he did this showed, as the narrative states, hesitation, alarm, increasing terror, and at last declared that the present hour was for him critical, perhaps fatal. Tiberius embraced him, and told him he was right in supposing he had been in danger but that he should escape it; and made him thenceforth a confidential counsellor. But Thrasyllus would have been but a sorry astrologer had he not foreseen such a question and prepared an answer which he thought fitted to the character of his patron.

A somewhat similar story is told[†] of Louis XI of France. He sent for a famous astrologer whose death he was meditating, and asked him to show his skill by foretelling his own future. The astrologer replied that his fate was uncertain, but it was so inseparably interwoven with that of his questioner that the latter would survive him but by a few hours, whereon the superstitious monarch not only dismissed him uninjured, but took steps to secure his subsequent safety. The same anecdote is also related of a Scotch student who, being captured by Algerian pirates, predicted to the Sultan that their fates were so involved that he should predecease the Sultan by only a few weeks. This may have been good enough for a barbarian, but with a civilized monarch it

---

[*] *Annales*, VI, 22: quoted by Whewell, *History of the Inductive Sciences*, vol. I, p. 313.

[†] *Personal Characteristics from French History*, by Baron F. Rothschild, London, 1896, p. 10. The story was introduced by Sir Walter Scott in Quentin Durward (chap. XV).

probably would in most cases be less effectual, as certainly it is less artistic, than the answer of Thrasyllus.

I may conclude by mentioning a few notable cases of horoscopy.

Among the most successful instances of horoscopy enumerated by Raphael* is one by W. Lilly, given in his *Monarchy or No Monarchy*, published in 1651, in which he predicted a plague in London so terrible that the number of deaths should exceed the number of coffins and graves, to be followed by "an exorbitant fire." The prediction was amply verified in 1665 and 1666. In fact Lilly's success was embarrassing, for the Committee of the House of Commons, which sat to investigate the causes of the fire and ultimately attributed it to the papists, thought that he must have known more about it than he chose to declare, and on Oct. 25, 1666, summoned him before them. I may add that Lilly proved himself a match for his questioners.

An even more curious instance of a lucky hit is told of Flamsteed†, the first astronomer royal. It is said that an old lady who had lost some property wearied Flamsteed by her perpetual requests that he would use his observatory to discover her property for her. At last, tired out with her importunities, he determined to show her the folly of her demand by making a prediction, and, after she had found it false, to explain again to her that nothing else could be expected. Accordingly he drew circles and squares round a point that represented her house and filled them with all sorts of mystical symbols. Suddenly striking his stick into the ground he said, "Dig there and you will find it." The old lady dug in the spot thus indicated, and found her property; and it may be conjectured that she believed in astrology for the rest of her life.

Perhaps the belief that the royal observatory was built for such purposes may be still held, for De Morgan, writing in 1850, says that "persons still send to Greenwich to have their fortunes told, and in one case a young gentleman wrote to know who his wife was to be, and what fee he was to remit."

It is easier to give instances of success in horoscopy than of failure. Not only are all ambiguous predictions esteemed to be successful, but it is notorious that prophecies which have been verified by the subsequent

---

* *Manual of Astrology*, p. 37.
† The story, though in a slightly different setting, is given in *The London Chronicle*, Dec. 3, 1771, and it is there stated that Flamsteed attributed the result to the direct action of the devil.

course of events are remembered and quoted, while the far more numerous instances in which the prophecies have been falsified are forgotten or passed over in silence.

As exceptionally well-authenticated instances of failures I may mention the twelve cases collected by Cardan in his *Geniturarum Exempla*. These are good examples because Cardan was not only the most eminent astrologer of his time, but was a man of science, and perhaps it is not too much to say was accustomed to accurate habits of thought; moreover, as far as I can judge, he was perfectly honest in his belief in astrology. To English readers the most interesting of these is the horoscope of Edward VI of England, the more so as Cardan has left a full account of the affair, and has entered into the reasons of his failure to predict Edward's death.

To show how Cardan came to be mixed up in the transaction I should explain that in 1552 Cardan went to Scotland to prescribe for John Hamilton, the archbishop of St Andrews, who was ill with asthma and dropsy and about whose treatment the physicians had disagreed[*]. On his return through London, Cardan stopped with Sir John Cheke, the Professor of Greek at Cambridge, who was tutor to the young king. Six months previously, Edward had been attacked by measles and small-pox which had made his health even weaker than before. The king's guardians were especially anxious to know how long he would live, and they asked Cardan to cast Edward's nativity with particular reference to that point.

The Italian was granted an audience in October, of which he wrote a full account in his diary, quoted in the *Geniturarum Exempla*. The king, says he[†], was "of a stature somewhat below the middle height, pale faced, with grey eyes, a grave aspect, decorous, and handsome. He

---

[*] Luckily they left voluminous reports on the case and the proper treatment for it. The only point on which there was a general agreement was that the phlegm, instead of being expectorated, collected in his Grace's brains, and that thereby the operations of the intellect were impeded. Cardan was celebrated for his success in lung diseases, and his remedies were fairly successful in curing the asthma. His fee was 500 crowns for travelling expenses from Pavia, 10 crowns a day, and the right to see other patients; the archbishop actually gave him 2300 crowns in money and numerous presents in kind; his fees from other persons during the same time must have amounted to about an equal sum (see Cardan's *De Libris Propriis*, ed. 1557, pp. 159–175; *Consilia Medica, Opera*, vol. IX, pp. 124–148; *De Vita Propria*, ed. 1557, pp. 138, 193 *et seq.*).

[†] I quote from Morley's translation, vol. II, p. 135 *et seq.*

was rather of a bad habit of body than a sufferer from fixed diseases, and had a somewhat projecting shoulder-blade." But, he continues, he was a boy of most extraordinary wit and promise. He was then but fifteen years old and he was already skilled in music and dialectics, a master of Latin, English, French, and fairly proficient in Greek, Italian,

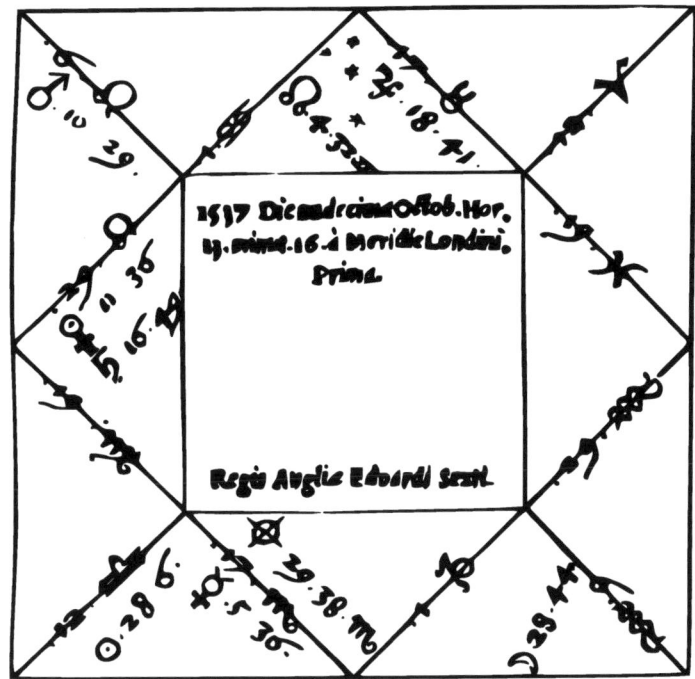

and Spanish. He "filled with the highest expectation every good and learned man, on account of his ingenuity and suavity of manners.... When a royal gravity was called for, you would think that it was an old man you saw, but he was bland and companionable as became his years. He played upon the lyre, took concern for public affairs, was liberal of mind, and in these respects emulated his father, who, while he studied to be [too] good, managed to seem bad." And in another place* he describes him as "that boy of wondrous hopes." At the close of the interview Cardan begged leave to dedicate to Edward a work on which he was then engaged. Asked the subject of the work, Cardan replied that he began by showing the cause of comets. The subsequent conversation, if it is reported correctly, shows good sense and considerable logical skill on the part of the young king.

---

* *De Rerum Varietate*, p. 285.

I have reproduced on the preceding page a facsimile of Cardan's original drawing of Edward's horoscope. The horoscope was cast and read with unusual care. I need not quote the minute details given about Edward's character and subsequent career, but obviously the predictions were founded on the impressions derived from the above-mentioned interview. The conclusion about his length of life was that he would certainly live past middle age, though after the age of 55 years 3 months and 17 days various diseases would fall to his lot[*].

In the following July the king died, and Cardan felt it necessary for his reputation to explain the cause of his error. The title of his dissertation is *Quae post consideravi de eodem*[†]. In effect his explanation is that a weak nativity can never be predicted from a single horoscope, and that to have ensured success he must have cast the nativity of every one with whom Edward had come intimately into contact; and, failing the necessary information to do so, the horoscope could be regarded only as a probable prediction.

This was the argument usually offered to account for non-success. A better defence would have been the one urged by Raphael[‡] and by Southey[§] that there might be other planets unknown to the astrologer which had influenced the horoscope, but I do not think that medieval astrologers assigned this reason for failure.

I have not alluded to the various adjuncts of the art, but astrologers so frequently claimed the power to be able to raise spirits that perhaps I may be pardoned for remarking that I believe some of the more important and elaborate of these deceptions were effected not infrequently by means of a magic lantern, the pictures being sometimes thrown on to a mirror, and at other times on to a thick cloud of smoke which caused the images to move and finally disappear in a fantastic way capable of many explanations[‖].

I would conclude by repeating again that though the practice of astrology was so often connected with impudent quackery, yet one ought not to forget that nearly every physician and man of science in medieval Europe was an astrologer. These observers did not consider that its

---

[*] *Geniturarum Exempla*, p. 19.
[†] *Ibid.*, p. 23.
[‡] *The Familiar Astrologer*, London, 1832, p. 248.
[§] *The Doctor*, chap. 92.
[‖] See *ex. gr.* the life of Cellini, chap. XIII, Roscoe's translation, pp. 144-146. See also Sir David Brewster's *Letters on Natural Magic*.

rules were definitely established, and they laboriously collected much of the astronomical evidence that was to crush their art. Thus, though there never was a time when astrology was not practised by knaves, there was a period of intellectual development when it was honestly accepted as a difficult but a real science.

CHAPTER XI.

CRYPTOGRAPHS AND CIPHERS.

THE art of constructing cryptographs or ciphers—intelligible to those who know the key and unintelligible to others—has been studied for centuries. Their usefulness on certain occasions, especially in time of war, is obvious, while it may be a matter of great importance to those from whom the key is concealed to discover it. But the romance connected with the subject, the not uncommon desire to discover a secret, and the implied challenge to the ingenuity of all from whom the key is hidden, have attracted to the subject the attention of many to whom its utility is a matter of indifference.

The leading authorities on the subject, few of which are less than a century old, are enumerated in an article by J.E. Bailey in the ninth edition of the *Encyclopaedia Britannica*, and references to various historic ciphers are there given. My knowledge of the subject, however, is limited to ciphers which I have met with in the course of casual reading, and I prefer to discuss the subject as it has presented itself to me, with no attempt to make it historically complete and no reference to other authorities. In fact the theory of the subject is not sufficiently important to make it worth while to try to deal with it historically or exhaustively.

Most writers use the words cryptograph and cipher as synonymous. I employ them, however, with different meanings, which I proceed to define.

A cryptograph may be defined as a manner of writing in which the letters or symbols employed are used in their normal sense, but are so arranged that the communication is intelligible only to those possessing the key. The word is sometimes used to denote the communication made. A simple example is a communication in which every word is spelt backwards. Thus:

ymene deveileb ot eb gniriter troper noitisop no ssorc daor.

A cipher may be defined as a manner of writing by characters arbitrarily invented or by an arbitrary use of letters, words, or characters in other than their ordinary sense, intelligible only to those possessing the key. The word is sometimes used to denote the communication made. A simple example is when each letter is replaced by the one that immediately follows it in the natural order of the alphabet, *a* being replaced by *b*, *b* by *c*, and so on, and finally *z* by *a*. In this cipher the above message would read:

fofnz cfmjfwfe up cf sfujsjoh sfqpsu qptjujpo po dsptt spbe.

In both cryptographs and ciphers the essential feature is that the communication may be freely given to all the world though it is unintelligible save to those who possess the key. The key must not be accessible to anyone, and if possible it should be known only to those using the cryptograph or cipher. The art of constructing a cryptograph lies in the concealment of the proper order of the essential letters or words: the art of constructing a cipher lies in concealing what letters or words are represented by the symbols used. In an actual communication cipher symbols may be arranged cryptographically, and thus further hinder a reading of the message. Thus the message given above would read in a cryptographic cipher as

znfof efwfjmfc pu fc hojsjufs uspqfs opjujtpq op ttpsd ebps.

If the message were sent in Latin or some foreign language it would further diminish the chance of it being read by a stranger through whose hands it passed. But I may confine myself to messages in English, and for the present to simple cryptographs and ciphers.

A communication in cryptograph or cipher must be in writing or in some permanent form. Thus to make small muscular movements—such, *ex. gr.*, as talking on the fingers, or breathing long and short in the Morse dot and dash system, or making use of pre-arranged signs by a fan or stick, or flashing signals by light—do not here concern us.

Again, the mere fact that the message is concealed or conveyed secretly does not make it a cryptograph or cipher. The majority of stories dealing with secret communications are concerned with the artfulness with which the message is concealed or conveyed and have nothing to do with cryptographs or ciphers. Many of the ancient instances of secret communication are of this type[*]. Illustrations are to be found in messages conveyed by pigeons, or wrapped round arrows shot over the head of a foe, or written on the paper wrapping of a cigarette, or the use of ink which becomes visible only when the recipient treats the paper on which it is written by some chemical or physical process.

Again, a communication in a foreign language or in any recognized notation like shorthand is not an instance of a cipher. A letter in Chinese or Polish or Russian might be often used for conveying a secret message from one part of England to another, but it fails to fulfil our test that if published to all the world it would be concealed from everyone, unless submitted to some special investigation. On the other hand, in practice, foreign languages or systems of shorthand which are but little known may serve to conceal a communication better than an easy cipher, for in the last case the key may be found with but little trouble, while in the other cases, though the key may be accessible, it is probable that there are only a few who know where to look for it. An illustration of this is afforded by the system used by Pepys in writing his Diary which is further alluded to below.

 I proceed to enumerate some of the better known types of cryptographs. There are at least three distinct types. The first type comprises those in which the order of the letters is changed in some pre-arranged manner. The second type comprises those in which the concealment is due to the introduction of non-significant letters. The third type comprises those in which the letters used are written in fragments. The types are not exclusive, and any particular cryptograph may comprise the distinctive feature of two or all the types.

A cryptograph of the first type is one in which the successive letters of the message are re-arranged in some pre-determined manner.

One of the most obvious cryptographs of this type is to write each word or the message itself backwards. He would, however, be a careless reader who could be deceived by this. Here is an instance in which the

---

[*] A long list of classical authorities for different devices used in ancient times for concealing messages is given in *Mercury* by J. Wilkins, London, 1641, pp. 27–36.

whole message is written backwards:

tsop yb tnes tnemeerga fo seniltuo smret ruo tpecca yeht.

In such a case it is unnecessary to indicate the division into words by leaving spaces between them, and we might divide the letters artificially, as thus:

Ts opybtne stne meer gafos eniltu osmret ruot peccaye ht.

Systems of this kind which depend on altering the places of letters or lines in some pre-arranged manner have always been common. I quote a couple of instances[*] from Wilkins's book to which I have already referred—it was a work which seems to have been studied diligently by many of those who took part in the civil disturbances of the 17th century, and gives an excellent account of some of the easier systems of cryptographs and ciphers.

The first example I take from him is where the letters which make up the communication are written vertically up or down. Thus the message: *The pestilence continues to increase* might be written thus:

e i o t n l i t
s n t i o e t h
a c s n c n s e
e r e u e c e p

Again, Wilkins says that the cryptograph may be yet further obscured by placing the letters which make up the message in any pre-arranged but discontinuate order. For instance if the message runs to four lines we may put the first letter at the beginning of the first line, the second at the beginning of the fourth line, the third at the end of the first line, the fourth at the end of the fourth line, the fifth at the beginning of the second line, the sixth at the beginning of the third line, the seventh at the end of the second line, the eighth at the end of the third line, and so on. Thus the message: *Wee shall make an Irruption upon the Enemie, from the North, at ten of the clock this*

---

[*] *Mercury, or the Secret and Swift Messenger*, by J. Wilkins, London, 1641, pp. 50–52.

night would read thus:

> *Wm rpeta hhs cteinpke*
> *haih fonoih kftoe nil*
> *anoerr ocgt tthmnu rl*
> *eauo mhtei nlen ettes,*

where, to obscure the message further, it is divided arbitrarily into what appear to be words.

Another instance of a cryptograph of this type may be constructed thus. First, by writing the message in lines of some arranged length, say, for instance, each containing seventeen letters—the letters in successive lines being arranged vertically under those corresponding to them in the upper line—and either leaving no spaces between the words or inserting some pre-arranged letter or letters or digits between them, such as j, q, z. The message can be then sent as a cryptograph by writing the letters in order in successive vertical lines. This only comes to saying that we write successively the 1st, 18th, 35th letters of the original message, and then the 2nd, 19th, 36th letters, and so on. To confuse the decipherer the final reading may be arbitrarily put into what might represent words. If, however, we know the clue number, say c, it is easy enough to read the communication. For if it divides into the number of letters n times with a remainder r it suffices to re-write the message in lines putting $n + 1$ letters in each of the first r lines, and n letters in each of the last $c - r$ lines, and then the communication can be read by reading the columns downwards. For instance, if the following communication, containing 270 letters, were received: *Ahtzeipqhgesoaeouazsesewaeqtmusfdtbenzcesjteottqizyczhtzjio arhqettjrfesftnzmroomohyearziaqneornbreotlennkaerwizesjuasjodezwjzz szjbrrittjnfjlweuzroqyfohtqayeizsleopjidihaloalhpepkrheanazsrvliimosiad ygtpekijscerqvvjqjqajqnyjintkaehsbhsnbgoaotqetqeuuesayqurntpebqstzam ztqrj*, and the clue number were *17* we should put *16* letters in each of the first *15* lines and *15* letters in each of the last *2* lines. The communication could then be discovered by reading the columns downwards: the letters j, q, and z marking the ends of words.

Another cryptograph of this type may be constructed by arranging the letters cyclically, and agreeing that the communication is to be made by selected letters, as, for instance, every seventh, second, seventh, second, and so on. Thus if the communication were *Ammunition too low to allow of a sortie*, which consists of 32 letters, the successive significant letters would come in the order 7, 9, 16, 18, 25, 27, 2, 4, 13, 15, 24, 28, 5, 8, 20, 22, 1, 6, 21, 26, 11, 14, 32, 10, 31, 12, 17, 23, 3, 29, 30, 19—the numbers being selected as in the decimation problem given above on pages 19–20, and being struck out from the 32 cycle as soon as they are determined. The above communication would then read *Ttriooalmolaoonmsueoawotnliotifw*. This is a good cryptograph, but it is troublesome to construct, especially if the message is long, and for that reason is not to be recommended.

A cryptograph of the second type is one in which the message is expressed in ordinary writing, but in it are introduced a number of dummies or non-significant letters or digits thus concealing which of the letters are relevant.

One way of picking out those letters which are relevant is by the use of a perforated card of the shape of (say) a sheet of note-paper, which when put over such a sheet permits only such letters as are on certain portions of it to be visible. Such a card is known as a *grille*. An example of a grille with four openings is figured below. A communication made in this way may be easily concealed from anyone who does not possess

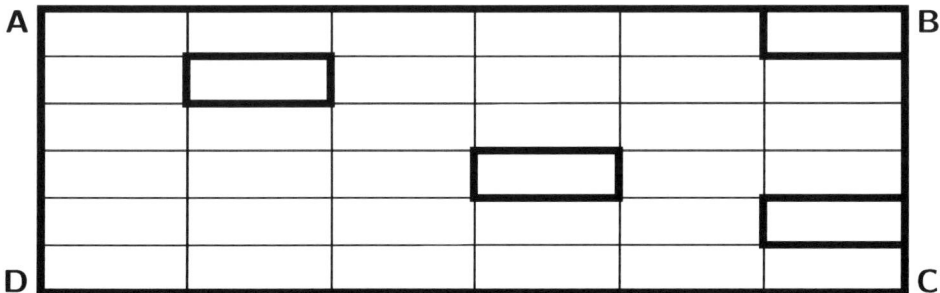

a card of the same pattern. If the recipient possesses such a card he has only to apply it in order to read the message.

The use of the grille may be rendered less easy to detect if it be used successively in different positions, for instance, with the edges AB and CD successively put along the top of the paper containing the message. On the next page, for instance, is a message which, with the aid of the grille figured above, is at once intelligible. On applying the grille to it with the line AB along the top HK we get the first half of the communication, namely, *1000 rifles se*. On applying the grille with the line CD along the top HK we get the rest of the message, namely, *nt to L to-day*. The other spaces in the paper are filled with non-significant letters or numerals in any way we please. Of course any one using such a grille would not divide the sheet of paper on which the communication was written into cells, but in the figure I have done so in order to render the illustration clearer.

H	981	264	070	523	479	100	K
	NTT	*ORI*	*SON*	*SON*	*AHY*	*DTC*	
	BFS	*PUM*	*OLT*	*KFE*	*LJO*	*EGX*	
	AEU	*QJT*	*EGO*	*FLE*	*HVE*	*WLA*	
	FML	*AES*	*REM*	*REM*	*ODA*	*SSE*	
	YZZ	*EPD*	*QJC*	*EKS*	*TIM*	*OEF*	

We can avoid the awkward expedient of having to use a perforated card, which may fall into undesired hands, by introducing a certain pre-arranged number of dummies or non-significant letters or symbols between those which make up the message. Thus, to take an extreme case, we might arrange that only every 101st letter should form our communication, and the intervening 100 letters should be written at random. But such a communication would be 101 times longer than the message, a nearly fatal objection if it had to be written in a hurry or telegraphed. A better method, and one which is not easily discovered by a stranger, is to arrange that (say) only every alternate second and third letter shall be relevant. Thus the first, third, sixth, eighth, eleventh, etc., letters are those that make up the message. Such a communication would be only two and a half times as long as the message, but even that might be a great disadvantage if time in sending the message was of importance. For a message written at leisure this need not matter much, and in such a code the introduction of a sufficient number of unnecessary letters in some pre-arranged manner gives an effectual means of conveying a message in secret.

We can also avoid the use of a perforated card if it be arranged that every n^{th} word shall give the message, the other words being non-significant, though of course inserted as far as possible so as to make the complete communication run as a whole. But the difficulty of composing a document of this kind and its great length render it unsuitable for any purpose except an occasional communication composed at leisure and sent in writing. This method is said to have been used by the Earl of Argyle when plotting against James II.

Similarly any system that rests on picking out certain letters in a document, which letters form a communication in ordinary writing, is a cryptograph. Thus a communication conveyed by a newspaper, in which the letters making up the message are indicated by pen dots or pin pricks or in some other agreed way, is a cryptograph.

A kind of secret writing which may perhaps be considered to constitute a third type of cryptograph is a communication on paper which is legible only when the paper is folded in a particular way. An example is a message written across the edges of a strip of paper wrapped spiralwise round a stick called a *scytale*. When the paper is unwound and taken off the stick the letters appear broken, and may seem to consist of arbitrary signs, but by wrapping the paper round a similar stick the message can be again read. This system is said to have been used by the Lacedemonians*. The concealment can never have been effectual against an intelligent reader who got possession of the paper.

The defect of the method is that the broken letters at once attract attention and suggest the system used. If the fact can be concealed that the visible symbols are parts of letters the cryptograph would be much improved. As an illustration take the appended communication which is said to have been given to the Young Pretender during his wanderings after Culloden. If it be creased along the lines BB and

CC (CC being along the second line of the second score), and then folded over, with B inside, so that the crease C lies over the line A (which is the second line of the first score) thus leaving only the top

---

* For references, see Wilkins, *supra*, p. 38.

and bottom of the piece of paper visible, it will be found to read *conceal yourself, your foes look for you*. I have seen what purports to be the original, but of the truth of the anecdote I know nothing, and the desirability of concealing himself must have been so patent that it was hardly necessary to communicate it by a cryptograph.

I proceed next to some of the more common types of ciphers. It is immaterial whether we invent characters to denote the various letters; or whether we employ special symbols to represent them, such as the symbol (for a, the symbol : for b, and so on; or whether we use the letters in a non-natural sense, such as the letter z for a, the letter y for b, and so on. The rules for reading the cipher will be the same in each case.

In early times it was a common practice to invent arbitrary symbols to represent the letters. If the symbols are invented for the purpose they provoke attention, hence it would seem that preferably we should use symbols which are not likely to attract special notice. For instance, the symbols may be musical notes, in which case the message would appear as a piece of music. Geometrical figures have also been used for the same purpose. It is not even necessary to employ written signs. Natural objects have often been used, as in a necklace of beads, or a bouquet of flowers, where the different shaped or coloured beads or different flowers stand for different letters or words. An even more subtle form of disguising the cipher is to make the different distances between consecutive knots or beads indicate the different letters.

Of all such systems we may say that a careful scrutiny shows that different symbols are being used, and as soon as the various symbols are distinguished one from the other no additional complication is introduced, while for practical purposes they are more trouble to send or receive than those written in symbols in current use. Accordingly I confine myself to ciphers written by the use of the current letters and numerals.

It is convenient to divide ciphers into four classes. The first class comprises ciphers in which the same letter or word is always represented by the same symbol, and this symbol always represents the same letter or word. The second class comprises ciphers in which the same letter or word is, in some or all cases, represented by more than one symbol, and this symbol always represents the same letter or word. The third class comprises ciphers in which the same symbol represents sometimes one letter or word and sometimes another. The fourth class comprises ciphers in which each letter or word is always represented by the same symbol, but more than one letter or word may be represented by the same symbol.

A cipher of the first type then is one in which the same letter or word is always represented by the same symbol, and this symbol always represents the same letter or word.

Perhaps the simplest illustration of a cipher of this type is to employ one language, but written as far as practical in the alphabet of another language. It is said that during the Indian Mutiny messages in English, but written in Greek characters, were used freely, and successfully baffled the ingenuity of the enemy, into whose hands they fell. If this is true, the intelligence of the Hindoos must have been much less than that with which they are usually credited. The device, however, is an old one, for we are told[*] that Edward VI was accustomed to make notes in cipher "with Greek characters, to the end that they who waited on him should not read them."

A common cipher of this type is made by using the actual letters of the alphabet, but in a non-natural sense as indicating other letters. Thus we may use each letter to represent the one immediately following it in the natural order of the alphabet—the letters being supposed to be cyclically arranged—a standing for b wherever it occurs, b standing for c, and so on, and finally z standing for a. Or more generally we may write the letters of the alphabet in a line, and under them re-write the letters in any order we like. For instance

$a\ b\ c\ d\ e\ f\ g\ h\ i\ j\ k\ l\ m\ n\ o\ p\ q\ r\ s\ t\ u\ v\ w\ x\ y\ z$
$o\ l\ k\ m\ a\ z\ s\ q\ x\ e\ u\ f\ y\ r\ t\ h\ c\ w\ b\ v\ n\ i\ d\ g\ j\ p$

In such a scheme, we must in our communication replace a by o, b by l, etc. The recipient will prepare a key by rearranging the letters in the second line in their natural order and placing under them the corresponding letter in the first line. Then wherever a comes in the message he receives he will replace it by e; similarly he will replace b by s, and so on.

A cipher of this kind is not uncommonly used in military signalling, the order of the letters being given by the use of a key-word. If, for instance, *Pretoria* is chosen as the key-word, we write the letters in

---

[*] Sir John Hayward, *Life of Edward VI.*, edition of 1636, p. 20.

this order, striking out any which occur more than once, and continue with the unused letters of the alphabet in their natural order, writing the whole in two lines thus:

$$p \quad r \quad e \quad t \quad o \quad i \quad a \quad b \quad c \quad d \quad f \quad g \quad h$$
$$z \quad y \quad x \quad w \quad v \quad u \quad s \quad q \quad n \quad m \quad l \quad k \quad j$$

Then in using the cipher p is replaced by z and *vice versâ*, r by y, and so on. A long message in such a cipher would be easily discoverable, but it is rapidly composed by the sender and read by the receiver, and for some purposes may be useful, especially if the discovery of the purport of the message is, after a few hours, immaterial.

A summary of the usual rules for reading ciphers of this type, whether written in English, French, German, Italian, Dutch, Latin, or Greek, was given by D.A. Conradus in 1742[*]; and similar rules have been given by various later writers. In English the letter which occurs most frequently is e. The next most common letters are said to be t, a, o, and i; then n; then r, s, and h; then d and l; then c, w, u, and m; then f, y, g, p, and b; then v and k; and then x, q, j, and z. The most common double letters are *ee*, *ll*, *oo*, and *ss*; while in more than half the cases of a double letter at the end of a word, the letter is either l or s. Also, t and h form a common conjunction. I need not, however, go here into further details of this kind.

Assuming that the division into words is given, that non-significant symbols are not introduced, and that the problem is not complicated by the avoidance of the use of common words, a communication of any considerable length can usually be read with but little difficulty. The hints given by Conradus will at once suggest certain hypotheses as to which letters stand for which. Taking one of these hypotheses we write the message, replacing the symbols by the letters we conjecture that they represent and replace the others by dots. If the hypothesis is tenable, the arrangement will probably suggest some of the missing letters. If, for example, we find two words emphs-all and empht-e where the missing letter is represented by the same symbol, the first word shows us that the missing letter is h, m, or t, and the second word shews us that it must be e, h, i, or o, hence it must be h. Every fresh letter so determined makes the hypothesis more probable and renders

---

[*] *Gentleman's Magazine*, 1742, vol. XII, pp. 133–135, 185–186, 241–242, 473–475. See also the *Collected Works of E.A. Poe* in 4 volumes, vol. I, p. 30 *et seq.*

it easier to guess what the remaining symbols represent. The chief difficulty is to get a working hypothesis for the first few letters—if it is the true solution, probably the puzzle will be readily solved—but to make up a working hypothesis for even a few letters requires patience.

Ciphers of this class in which the division between the words is given are to be avoided. If we leave a space between such words a would-be decipherer is given an immense help. He will naturally try if a word denoted by a single symbol can be an i or an a, while the words of two or three letters will often stand revealed and so provide a definite groundwork on which he can construct the key. A long word may also betray the secret. For instance, if the decipherer has reason to suspect that the message related to something connected with Birmingham, and he found that a particular word of ten letters had its second and fifth letters alike, as also its fourth and tenth letters, he would naturally see how the key would work if the word represented Birmingham, and on this hypothesis would at once know the letters represented by eight symbols. With reasonable luck this should suffice to enable him to tell if the hypothesis was tenable. The effect of this can be avoided by leaving no spaces between the words, but this might lead to confusion and is not to be recommended. We can also use letters which occur but rarely, such as j, q, x, z, to separate words, and probably this is the best method.

Ciphers of this type suggest themselves naturally to those approaching the subject for the first time, and are commonly made by merely shifting the letters a certain number of places forward. If this is done we may decrease the risk of detection by altering the amount of shifting at short (and preferably irregular) intervals. Thus it may be agreed that if initially we shift every letter one place forward then whenever we come to the letter (say) n we shall shift every letter one more place forward. In this way the cipher changes continually, and is essentially changed to one of the third class; but even with this improvement it is probable that an expert would decode a fairly long message without much difficulty.

We can have ciphers for numerals as well as for letters: such ciphers are common in many shops. Any word or sentence containing ten different letters will answer the purpose. Thus, an old tradesman of my acquaintance used the excellent precept *Be just O Man*—the first letter representing 1, the second 2, and so on. In this cipher the price 10/6 would be marked bn/t. This is an instance of a cipher of the first type.

A cipher of the second type is one in which the same letter or word is, in some or all cases, represented by more than one symbol, and this symbol always represents the same letter or word. Such ciphers were uncommon before the Renaissance, but the fact that to those who held the key they were not more difficult to write or read than ciphers of the first type, while the key was not so easily discovered, led to their common adoption in the seventeenth century.

A simple instance of such a cipher is given by the use of numerals to denote the letters of the alphabet. Thus a may be represented by 11 or by 37 or by 63, b by 12 or by 38 or by 64, and so on, and finally z by 36 or by 62 or by 88, while we can use 89 or 90 to signify the end of a word and the numbers 91 to 99 to denote words or sentences which constantly occur. Of course in practice no one would employ the numbers in an order like this, which suggests their meaning, but it will serve to illustrate the principle. I have deliberately used numbers of only two digits, as the recipient can then point off the symbols used in twos, and will know that each pair of symbols represents a letter, word, or sentence in the message. A disadvantage of this cipher is that since each letter is denoted by two symbols the length of the message is doubled by putting it in cipher.

The cipher can be improved by introducing after every (say) eleventh digit a non-significant digit. If this is done the recipient of the message must erase every twelfth digit before he begins to read the message. With this addition the difficulty of discovering the key is considerably increased.

The same principle is sometimes applied with letters instead of numbers. For instance, if we take a word (say) of n letters, preferably all different, and construct a table as shown below of n^2 cells, each cell is defined by two letters of the key-word. Thus, if we choose the word *smoking-cap* we shall have 100 cells, and each cell is determined uniquely by the two letters denoting its row and column. If we fill these cells in order with the letters of the alphabet we shall have a system similar to that explained above, where a will be denoted by *ss* or *og* or *no*, and so for the other letters. The last 22 cells may be used to denote the first 22 letters of the alphabet, or better, three or four of them may be used after the end of a word to show that it is ended, and the rest may be used to denote words or sentences which are likely

	S	M	O	K	I	N	G	C	A	P
S	a	b	c	d	e	f	g	h	i	j
M	k	l	m	n	o	p	q	r	s	t
O	u	v	w	x	y	z	a	b	c	d
K	e	f	g	h	i	j	k	l	m	n
I	o	p	q	r	s	t	u	v	w	x
N	y	z	a	b	c	d	e	f	g	h
G	i	j	k	l	m	n	o	p	q	r
C	s	t	u	v	w	x	y	z		
A										
P										

to occur frequently.

Like the similar cipher with numbers this can be improved by introducing after every mth letter any single letter which it is agreed shall be non-significant. To decipher a communication so written it is necessary to know the clue-word and the clue-number.

Here for instance is a communication written in the above cipher with the clue-word *smoking-cap*, and with 7 as the clue-number: *ngmk sigrioicpssamckscakqignassnxmigpoasuiamnocmpaminscnogcpncisyiksk amsssgnncaekknoomkhscpcmscbgpngsiawssgiggndiica*[1]. In this sentence the letters denoting the 79th, 80th, 81st, and 82nd cells have been used to denote the end of a word, and no use has been made of the last 18 cells.

Another cipher of this type is made as follows[*]. The sender and recipient of the message furnish themselves with identical copies of some book. In the cipher only numerals are used, and these numerals indicate the locality of the letters in the book. For example, the first letter in the communication might be indicated by 79–8–5, meaning that it is the 5th letter in the 8th line of the 79th page. But though secrecy might

---

[*] The method is well known. It is mentioned by E.A. Poe, *Collected Works*, vol. III, pp. 338–9, but is much older.

---

1. The original text read ... *sssgnnn*..., which leads to gobbledegook in the deciphered message.

be secured, it would be very tedious to prepare or decode a message, and the method is not as safe as some of those described below.

Another cipher of this type is for the sender and receiver to agree on some common book of reference and to agree further on a number which, if desired, may be communicated as part of the message. To employ this cipher the page of the book indicated by the given number must be used. The first letter in it is taken to signify a, the next b, and so on–any letter which occurs a second time or more frequently being neglected. It may be also arranged that after n letters of the message have been ciphered, the next n letters shall be written in a similar cipher taken from the pth following page of the book, and so on. Thus the possession of the code-book would be of little use to anyone who did not also know the numbers employed. It is so easy to conceal the clue number that with ordinary prudence it would be almost impossible for an unauthorized person to discover a message sent in this cipher. The clue number may be communicated indirectly in many ways. For instance, it may be arranged that the number to be used shall be the number sent, plus (say) q, or that the number to be used shall be an agreed multiple of the number actually sent.

A cipher of the third type is one in which the same symbol represents sometimes one letter or word and sometimes another. Usually such ciphers are easily made or read by those who have the key, but are difficult to discover by those who do not possess it.

A simple example is the employment of pre-arranged numbers in shifting forward the letters that make the communication. For instance, if we agree on the clue number (say) 4276, then the first letter in the communication is replaced by the fourth letter which follows it in the natural order of the alphabet: for instance, if it were an a it would be replaced by e. The next letter is replaced by the second letter which follows it in the natural order of the alphabet: for instance, if it were an a it would be replaced by c. The next letter is replaced by the seventh after it. The next by the sixth after it. The next by the fourth, and so on to the end of the message. Of course to read the message the recipient would reverse the process. If the letters of the alphabet are written at uniform intervals along a ruler, and another ruler similarly marked with the digits can slide along it, the letter corresponding to the shifting of any given number of places can be read at once.

It would be undesirable to allow the division into words to appear in the message, and either the words must be run on continuously, or preferably the less common letters *j*, *q*, *z* may be used to mark the division of words. It is also well to conceal the number of digits in the clue-number. This can be done and the cipher much improved by inserting after every (say) *m*th letter a non-significant letter.

Here for instance is a communication written in this cipher with the clue-numbers 4276 and 7: *atpznhvaxuxhiepxafwghzniyprpsikbdkzy ygkqprgezuytlkobldifebzmxlpogquyitcmgxkckuexvsqkaziaggsigaytnvvsstyv uaslywgjuzmcsfctqbpwjvaepfxhibwpxiultxlavvtqzoxwkvtuvvfheqbxnpvismp hzmqtuwxjykeevltif*. The recipient would begin by striking out every eighth letter. He would then shift back every letter 4, 2, 7, 6, 4, 2, &c., places respectively, and in reading it would leave out the letters *j*, *q*, and *z* as only marking the ends of words.

This is an excellent cipher, and it has the additional merit of not materially lengthening the message. It can be rendered still more difficult by arranging that either or both the clue-numbers shall be changed according to some definite scheme, and it may be further agreed that they shall change automatically every day or week.

A somewhat similar system was proposed by Wilkins[*]. He took a key-word, such as *prudentia*, and constructed as many alphabets as there were letters in it, each alphabet being arranged cyclically and beginning respectively with the letters *p*, *r*, *u*, *d*, *e*, *n*, *t*, *i*, and *a*. He thus got a table like the following, giving nine possible letters which might stand for any letter of the alphabet. Using this we may vary the cipher in successive words or letters of the communication. Thus

a	b	c	d	e	f	g	h	i	k	l	m	n	o	p	q	r	s	t	u	v	w	x	y	z
p	q	r	s	t	u	v	w	x	y	z	a	b	c	d	e	f	g	h	i	k	l	m	n	o
r	s	t	u	v	w	x	y	z	a	b	c	d	e	f	g	h	i	k	l	m	n	o	p	q
u	v	w	x	y	z	a	b	c	d	e	f	g	h	i	k	l	m	n	o	p	q	r	s	t
d	e	f	g	h	i	k	l	m	n	o	p	q	r	s	t	u	v	w	x	y	z	a	b	c
e	f	g	h	i	k	l	m	n	o	p	q	r	s	t	u	v	w	x	y	z	a	b	c	d
n	o	p	q	r	s	t	u	v	w	x	y	z	a	b	c	d	e	f	g	h	i	k	l	m
t	u	v	w	x	y	z	a	b	c	d	e	f	g	h	i	k	l	m	n	o	P	q	r	s
i	k	l	m	n	o	p	q	r	s	t	u	v	w	x	y	z	a	b	c	d	e	f	g	h
a	b	c	d	e	f	g	h	i	k	l	m	n	o	p	q	r	s	t	u	v	w	x	y	z

[*] *Mercury*, by J. Wilkins, London, 1641, pp. 59, 60.

the message *The prisoners have mutinied and seized the railway station* would, according as the cipher changes in successive words or letters, read as *Hwt fhziedvhi bupy pxwmqmhg erh ervmrq max zirteig station* or as *Hyy svvlwnthm lehx uukzgmiq tvd gvcciq mqe frcoanr atpkcrr*. I have taken Wilkins's key-word, but it is obvious that it would be desirable to omit *a* wherever it appears in it, since otherwise, if the cipher changes in successive words, some of the words may appear unaltered in the cipher, as is shown in the first of the examples given above.

A cipher of the fourth type is one in which each letter is always represented by the same symbol, but more than one letter may be represented by the same symbol. Such ciphers were not uncommon at the beginning of the nineteenth century, and were usually framed by means of a key-sentence containing about as many letters as there are letters in the alphabet.

Thus if the key-phrase is *The fox jumped over the garden gate*, we write under it the letters of the alphabet in their usual sequence as shown below:

T h e   f o x   j u m p e d   o v e r   t h e   g a r d e n   g a t e.
a b c   d e f   g h i j k l   m n o p   q r s   t u v w x y   z a b c.

Then we write the message replacing *a* by *t* or *a*, *b* by *h* or *t*, *c* by *e*, *d* by *f*, and so on. Here is such a message. *M foemho nea ge eoo jmdhohg avf teg ev ume afrmeo*. But it will be observed that in the cipher *a* may represent *a* or *u*, *d* may represent *l* or *w*, *e* may represent *c* or *k* or *o* or *s* or *x*, *g* may represent *t* or *z*, *h* may represent *b* or *r*, *o* may represent *e* or *m*, *r* may represent *p* or *v*, and *t* may represent *a* or *b* or *q*. And the recipient, in deciphering it, must judge as best he can which is the right meaning to be assigned to these letters when they appear.

An instance of a cipher of the fourth type is afforded by a note sent by the Duchess de Berri to her adherents in Paris, in which she employed the key phrase

l e   g o u v e r n e m e n t   p r o v i s o i r e.
a b   c d e f g h i j k l m n   o p q r s t u v x y.

Hence in putting her message into cipher she replaced *a* by *l*, *b* by *e*, *c* by *g*, and so on. She forgot however to supply the key to the recipients of the message, but her friend Berryer had little difficulty in reading it by the aid of the rules I have indicated, and thence deduced the key-phrase she had employed.

Having considered various classes of cryptographs and ciphers I may now consider what features we should regard as important in choosing a cipher intended for practical use.

In the first place, it is obvious that the means employed should be such as not to excite suspicion if the communication falls into unauthorized hands. But this is a counsel of perfection, and almost impossible to attain.

In the second place, we may say that, under modern conditions in war, finance, or diplomacy, a cipher may be useless unless it can be telegraphed or telephoned. If this is deemed important, it will practically restrict us to the use of the 26 letters of the alphabet, the 10 numerical symbols for the digits, to which if we like we may add a few additional marks such as punctuation stops, brackets, &c. The same condition will require that the message should not be unduly lengthened by being turned into cipher. Hence any considerable use of non-significant symbols is to be deprecated.

In the third place, the key to the cipher should be such that it can be easily reproduced from memory. For, if the key is so elaborate that those who use it are obliged to preserve it in some tangible and accessible form, unauthorized persons may obtain the power of reading messages. Hence the key should be reproducible at will. Further, it is desirable that the key should be of such a character that it (or a change of it) can be telegraphed or otherwise communicated without the probability of exciting suspicion.

In the fourth place, a cipher should be capable of change at short intervals. So that if the reading of one message in it be discovered subsequent messages may be undecipherable even though the system used is unaltered.

Lastly, no ambiguity should be possible in deciphering the communication. This will exclude ciphers of the fourth type.

Accordingly in choosing a good cipher we should seek for one in which only current letters, symbols, or words are employed; such that its use does not unduly lengthen the message; such that the key to it can be reproduced at will and need not be kept in a form which might betray the secret to an unauthorized person; such that the key to it changes or can be changed at short intervals; and such that it is not ambiguous. Many ciphers of the second and third types fulfil these conditions, but it is generally desirable to avoid ciphers of the first type unless circumstances permit of the free use of a code-book.

The use of instruments giving a cipher, which is or can be varied constantly and automatically, has been often recommended. Several have been constructed on the lines of the well-known letter-locks*. The possession of the key of the instrument as well as a knowledge of the clue-word is necessary to enable anyone to read a message, but the risk of some instrument, when set, falling into unauthorized hands must be taken into account. Since equally good ciphers can be constructed without the use of mechanical devices I do not think their employment can be recommended.

This chapter has already run to such a length that I cannot find space to describe more than one or two ciphers that appear in history or fiction, but, we may say that until recently most of the historical ciphers are not difficult to read.

It is said that Julius Caesar in making secret memoranda was accustomed to move every letter four places forward, writing *d* for *a*, *e* for *b*, &c. This would be a very easy instance of a cipher of the first type, but it may have been effective at that time. His nephew Augustus sometimes used a similar cipher, in which each letter was moved forward one place†.

Bacon proposed a cipher in which each letter was denoted by a group of five letters consisting of *A* and *B* only. Since there are 32 such groups, he had 6 symbols to spare, which he could use to separate words or to which he could assign special meanings. A message in this cipher would be five times as long as the original message. This may be compared with the far superior system of the five (or four) digit codebook system in use at the present time.

Charles I used ciphers freely in important correspondence—the majority being of the second type. He was foolish enough to take a cabinet containing many confidential letters in cipher, to some of which their

---

* See, for instance, the descriptions of those devised by Sir Charles Wheatstone, given in his *Scientific Papers*, London, 1879, pp. 342–347; and by Capt. Bazeries in *Comptes Rendus, Association Français pour l'avancement des sciences*, vol. XX (Marseilles), 1891, p. 160, *et seq.*

† Of some of Caesar's correspondence, Suetonius says (cap. 56) *si quis investigare et persequi velit, quartam elementorum literam, id est, d pro a, et perinde reliquas commutet.* And of Augustus he says (cap. 88) *quoties autem per notas scribit, b pro a, c pro b, ac deinceps eadem ratione, sequentes literas ponit; pro x autem duplex a.*

readings were appended, on the field of Naseby, where they fell into the hands of Fairfax[*]. The House of Commons sent them to a committee presided over by a Mr Tate. It is commonly believed that the Committee referred the papers to J. Wallis[†], then Fellow of Queens' College, Cambridge, and subsequently Savilian Professor at Oxford, who discovered the key to them. At any rate the letters were read.

In these ciphers each letter was represented by a number. The clues to some of the ciphers were provided by the King who had written over the number the letter which it represented, as shown in the following quotation:

c	a	t	o	l	i	c	k	s	i	n	F		
11	18	45	35	23	27	11	25	47	28	40	148	haue	layed

t	h	i	r	p	u	r	s	e	s	t	o	g	e	t	h	e	r
45	31	27	51	33	62	50	47	7	48	45	35	21	7	46	32	7	51

f	o	r		s	u	p	l	y	of	a	r	m	e	s.
15	35	50	a	47	62	33	23	74	k1	17	51	42	7	47.

The published letters show that the King used different ciphers at different times, though perhaps he used the same one in all correspondence with any particular person, but the general character of those he employed is the same. The sentence quoted above is taken from a letter from Queen Henrietta Maria of January 26, 1643. In this and another letter a few months later a is represented by 17 or 18, b by 13, c by 11 or 12, d by 5, e by 7 or 8 or 9 or 10, f by 15 or 16, g by 21, h by 31 or 32, i by 27 or 28, k by 25, l by 23 or 24, m by 42 or 44, n by 39 or 40 or 41, o by 35 or 36 or 37 or 38, p by 33 or 34, r by 50 or 51 or 52, s by 47 or 48, t by 45 or 46, u by 62 or 63, w by 58, and y by 74 or 77. Numbers of three digits were used to represent particular people or places. Thus 148 stood for *France*, 189 for the *King*, 260 for the *Queen*, 354 for *Prince Rupert*, and so on. Further, there were a few special symbols, thus k1 stood for *but*, n1 for *to*, and f1 for *is*. The numbers 2 to 4 and 65 to 72 were non-significant, and were to be struck out or neglected by the recipient of the message. Each symbol is separated from that which follows it by a full-stop.

---

[*] *First Report of the Royal Commission on Historical Manuscripts*, 1870, pp. 2, 4.

[†] See his letters on Cryptography, *Opera*, vol. III, pp. 659–672.

The Queen seems to have found writing in cipher a great trouble. In the letter from which I have already quoted a sentence she says ... *que je suis extrement tourmantee du mal de teete qui fait que je mesteray en syfre par un autre se qui jovois fait moy mesme*, and she uses the cipher only for the particular words it was desired to conceal. Thus she writes *Mr Capell nous a fait voir que sy* 27, 23, &c., &c. If by this she saved herself trouble, she did it at the cost of rendering the cipher much easier to read.

The system used by Charles was in considerable repute during the seventeenth century, but even without extraneous help it is possible for a diligent student to discover the key if the message is fairly long. An excellent illustration of this fact is to be found in the writings of the late Sir Charles Wheatstone. A paper in cipher, every page of which was initialled by Charles I, and countersigned by Lord Digby, was purchased some years ago by the British Museum. It was believed to be a state paper of importance. It consists of a series of numbers (about 150 different symbols being used) without any clue to their meaning, or any indication of a division between the words employed. The task of reading it was rendered the more difficult by the supposition, which proved incorrect, that the document was in English; but notwithstanding this, Sir Charles Wheatstone discovered the key[*]. In this cipher a was represented by any of the numbers 12, 13, 14, 15, 16, or 17, b by 18 or 19, and so on, while some 65 special words were represented by particular numbers.

I may note in passing that Charles also used a species of shorthand, in which the letters were represented by four strokes varying in length and position. Essentially the system is simple, though it is troublesome to read or write.

The famous diary of Samuel Pepys is commonly said to have been written in cipher, but in reality it is written in shorthand according to a system invented by T. Shelton[†]. It is however somewhat difficult to read, for the vowels are usually omitted, and Pepys used some arbitrary signs for terminations, particles, and certain words—so far turning it into a cipher. Further, in certain places, when the matter

---

[*] The document, its translation, and the key used are given in Wheatstone's *Scientific Papers*, London, 1879, pp. 321–341.

[†] *Tachy-graphy* by T. Shelton. The earliest edition I have seen is dated 1641. A somewhat similar system by W. Cartwright was issued by J. Rich under the title *Semographie*, London, 1644.

is such that it can hardly be expressed with decency, he changed from English to a foreign language, or inserted non-significant letters. Shelton's system had been forgotten when attention was first attracted to the diary. Accordingly we may say that, to those who first tried to read it, it was written in cipher, but Pepys's contemporaries would have properly described it as being written in shorthand, though with a few modifications of his own invention.

A system of shorthand specially invented for the purpose is a true cipher. One such system in which each letter is represented either by a dot or by a line of constant length was used by the Earl of Glamorgan, better known by his subsequent title as Marquis of Worcester, in 1645, as also by Charles I. in some of his private correspondence. It is a cipher of the first type and has the defects inherent in almost every cipher of this kind: in fact Glamorgan's letter was deciphered, and the system discovered by Mr Dircks[*]. Obsolete systems of shorthand[†] might be thus used to form an effective cipher.

It is always difficult to read a very short message in cipher, since necessarily the clues are few in number. When the Chevalier de Rohan was sent to the Bastille, on suspicion of treason, there was no evidence against him except what might be extracted from Monsieur Latruaumont. The latter died without making any admission. De Rohan's friends had arranged with him to communicate the result of Latruaumont's examination, and accordingly in sending him some fresh body linen they wrote on one of the shirts *Mg dulhxcclgu ghj yxuj, lm ct ulgc alj*. For twenty-four hours de Rohan pored over the message, but, failing to read it, he admitted his guilt, and was executed November 27, 1674.

The cipher is a very simple one of the first type, but the communication is so short that unless the key were known it would not be easy to read it. Had de Rohan suspected that the second word was *prisonnier*, it would have given him 7 out of the 12 letters used, and as the first and third words suggest the symbols used for l and t, he could hardly have failed to read the message.

The cipher on the facing page is said to have been employed by

---

[*] *Life of the Marquis of Worcester* by H. Dircks, London, 1865. Worcester's system of shorthand was described by him in his *Century of Inventions*, London, 1663, sections 3, 4, 5.

[†] Various systems, including those used in classical and medieval times, are described in the *History of Shorthand* by T. Anderson, London, 1882.

Marie Antoinette*. I take it that it was used in the method indicated

AB	A O	B P	C Q	D R	E S	F T	G V	H X	I Y	L Z	M N
CD	M Z	A N	B O	C P	D Q	E R	F S	G T	H V	I X	L Y
EF	L N	M O	A P	B Q	C R	D S	E T	F V	G X	H Y	I Z
GH	I N	L O	M P	A Q	B R	C S	D T	E V	F X	G Y	H Z
IL	H N	I O	L P	M Q	A R	B S	C T	D V	E X	F Y	G Z
MN	G N	H O	I P	L Q	M R	A S	B T	C V	D X	E Y	F Z
OP	F N	G O	H P	I Q	L R	M S	A T	B V	C X	D Y	E Z
QR	E N	F O	G P	H Q	I R	L S	M T	A V	B X	C Y	D Z
ST	D N	E O	F P	G Q	H R	I S	L T	M V	A X	B Y	C Z
VX	C N	D O	E P	F Q	G R	H S	I T	L V	M X	A Y	B Z
YZ	B N	C O	D P	E Q	F R	G S	H T	I V	L X	M Y	A Z

on page 266 above. If so, the first word in the communication would be rewritten according to the scheme given in the first line, *a* being replaced by *o*, and *vice versâ*, *b* by *p*, and so on. The second word would be rewritten according to the scheme in the second line, and so on.

---

* The key is given, but without explanation, in *Juniper Hall*, by C. Hill, London, 1904, p. 13.

One of the modern systems is the five digit code-book cipher, to which I have already alluded. According to the general belief, it is frequently employed in certain official communications at the present day. A code dictionary is prepared in which every word likely to be used is printed, and the words are numbered consecutively 00000, 00001, ... up, if necessary, to 99999. Thus each word is represented by a number of five digits, and there are 10^5 such numbers available. The message is first written down in words. Below that it is written in numbers, each word being replaced by the number corresponding to it. To each of these numbers is added some definite prearranged clue-number—the words in the dictionary being assumed to be arranged cyclically, so that if the resulting number exceeds 10^5 it is denoted only by the excess above 10^5. The resulting numbers are sent as a message. On receipt of a message it is divided into consecutive groups of five numbers, each group representing a word. From each number is subtracted the prearranged clue-number, and then the message can be read off by the code dictionary. When a code message is published by the Government receiving it, the construction of the sentences is usually altered before publication, so that the key may not be discoverable by anyone in possession of the code-book or who has seen the cipher message. This is a rule applicable to all cryptographs and ciphers.

This is a cipher with 10^5 symbols, and as each symbol consists of five digits, a message of n words is denoted by $5n$ digits, and probably is not longer than the message when written in the ordinary way. Since however the number of words required is less than 10^5, the spare numbers may be used to represent collocations of words which constantly occur, and if so the cipher message may be slightly shortened.

If the clue number is the same all through the message it would be possible by not more than 10^5 trials to discover the message. This is not a serious risk, but, slight though it is, it can be avoided if the clue number is varied; the clue number might, for instance, be 781 for the first three words, 791 for the next five words, 801 for the next seven words, and so on. Further it may be arranged that the clue numbers shall be changed every day; thus on the seventh day of the month they might be 781, 791, &c., and on the eighth day 881, 891, &c., and so on.

This cipher can however be further improved by inserting at some step, say after each mth digit, an unmeaning digit. For example, if, in the original message written in numbers, we insert a 9 after every seven digits we shall get a collection of words (each represented by five digits), most of which would have no connection with the original message, and probably the number of digits used in the message itself would no longer be a multiple of 5. Of course the receiver has only to reverse the process in order to read the message.

It is however unnecessary to use five symbols for each word. For if we make a similar code with the twenty-six letters of the alphabet instead of the ten digits, four letters for each word or phrase would give us 26^4, that is, 456976 possible variations. Thus the message would be shorter and the power of the code increased. Further, if we like to use the ten digits and the twenty-six letters of the alphabet—all of which are easily telegraphed—we could, by only using three symbols, obtain 36^3, that is, 46656 possible words, which would be sufficient for all practical purposes.

This code, at any rate with these modifications, is undecipherable by strangers, but it has the disadvantages that those who use it must always have the code dictionary available, and that it takes a considerable time to code or decode a communication. For practical purposes its use would be confined to communications which could be deciphered at leisure in an office, It is especially suitable in the case of communications between officials, each supplied with a competent staff of secretaries or clerks—as from an ambassador to his chief, or a commander in the field to his war office. It is an excellent example of a cipher of the first type, but it is not clear that it possesses any superiority over some of the simple ciphers of the third type.

One of the best known writers on the subject of cryptographs and ciphers is E.A. Poe, indeed probably a good many readers have made their first acquaintance with a cipher in his story of *The Gold Bug*, the interest of which turns on reading a simple cipher of the first type. In earlier times J. Tritheim of Spanheim, G. Porta of Naples, Cardan, Niceron, and J. Wilkins occupied much the same position, while whenever ciphers were freely used skilful decipherers seem to have arisen.

Poe wrote an essay on cryptography in which he said that it may be roundly asserted that human ingenuity cannot concoct a cipher which human ingenuity cannot resolve—a conclusion which is hardly justified by the known facts. In an earlier article he once made a similar remark so far as ciphers of the first class are concerned, with the implied limitation that only 26 symbols may be used. In this sense the observation is correct. His assertion excited some attention, and numerous communications in cipher were sent to him. More than one of his correspondents did not play the game fairly, not only employing foreign languages, but using several different ciphers in the same communication. Nevertheless he resolved all except one; and he proved that this last was a fraud, being merely a jargon of random characters, having no meaning whatever.

CHAPTER XII.

HYPER-SPACE[*].

I PROPOSE to devote the remaining pages to the consideration, from the point of view of a mathematician, of certain properties of space, time, and matter, and to a sketch of some hypotheses as to their nature. I shall not discuss the metaphysical theories that profess to

---

[*] On the possibility of the existence of space of more than three dimensions see C.H. Hinton, *Scientific Romances*, London, 1886, a most interesting work, from which I have derived much assistance in compiling the earlier part of this chapter; his later work, *The Fourth Dimension*, London, 1904, may be also consulted. See also G.F. Rodwell, *Nature*, May 1, 1873, vol. VIII, pp. 8, 9; and E.A. Abbott, *Flatland*, London, 1884.

The theory of Non-Euclidean geometry is due primarily to Lobatschewsky, *Geometrische Untersuchungen zur Theorie der Parallellinien*, Berlin, 1840 (originally given in a lecture in 1826); to Gauss (*ex. gr.* letters to Schumacher, May 17, 1831, July 12, 1831, and Nov. 28, 1846, printed in Gauss's collected works); and to J. Bolyai, Appendix to the first volume of his father's *Tentamen*, Maros-Vásárkely, 1832; though the subject had been discussed by J. Saccheri as long ago as 1733: its development was mainly the work of G.F.B. Riemann, *Ueber die Hypothesen welche der Geometrie zu Grunde liegen*, written in 1854, *Göttinger Abhandlungen*, 1866–7, vol. XIII, pp. 131–152 (translated in *Nature*, May 1 and 8, 1873, vol. VIII, pp. 14–17, 36–37); H.L.F. von Helmholtz, *Göttinger Nachrichten*, June 3, 1868, pp. 193–221; and E. Beltrami, *Saggio di Interpretazione della Geometria non-Euclidea*, Naples, 1868, and the *Annali di Matematica*, series 2, vol. II, pp. 232–255: see an article by von Helmholtz in the *Academy*, Feb. 12, 1870, vol. I, pp. 128–131. Within the last twenty-five years the theory has been treated by several mathematicians.

A bibliography of hyper-space, compiled by G.B. Halsted, appeared in the *American Journal of Mathematics*, vol. I (1878), pp. 261–276, 384–385; and vol. II (1879), pp. 65–70.

account for the origin of our conceptions of them, for these theories lead to no practical result and rest on assertions which are incapable of definite proof—a foundation which does not commend itself to a scientific student. Space, time, and matter cannot be defined; but the means of measuring them and the investigation of their properties fall within the domain of mathematics.

I devote this chapter to considerations connected with space, leaving the subjects of time and mass to the following two chapters.

I shall confine my remarks on the properties of space to two speculations which recently have attracted considerable attention. These are (i) the possibility of the existence of space of more than three dimensions, and (ii) the possibility of kinds of geometry, especially of two dimensions, other than those which are treated in the usual text-books. These problems are related. The term hyper-space was used originally of space of more than three dimensions, but now it is often employed to denote also any Non-Euclidean space. I attach the wider meaning to it, and it is in that sense that this chapter is on the subject of hyper-space.

In regard to the first of these questions, the conception of a world of more than three dimensions is facilitated by the fact that there is no difficulty in imagining a world confined to only two dimensions—which we may take for simplicity to be a plane, though equally well it might be a spherical or other surface. We may picture the inhabitants of flatland as moving either on the surface of a plane or between two parallel and adjacent planes. They could move in any direction along the plane, but they could not move perpendicularly to it, and would have no consciousness that such a motion was possible. We may suppose them to have no thickness, in which case they would be mere geometrical abstractions: or, preferably, we may think of them as having a small but uniform thickness, in which case they would be realities.

Several writers have amused themselves by expounding and illustrating the conditions of life in such a world. To take a very simple instance, in flatland—or any even dimensional space—a knot is impossible, a simple alteration which alone would make some difference in the experience of the inhabitants as compared with our own.

If an inhabitant of flatland was able to move in three dimensions, he would be credited with supernatural powers by those who were unable so to move; for he could appear or disappear at will, could (so far as they could tell) create matter or destroy it, and would be free from so many constraints to which the other inhabitants were subject that his actions would be inexplicable by them.

We may go one step lower, and conceive of a world of one dimension—like a long tube—in which the inhabitants could move only forwards and backwards. In such a universe there would be lines of varying lengths, but there could be no geometrical figures. To those who are familiar with space of higher dimensions, life in line-land would seem somewhat dull. It is commonly said that an inhabitant could know only two other individuals; namely, his neighbours, one on each side. If the tube in which he lived was itself of only one dimension, this is true; but we can conceive an arrangement of tubes in two or three dimensions, where an occupant would be conscious of motion in only one dimension, and yet which would permit of more variety in the number of his acquaintances and conditions of existence.

Our conscious life is in three dimensions, and naturally the idea occurs whether there may not be a fourth dimension. No inhabitant of flatland could realize what life in three dimensions would mean, though, if he evolved an analytical geometry applicable to the world in which he lived, he might be able to extend it so as to obtain results true of that world in three dimensions which would be to him unknown and inconceivable. Similarly we cannot realize what life in four dimensions is like, though we can use analytical geometry to obtain results true of that world, or even of worlds of higher dimensions. Moreover the analogy of our position to the inhabitants of flatland enables us to form some idea of how inhabitants of space of four dimensions would regard us.

Just as the inhabitants of flatland might be conceived as being either mere geometrical abstractions, or real and of a uniform thickness in the third dimension, so, if there is a fourth dimension, we may be regarded either as having no thickness in that dimension, in which event we are mere (geometrical) abstractions—as indeed idealist philosophers have asserted to be the case—or as having a uniform thickness in that dimension, in which event we are living in four dimensions although we are not conscious of it. In the latter case it is reasonable to suppose that the thickness in the fourth dimension of bodies in our world is small and possibly constant; it has been conjectured also that it is comparable with the other dimensions of the molecules of matter, and if so it is possible that the constitution of matter and its fundamental properties may supply experimental data which will give a physical basis for proving or disproving the existence of this fourth dimension.

If we could look down on the inhabitants of flatland we could see their anatomy and what was happening inside them. Similarly an inhabitant of four-dimensional space could see inside us.

An inhabitant of flatland could get out of a room, such as a rectangle, only through some opening, but, if for a moment he could step into three dimensions, he could reappear on the other side of any boundaries placed to retain him. Similarly, if we came across persons who could move out of a closed prison-cell without going through any of the openings in it, there might be some reason for thinking that they did it by passing first in the direction of the fourth dimension and then back again into our space. This however is unknown.

Again, if a finite solid was passed slowly through flatland, the inhabitants would be conscious only of that part of it which was in their plane. Thus they would see the shape of the object gradually change and ultimately vanish. In the same way, if a body of four dimensions was passed through our space, we should be conscious of it only as a solid body, namely, the section of the body by our space, whose form and appearance gradually changed and perhaps ultimately vanished. It has been suggested that the birth, growth, life, and death of animals may be explained thus as the passage of finite four-dimensional bodies through our three-dimensional space. I believe that this idea is due to Mr Hinton.

The same argument is applicable to all material bodies. The impenetrability and inertia of matter are necessary consequences; the conservation of energy follows, provided that the velocity with which the bodies move in the fourth dimension is properly chosen: but the indestructibility of matter rests on the assumption that the body does not pass completely through our space. I omit the details connected with change of density as the size of the section by our space varies.

We cannot prove the existence of space of four dimensions, but it is interesting to enquire whether it is probable that such space actually exists. To discuss this, first let us consider how an inhabitant of flatland might find arguments to support the view that space of three dimensions existed, and then let us see whether analogous arguments apply to our world. I commence with considerations based on geometry and then proceed to those founded on physics.

Inhabitants of flatland would find that they could have two triangles of which the elements were equal, element to element, and yet which could not be superposed. We know that the explanation of this fact is that, in order to superpose them, one of the triangles would have to be turned over so that its undersurface came on to the upper side, but of course such a movement would be to them inconceivable. Possibly however they might have suspected it by noticing that inhabitants of one-dimensional space might experience a similar difficulty in comparing the equality of two lines, ABC and $CB'A'$, each defined by a set of three points. We may suppose that the lines are equal and such that corresponding points in them could be superposed by rotation round C—a movement inconceivable to the inhabitants—but an inhabitant of such a world in moving along from A to A' would not arrive at the corresponding points in the two lines in the same relative order, and thus might hesitate to believe that they were equal. Hence inhabitants of flatland might infer by analogy that by turning one of the triangles over through three-dimensional space they could make them coincide.

We have a somewhat similar difficulty in our geometry. We can construct triangles in three dimensions—such as two spherical triangles—whose elements are equal respectively one to the other, but which cannot be superposed. Similarly we may have two spirals whose elements are equal respectively, one having a right-handed twist and the other a left-handed twist, but it is impossible to make one fill exactly the same parts of space as the other does. Again, we may conceive of two solids, such as a right hand and a left hand, which are exactly similar and equal but of which one cannot be made to occupy exactly the same position in space as the other does. Those are difficulties similar to those which would be experienced by the inhabitants of flatland in comparing triangles; and it may be conjectured that in the same way as such difficulties in the geometry of an inhabitant in space of one dimension are explicable by temporarily moving the figure into space of two dimensions by means of a rotation round a point, and as such difficulties in the geometry of flatland are explicable by temporarily moving the figure into space of three dimensions by means of a rotation round a line, so such difficulties in our geometry would disappear if we could temporarily move our figures into space of four dimensions by means of a rotation round a plane—a movement which of course is inconceivable to us.

Next we may enquire whether the hypothesis of our existence in a space of four dimensions affords an explanation of any difficulties or apparent inconsistencies in our physical science*. The current conception of the luminiferous ether, the explanation of gravity, and the fact that there are only a finite number of kinds of matter, all the atoms of each kind being similar, present such difficulties and inconsistencies. To see whether the hypothesis of a four-dimensional space gives any aid to their elucidation, we shall do best to consider first the analogous problems in two dimensions.

We live on a solid body, which is nearly spherical, and which moves round the sun under an attraction directed to it. To realize a corresponding life in flatland we must suppose that the inhabitants live on the rim of a (planetary) disc which rotates round another (solar) disc under an attraction directed towards it. We may suppose that the planetary world thus formed rests on a smooth plane, or other surface of constant curvature; but the pressure on this plane and even its existence would be unknown to the inhabitants, though they would be conscious of their attraction to the centre of the disc on which they lived. Of course they would be also aware of the bodies, solid, liquid, or gaseous, which were on its rim, or on such points of its interior as they could reach.

Every particle of matter in such a world would rest on this plane medium. Hence, if any particle was set vibrating, it would give up a part of its motion to the supporting plane. The vibrations thus caused in the plane would spread out in all directions, and the plane would communicate vibrations to any other particles resting on it. Thus any form of energy caused by vibrations, such as light, radiant heat, electricity, and possibly attraction, could be transmitted from one point to another without the presence of any intervening medium which the inhabitants could detect.

If the particles were supported on a uniform elastic plane film, the intensity of the disturbance at any other point would vary inversely as the distance of the point from the source of disturbance; if on a uniform elastic solid medium, it would vary inversely as the square of that distance. But, if the supporting medium was vibrating, then, wherever a particle rested on it, some of the energy in the plane would

---

* See a note by myself in the *Messenger of Mathematics*, Cambridge, 1891, vol. XXI, pp. 20–24, from which the above argument is extracted. The question has been treated by Mr Hinton on similar lines.

be given up to that particle, and thus the vibrations of the intervening medium would be hindered when it was associated with matter.

If the inhabitants of this two-dimensional world were sufficiently intelligent to reason about the manner in which energy was transmitted they would be landed in a difficulty. Possibly they might be unable to explain gravitation between two particles—and therefore between the solar disc and their disc—except by supposing vibrations in a rigid medium between the two particles or discs. Again, they might be able to detect that radiant light and heat, such as the solar light and heat, were transmitted by vibrations transverse to the direction from which they came, though they could realize only such vibrations as were in their plane, and they might determine experimentally that in order to transmit such vibrations a medium of great rigidity (which we may call ether) was necessary. Yet in both the above cases they would have also distinct evidence that there was no medium capable of resisting motion in the space around them, or between their disc and the solar disc. The explanation of these conflicting results lies in the fact that their universe was supported by a plane, of which they were necessarily unconscious, and that this rigid elastic plane was the ether which transmitted the vibrations.

Now suppose that the bodies in our universe have a uniform thickness in the fourth dimension, and that in that direction our universe rests on a homogeneous elastic body whose thickness in that direction is small and constant. The transmission of force and radiant energy, without the intervention of an intervening medium, may be explained by the vibrations of the supporting space, even though the vibrations are not themselves in the fourth dimension. Also we should find, as in fact we do, that the vibrations of the luminiferous ether are hindered when it is associated with matter. I have assumed that the thickness of the supporting space is small and uniform, because then the intensity of the energy transmitted from a source to any point would vary inversely as the square of the distance, as is the case; whereas if the supporting space was a body of four dimensions, the law would be that of the inverse cube of the distance.

The application of this hypothesis to the third difficulty mentioned above—namely, to show why there are in our universe only a finite number of kinds of atoms, all the atoms of each kind having in common a number of sharply defined properties—will be given later*.

Thus the assumption of the existence of a four-dimensional homogeneous elastic body on which our three-dimensional universe rests, affords an explanation of some difficulties in our physical science.

It may be thought that it is hopeless to try to realize a figure in four dimensions. Nevertheless attempts have been made to see what the sections of such a figure would look like.

If the boundary of a solid is $\varphi(x, y, z) = 0$, we can obtain some idea of its form by taking a series of plane sections by planes parallel to $z = 0$, and mentally superposing them. In four dimensions the boundary of a body would be $\varphi(x, y, z, \omega) = 0$, and attempts have been made to realize the form of such a body by making models of a series of solids in three dimensions formed by sections parallel to $\omega = 0$. Again, we can represent a solid in perspective by taking sections by three co-ordinate planes. In the case of a four-dimensional body the section by each of the four co-ordinate solids will be a solid, and attempts have been made by drawing these to get an idea of the form of the body. Of course a four-dimensional body will be bounded by solids.

The possible forms of regular bodies in four dimensions, analogous to polyhedrons in space of three dimensions, have been discussed by Mr Stringham[†].

I now turn to the second of the two problems mentioned at the beginning of the chapter: namely, the possibility of there being kinds of geometry other than those which are treated in the usual elementary text-books. This subject is so technical that in a book of this nature I can do little more than give a sketch of the argument on which the idea is based.

The Euclidean system of geometry, with which alone most people are acquainted, rests on a number of independent axioms and postulates. Those which are necessary for Euclid's geometry have, within recent years, been investigated and scheduled. They include not only those explicitly given by him, but some others which he unconsciously used. If these are varied, or other axioms are assumed, we get a different series of propositions, and any consistent body of such propositions

---

* See below, p. 318 (3).
† *American Journal of Mathematics*, 1880, vol. III, pp. 1–14.

constitutes a system of geometry. Hence there is no limit to the number of possible Non-Euclidean geometries that can be constructed.

Among Euclid's axioms and postulates is one on parallel lines, which is usually stated in the form that if a straight line meets two straight lines, so as to make the two interior angles on the same side of it taken together less than two right angles, then these straight lines being continually produced will at length meet upon that side on which are the angles which are less than two right angles. Expressed in this form the axiom is far from obvious, and from early times numerous attempts have been made to prove it[*]. All such attempts failed, and it is now known that the axiom cannot be deduced from the other axioms assumed by Euclid. It can be replaced by other statements about parallel lines, such as that the distance between two parallel lines is always the same, but such alternative statements, though perhaps *primâ facie* more axiomatic, are not to be preferred to Euclid's form, since his statement brings out prominently a characteristic feature of the space with which he is concerned.

The earliest conception of a body of Non-Euclidean geometry was due to the discovery, made independently by Saccheri, Lobatschewsky, and John Bolyai, that a consistent system of geometry of two dimensions can be produced on the assumption that the axiom on parallels is not true, and that through a point a number of straight (that is, geodetic) lines can be drawn parallel to a given straight line. The resulting geometry is called *hyperbolic*.

Riemann later distinguished between boundlessness of space and its infinity, and showed that another consistent system of geometry of two dimensions can be constructed in which all straight lines are of a finite length, so that a particle moving along a straight line will return to its original position. This leads to a geometry of two dimensions, called *elliptic geometry*, analogous to the hyperbolic geometry, but characterized by the fact that through a point no straight line can be drawn which, if produced far enough, will not meet any other given straight line. This can be compared with the geometry of figures drawn on the surface of a sphere.

Thus according as no straight line, or only one straight line, or a pencil of straight lines can be drawn through a point parallel to a given

---

[*] Some of the more interesting and plausible attempts have been collected by J. Richard in his *Philosophie de Mathématiques*, Paris, 1903.

straight line, we have three systems of geometry of two dimensions known respectively as elliptic, parabolic or homaloidal or Euclidean, and hyperbolic.

In the parabolic and hyperbolic systems straight lines are infinitely long. In the elliptic they are finite. In the hyperbolic system there are no similar figures of unequal size; the area of a triangle can be deduced from the sum of its angles, which is always less than two right angles; and there is a finite maximum to the area of a triangle. In the elliptic system all straight lines are of the same finite length; any two lines intersect; and the sum of the angles of a triangle is greater than two right angles. In the elliptic system it is possible to get from one point to a point on the other side of a plane without passing through the plane, namely, by going the other way round the straight line joining the two points; thus a watch-dial moving face upwards continuously forward in a plane in a straight line in the direction from the mark VI to the mark XII will finally appear to a stationary observer with its face downwards; and if originally the mark III was to the right of the observer it will finally be on his left hand.

In spite of these and other peculiarities of hyperbolic and elliptical geometries, it is impossible to prove by observation that one of them is not true of the space in which we live. For in measurements in each of these geometries we must have a unit of distance; and if we live in a space whose properties are those of either of these geometries, and such that the greatest distances with which we are acquainted (*ex. gr.* the distances of the fixed stars) are immensely smaller than any unit, natural to the system, then it may be impossible for our observations to detect the discrepancies between the three geometries. It might indeed be possible by observations of the parallaxes of stars to prove that the parabolic system and either the hyperbolic or elliptic system were false, but never can it be proved by measurements that Euclidean geometry is true. Similar difficulties might arise in connection with excessively minute quantities. In short, though the results of Euclidean geometry are more exact than present experiments can verify for finite things, such as those with which we have to deal, yet for much larger things or much smaller things or for parts of space at present inaccessible to us they may not be true.

If however we go a step further and ask what is meant by saying that a geometry is true or false, I can only quote the remark of Poincaré, that the selection of a geometry is really a matter of convenience, and that that geometry is the best which enables us to state the physical laws in the simplest form. This opinion has been strongly controverted, but at any rate it expresses one view of the question.

The above refers only to hyper-space of two dimensions. Naturally there arises the question whether there are different kinds of Non-Euclidean space of three or more dimensions. Riemann showed that there are three kinds of Non-Euclidean space of three dimensions having properties analogous to the three kinds of Non-Euclidean space of two dimensions already discussed. These are differentiated by the test whether at every point no geodetical surface, or one geodetical surface, or a fasciculus of geodetical surfaces can be drawn parallel to a given surface: a geodetical surface being defined as such that every geodetic line joining two points on it lies wholly on the surface. It may be added that each of the three systems of geometry of two dimensions described above may be deduced as properties of a surface in each of these three kinds of Non-Euclidean space of three dimensions.

It is evident that the properties of Non-Euclidean space of three dimensions are deducible only by the aid of mathematics, and cannot be illustrated materially, for in order to realize or construct surfaces in Non-Euclidean space of two dimensions we think of or use models in space of three dimensions; similarly the only way in which we could construct models illustrating Non-Euclidean space of three dimensions would be by utilizing space of four dimensions.

We may proceed yet further and conceive of Non-Euclidean geometries of more than three dimensions, but this remains, as yet, an unworked field.

Returning to the former question of Non-Euclidean geometries, I wish again to emphasize the fact that, if the axioms enunciated by Euclid are replaced by others, it is possible to construct other consistent systems of geometry. Some of these are interesting, but those which have been mentioned above have a special importance, from the somewhat sensational fact that they lead to no results necessarily inconsistent with the properties, as far as we can observe them, of the space in which we live; we are not at present acquainted with any other systems which are consistent with our experience.

CHAPTER XIII.

TIME AND ITS MEASUREMENT.

THE problems connected with time are totally different in character from those concerning space which I discussed in the last chapter. I there stated that the life of people living in space of one dimension would be uninteresting, and that probably they would find it impossible to realize life in space of higher dimensions. In questions connected with time we find ourselves in a somewhat similar position. Mentally, we can realize a past and a future—thus going backwards and forwards—actually we go only forwards. Hence time is analogous to space of one dimension. Were our time of two dimensions, the conditions of our life would be infinitely varied, but we can form no conception of what such a phrase means, and I do not think that any attempts have been made to work it out.

I shall concern myself here mainly with questions concerning the measurement of time, and shall treat them rather from a historical than from a philosophical point of view.

In order to measure anything we must have an unalterable unit of the same kind, and we must be able to determine how often that unit is contained in the quantity to be measured. Hence only those things can be measured which are capable of addition to things of the same kind.

Thus to measure a length we may take a foot-rule, and by applying it to the given length as often as is necessary, we shall find how many feet the length contains. But in comparing lengths we assume as the result of experience that the length of the foot-rule is constant, or rather that any alteration in it can be determined; and, if this assumption was denied, we could not prove it, though, if numerous repetitions of the experiment under varying conditions always gave the same result, probably we should feel no doubt as to the correctness of our method.

It is evident that the measurement of time is a more difficult matter. We cannot keep a unit by us in the same way as we can keep a foot-rule; nor can we repeat the measurement over and over again, for time once passed is gone for ever. Hence we cannot appeal directly to our sensations to justify our measurement. Thus, if we say that a certain duration is four hours, it is only by a process of reasoning that we can show that each of the hours is of the same duration.

The establishment of a scientific unit for measuring durations has been a long and slow affair. The process seems to have been as follows. Originally man observed that certain natural phenomena recurred after the interval of a day, say from sunrise to sunrise. Experience—for example, the amount of work that could be done in it—showed that the length of every day was about the same, and, assuming that this was accurately so, man had a unit by which he could measure durations. The present subdivision of a day into hours, minutes, and seconds is artificial, and apparently is derived from the Babylonians.

Similarly a month and a year are natural units of time though it is not easy to determine precisely their beginnings and endings.

So long as men were concerned merely with durations which were exact multiples of these units or which needed only a rough estimate, this did very well; but as soon as they tried to compare the different units or to estimate durations measured by part of a unit they found difficulties. In particular it cannot have been long before it was noticed that the duration of the same day differed in different places, and that even at the same place different days differed in duration at different times of the year, and thus that the duration of a day was not an invariable unit.

The question then arises as to whether we can find a fixed unit by which a duration can be measured, and whether we have any assurance that the seconds and minutes used to-day for that purpose are all of equal duration. To answer this we must see how a mathematician would define a unit of time. Probably he would say that experience leads us to believe that, if a rigid body is set moving in a straight line without any external force acting on it, it will go on moving in that line; and those times are taken to be equal in which it passes over equal spaces: similarly, if it is set rotating about a principal axis passing through its centre of mass, those times are taken to be equal in which it turns through equal angles. Our experiences are consistent with this statement, and that is as high an authority as a mathematician hopes to get.

The spaces and the angles can be measured, and thus durations can be compared. Now the earth may be taken roughly as a rigid body rotating about a principal axis passing through its centre of mass, and subject to no external forces affecting its rotation: hence the time it takes to turn through four right angles, *i.e.* through $360°$, is always the same; this is called a sidereal day: the time to turn through one twenty-fourth part of $360°$, *i.e.* through $15°$, is an hour: the time to turn through one-sixtieth part of $15°$, *i.e.* through $15'$, is a minute: and so on.

If, by the progress of astronomical research, we find that there are external forces affecting the rotation of the earth, mathematics would have to be invoked to find what the time of rotation would be if those forces ceased to act, and this would give us a correction to be applied to the unit chosen. In the same way we may say that although an increase of temperature affects the length of a foot-rule, yet its change of length can be determined, and thus applied as a correction to the foot-rule when it is used as the unit of length. As a matter of fact there is reason to think that the earth takes about one sixty-sixth of a second longer to turn through four right angles now than it did 2500 years ago, and thus the duration of a second is just a trifle longer to-day than was the case when the Romans were laying the foundations of the power of their city.

The sidereal day can be determined only by refined astronomical observations and is not a unit suitable for ordinary purposes. The relations of civil life depend mainly on the sun, and he is our natural time keeper. The true solar day is the time occupied by the earth in making one revolution on its axis relative to the sun; it is true noon when the sun is on the meridian. Owing to the motion of the sun relative to the earth, the true solar day is about four minutes longer than a sidereal day.

The true solar day is not however always of the same duration. This is inconvenient if we measure time by clocks (as now for nearly two centuries has been usual in Western Europe) and not by sun-dials, and therefore we take the average duration of the true solar day as the measure of a day: this is called the mean solar day. Moreover to define the noon of a mean solar day we suppose a point to move uniformly round the ecliptic coinciding with the sun at each apse, and further we suppose a fictitious sun, called the mean sun, to move in the celestial equator so that its distance from the first point of Aries is the same as that of this point: it is mean noon when this mean sun is on the meridian. The mean solar day is divided into hours, minutes, and seconds; and these are the usual units of time in civil life.

The time indicated by our clocks and watches is mean solar time; that marked on ordinary sun-dials is true solar time. The difference between them is the equation of time: this may amount at some periods of the year to a little more than a quarter of an hour. In England we take the Greenwich meridian as our origin for longitudes, and instead of local mean solar time we take Greenwich mean solar time as the civil standard.

Of course mean time is a comparatively recent invention. The French were the last civilized nation to abandon the use of true time: this was in 1816.

Formerly there was no common agreement as to when the day began. In parts of ancient Greece and in Japan the interval from sunrise to sunset was divided into 12 hours, and that from sunset to sunrise into 12 hours. The Jews, Chinese, Athenians, and, for a long time, the Italians, divided their day into 24 hours, beginning at the hour of sunset, which of course varies every day: this method is said to be still used in certain villages near Naples, except that the day begins half-an-hour after sunset—the clocks being re-set once a week. Similarly the Babylonians, Assyrians, Persians, and until recently the modern Greeks and the inhabitants of the Balearic Islands counted the twenty-four hours of the day from sunrise. Until the middle of last century, the inhabitants of Basle reckoned the twenty-four hours from our 11.0 p.m. The ancient Egyptians and Ptolemy counted the twenty-four hours from noon: this is the practice of modern astronomers. In Western Europe the day is taken to begin at midnight—as was first suggested by Hipparchus—and is divided into two equal periods of twelve hours each.

The week of seven days is an artificial unit of time. It had its origin in the East, and was introduced into the West by the Roman emperors, and, except during the French Revolution, has been subsequently in general use among civilized races. The names of the days are derived from the seven astrological planets, arranged, as was customary, in the order of their apparent times of rotation round the earth, namely, Saturn, Jupiter, Mars, the Sun, Venus, Mercury, and the Moon. The twenty-four hours of the day were dedicated successively to these planets: and the day was consecrated to the planet of the first hour.

Thus if the first hour was dedicated to Saturn, the second would be dedicated to Jupiter, and so on; but the day would be Saturn's day. The twenty-fourth hour of Saturn's day would be dedicated to Mars, thus the first hour of the next day would belong to the Sun; and the day would be Sun's day. Similarly the next day would be Moon's day; the next, Mars's day; the next, Mercury's day: the next, Jupiter's day; and the next, Venus's day.

The astronomical month is a natural unit of time depending on the motion of the moon, and containing about $29\frac{1}{2}$ days. The months of the calendar have been evolved gradually as convenient divisions of time, and their history is given in numerous astronomies. In the original Julian arrangement the months in a leap year contained alternately 31 and 30 days, while in other years February had 29 days. This was altered by Augustus in order that his month should not be inferior to one named after his uncle.

The solar tropical year is another natural unit of time. According to a recent determination, it contains 365.242216 days, that is, $365^{\text{d.}}$ $5^{\text{h.}}$ $48^{\text{m.}}$ $47^{\text{s.}}.4624$.

The Egyptians knew that it contained between 365 and 366 days, but the Romans did not profit by this information, for Numa is said to have reckoned 355 days as constituting a year—extra months being occasionally intercalated, so that the seasons might recur at about the same period of the year.

In 46 B.C. Julius Caesar decreed that thenceforth the year should contain 365 days, except that in every fourth or leap year one additional day should be introduced. He ordered this rule to come into force on January 1, 45 B.C. The change was made on the advice of Sosigenes of Alexandria.

It must be remembered that the year 1 A.D. follows immediately 1 B.C., that is, there is no year 0, and thus 45 B.C. would be a leap year. All historical dates are given now as if the Julian calendar was reckoned backwards as well as forwards from that year[*]. As a matter of fact, owing to a mistake in the original decree, the Romans, during the

---

[*] Herschel, *Astronomy*, London, 11th ed. 1871, arts. 916–919.

first 36 years after 45 B.C., intercalated the extra day every third year, thus producing an error of 3 days. This was remedied by Augustus, who directed that no intercalation of an extra day should be made in any of the twelve years A.U.C. 746 to 757 inclusive, but that the intercalation should be again made in the year A.U.C. 761 (that is, 8 A.D.) and every succeeding fourth year.

The Julian calendar made the year, on an average, contain 365.25 days. The actual value is, very approximately, 365.242216 days. Hence the Julian year is too long by about $11\frac{1}{4}$ minutes: this produces an error of nearly one day in 128 years. If the extra day in every thirty-second leap year had been omitted—as was suggested by some unknown Persian astronomer—the error would have been less than one day in 100,000 years. It may be added that Sosigenes was aware that his rule made the year slightly too long.

The error in the Julian calendar of rather more than eleven minutes a year gradually accumulated, until in the sixteenth century the seasons arrived some ten days earlier than they should have done. In 1582 Gregory XIII corrected this by omitting ten days from that year, which therefore contained only 355 days. At the same time he decreed that thenceforth every year which was a multiple of a century should be or not be a leap year according as the multiple was or was not divisible by four.

The fundamental idea of the reform was due to Lilius, who died before it was carried into effect. The work of framing the new calendar was entrusted to Clavius, who explained the principles and necessary rules in a prolix but accurate work[*] of over 700 folio pages. The plan adopted was due to a suggestion of Pitatus made in 1552 or perhaps 1537: the alternative and more accurate proposal of Stöffler, made in 1518, to omit one day in every 134 years being rejected by Lilius and Clavius for reasons which are not known.

Clavius believed the year to contain 365.2425432 days, but he framed his calendar so that a year, on the average, contained 365.2425 days, which he thought to be wrong by one day in 3323 years: in reality it is a trifle more accurate than this, the error amounting to one day in about 3600 years.

The change was unpopular, but Riccioli[†] tells us that, as those

---

[*] *Romani Calendarii a Greg. XIII, restituti Explicatio*, Rome, 1603.
[†] *Chronologia Reformata*, Bonn, 1669, vol. II, p. 206.

miracles which take place on fixed dates—*ex. gr.* the liquefaction of the blood of S. Januarius—occurred according to the new calendar, the papal decree was presumed to have a divine sanction—Deo ipso huic correctioni Gregorianae subscribente—and was accepted as a necessary evil.

In England a bill to carry out the same reform was introduced in 1584, but was withdrawn after being read a second time; and the change was not finally effected till 1752, when eleven days were omitted from that year. In Roman Catholic countries the new style was adopted in 1582. In Scotland the change was made in 1600. In the German Lutheran States it was made in 1700. In England, as I have said above, it was introduced in 1752; and in Ireland it was made in 1782. It is well known that the Greek Church still adheres to the Julian calendar.

The Mohammedan year contains 12 lunar months, or $354\frac{1}{3}$ days, and thus has no connection with the seasons.

The Gregorian change in the calendar was introduced in order to keep Easter at the right time of year. The date of Easter depends on that of the vernal equinox, and as the Julian calendar made the year of an average length of 365.25 days instead of 365.242216 days, the vernal equinox came earlier and earlier in the year, and in 1582 had regreded to within about ten days of February.

The rule for determining Easter is as follows[*]. In 325 the Nicene Council decreed that the Roman practice should be followed; and after 463 (or perhaps, 530) the Roman practice required that Easter-day should be the first Sunday after the full moon which occurs on or next following the vernal equinox—full moon being assumed to occur on the fourteenth day from the day of the preceding new moon (though as a matter of fact it occurs on an average after an interval of rather more than $14\frac{3}{4}$ days), and the vernal equinox being assumed to fall on March 21 (though as a matter of fact it sometimes falls on March 22).

This rule and these assumptions were retained by Gregory on the ground that it was inexpedient to alter a rule with which so many traditions were associated; but, in order to save disputes as to the exact instant of the occurrence of the new moon, a mean sun and a mean moon defined by Clavius were used in applying the rule. One consequence of using this mean sun and mean moon and giving an

---

[*] De Morgan, *Companion to the Almanac*, London, 1845, pp. 1–36; *Ibid.*, 1846, pp. 1–10.

artificial definition of full moon is that it may happen, as it did in 1818 and 1845, that the actual full moon occurs on Easter Sunday. In the British Act, 24 Geo. II. cap. 23, the explanatory clause which defines full moon is omitted, but practically full moon has been interpreted to mean the Roman ecclesiastical full moon; hence the Anglican and Roman rules are the same. Until 1774 the German Lutheran States employed the actual sun and moon. Had full moon been taken to mean the fifteenth day of the moon, as is the case in the civil calendar, then the rule might be given in the form that Easter-day is the Sunday on or next after the calendar full moon which occurs next after March 21.

Assuming that the Gregorian calendar and tradition are used, there still remains one point in this definition of Easter which might lead to different nations keeping the feast at different times. This arises from the fact that local time is introduced. For instance the difference of local time between Rome and London is about 50 minutes. Thus the instant of the first full moon next after the vernal equinox might occur in Rome on a Sunday morning, say at 12.30 a.m., while in England it would still be Saturday evening, 11.40 p.m., in which case our Easter would be one week earlier than at Rome. Clavius foresaw the difficulty, and the Roman Communion all over the world keep Easter on that day of the month which is determined by the use of the rule at Rome. But presumably the British Parliament intended time to be determined by the Greenwich meridian, and if so the Anglican and Roman dates for Easter might differ by a week; whether such a case has ever arisen or been discussed I do not know, and I leave to ecclesiastics to say how it should be settled.

The usual method of calculating the date on which Easter-day falls in any particular year is involved, and possibly the following simple rule* may be unknown to some of my readers.

Let m and n be numbers as defined below. (i) Divide the number of the year by 4, 7, 19; and let the remainders be a, b, c, respectively. (ii) Divide $19c + m$ by 30, and let d be the remainder. (iii) Divide $2a + 4b + 6d + n$ by 7, and let e be the remainder. (iv) Then the Easter full moon occurs d days after March 21; and Easter-day is the $(22 + d + e)$th of March or the $(d + e - 9)$th day of April, except that if the calculation gives $d = 29$ and $e = 6$ (as happens in 1981) then

---

* It is due to Gauss, and his proof is given in Zach's *Monatliche Correspondenz*, August, 1800, vol. II, pp. 221–230.

Easter-day is on April 19 and not on April 26, and if the calculation gives $d = 28$, $e = 6$, and also $c > 10$ (as happens in 1954) then Easter-day is on April 18 and not on April 25, that is, in these two cases Easter falls one week earlier than the date given by the rule. These two exceptional cases cannot occur in the Julian calendar, and in the Gregorian calendar they occur only very rarely. It remains to state the values of m and n for the particular period. In the Julian calendar we have $m = 15$, $n = 6$. In the Gregorian calendar we have, from 1582 to 1699 inclusive, $m = 22$, $n = 2$; from 1700 to 1799, $m = 23$, $n = 3$; from 1800 to 1899, $m = 23$, $n = 4$; from 1900 to 2099, $m = 24$, $n = 5$; from 2100 to 2199, $m = 24$, $n = 6$; from 2200 to 2299, $m = 25$, $n = 0$; from 2300 to 2399, $m = 26$, $n = 1$; and from 2400 to 2499, $m = 25$, $n = 1$. Thus for the year 1908 we have $m = 24$, $n = 5$; hence $a = 0$, $b = 4$, $c = 8$; $d = 26$; and $e = 2$: therefore Easter Sunday will be on the 19th of April. After the year 4200 the form of the rule will have to be slightly modified.

The dominical letter and the golden number of the ecclesiastical calendar can be at once determined from the values of b and c. The epact, that is, the moon's age at the beginning of the year, can be also easily calculated from the above data in any particular case; the general formula was given by Delambre, but its value is required so rarely by any but professional astronomers and almanack-makers that it is unnecessary to quote it here.

We can evade the necessity of having to recollect the values of m and n by noticing that, if N is the given year, and if $\{N/x\}$ denotes the integral part of the quotient when N is divided by x, then m is the remainder when $15 + \xi$ is divided by 30, and n is the remainder when $6 + \eta$ is divided by 7: where, in the Julian calendar, $\xi = 0$, and $\eta = 0$; and, in the Gregorian calendar, $\xi = \{N/100\} - \{N/400\} - \{N/300\}$, and $\eta = \{N/100\} - \{N/400\} - 2$.

If we use these values of m and n, and if we put for a, b, c, their values, namely, $a = N - 4\{N/4\}$, $b = N - 7\{N/7\}$, $c = N - 19\{N/19\}$, the rule given on the previous page takes the following form. "Divide $19N - \{N/19\} + 15 + \xi$ by 30, and let the remainder be d. Next divide $6(N + d + 1) - \{N/4\} + \eta$ by 7, and let the remainder be e. Then Easter full moon is on the dth day after March 21, and Easter-day is on the $(22 + d + e)$th of March or the $(d + e - 9)$th of April as the case may be; except that if the calculation gives $d = 29$, and $e = 6$, or if it gives $d = 28$, $e = 6$, and $c > 10$, then Easter-day is on the $(d + e - 16)$th of April."

Thus, if $N = 1899$, we divide $19(1899) - 99 + 15 + (18 - 4 - 6)$ by 30, which gives $d = 5$, and then we proceed to divide $6(1899 + 5 + 1) - 474 + (18 - 4 - 2)$ by 7, which gives $e = 6$: therefore Easter-day is on April 2.

The above rules cover all the cases worked out with so much labour by Clavius and others*.

I may add here a rule, quoted by Zeller, for determining the day of the week corresponding to any given date. Suppose that the pth day of the qth month of the year N *anno domini* is the rth day of the week, reckoned from the preceding Saturday. Then r is the remainder when $p + 2q + \{3(q + 1)/5\} + N + \{N/4\} - \eta$ is divided by 7; provided January and February are reckoned respectively as the 13th and 14th months of the preceding year.

For instance, Columbus first landed in the New World on Oct. 12, 1492. Here $p = 12$, $q = 10$, $N = 1492$, $\eta = 0$. If we divide $12 + 20 + 6 + 1492 + 373$ by 7 we get $r = 6$; hence it was on a Friday. Again, Charles I was executed on Jan. 30, 1649. Here $p = 30$, $q = 13$, $N = 1648$, $\eta = 0$, and we find $r = 3$; hence it was on a Tuesday. As another example, the battle of Waterloo was fought on June 18, 1815. Here $p = 18$, $q = 6$, $N = 1815$, $\eta = 12$, and we find $r = 1$; hence it took place on a Sunday.

I proceed now to give a short account of some of the means of measuring time which were formerly in use.

Of devices for measuring time, the earliest of which we have any positive knowledge are the *styles* or *gnomons* erected in Egypt and Asia Minor. These were sticks placed vertically in a horizontal piece of ground, and surrounded by three concentric circles, such that every two hours the end of the shadow of the stick passed from one circle to another. Some of these have been found at Pompeii and Tusculum.

The *sun-dial* is not very different in principle. It consists of a rod or style fixed on a plate or dial; usually, but not necessarily, the style is placed so as to be parallel to the axis of the earth. The shadow of the

---

* Most of the above-mentioned facts about the calendar are taken from Delambre's *Astronomie*, Paris, 1814, vol. III, chap. xxxviii; and his *Histoire de l'astronomie moderne*, Paris, 1821, vol. I, chap. i: see also A. De Morgan, *The Book of Almanacs*, London, 1851; S. Butcher, *The Ecclesiastical Calendar*, Dublin, 1877; and C. Zeller, *Acta Mathematica*, Stockholm, 1887, vol. IX, pp. 131–136: on the chronological details see J.L. Ideler, *Lehrbuch der Chronologie*, Berlin, 1831.

style cast on the plate by the sun falls on lines engraved there which are marked with the corresponding hours.

The earliest sun-dial, of which I have read, is that made by Berosus in 540 B.C. One was erected by Meton at Athens in 433 B.C. The first sun-dial at Rome was constructed by Papirius Cursor in 306 B.C. Portable sun-dials, with a compass fixed in the face, have been long common in the East as well as in Europe. Other portable instruments of a similar kind were in use in medieval Europe, notably the sun-rings, hereafter described, and the sun-cylinders*.

I believe it is not generally known that a sun-dial can be so constructed that the shadow will, for a short time near sunrise and sunset, move backwards on the dial†. This was discovered by Nonez. The explanation is as follows. Every day the sun appears to describe a circle round the pole, and the line joining the point of the style to the sun describes a right cone whose axis points to the pole. The section of this cone by the dial is the curve described by the extremity of the shadow, and is a conic. In our latitude the sun is above the horizon for only part of the twenty-four hours, and therefore the extremity of the shadow of the style describes only a part of this conic. Let QQ' be the arc described by the extremity of the shadow of the style from sunrise at Q to sunset at Q', and let S be the point of the style and F the foot of the style, *i.e.* the point where the style meets the plane of the dial. Suppose that the dial is placed so that the tangents drawn from F to the conic QQ' are real, and that P and P', the points of contact of these tangents, lie on the arc QQ'. If these two conditions are fulfilled, then the shadow will regrede through the angle QFP as its extremity moves from Q to P, it will advance through the angle PFP' as its extremity moves from P to P', and it will regrede through the angle $P'FQ'$ as its extremity moves from P' to Q'.

If the sun's apparent diurnal path crosses the horizon—as always happens in temperate and tropical latitudes—and if the plane of the dial is horizontal, the arc QQ' will consist of the whole of one branch of a hyperbola, and the above conditions will be satisfied if F is within the space bounded by this branch of the hyperbola and its asymptotes.

---

* Thus Chaucer in the *Shipman's Tale*, "by my chilindre it is prime of day," and Lydgate in the *Siege of Thebes*, "by my chilyndre I gan anon to see... that it drew to nine."

† Ozanam, 1803 edition, vol. III, p. 321; 1840 edition, p. 529.

As a particular case, in a place of latitude 12° N. on a day when the sun is in the northern tropic (of Cancer) the shadow on a dial whose face is horizontal and style vertical will move backwards for about two hours between sunrise and noon.

If, in the case of a given sun-dial placed in a certain position, the conditions are not satisfied, it will be possible to satisfy them by tilting the sun-dial through an angle properly chosen. This was the rationalistic explanation, offered by the French encyclopaedists, of the miracle recorded in connection with Isaiah and Hezekiah[*]. Suppose, for instance, that the style is perpendicular to the face of the dial. Draw the celestial sphere. Suppose that the sun rises at M and culminates at N, and let L be a point between M and N on the sun's diurnal path. Draw a great circle to touch the sun's diurnal path MLN at L, let this great circle cut the celestial meridian in A and A', and of the arcs AL, $A'L$ suppose that AL is the less and therefore is less than a quadrant. If the style is pointed to A, then, while the sun is approaching L, the shadow will regrede, and after the sun passes L the shadow will advance. Thus if the dial is placed so that a style which is normal to it cuts the meridian midway between the equator and the tropic, then between sunrise and noon on the longest day the shadow will move backwards through an angle

$$\sin^{-1}(\cos\omega \sec \tfrac{1}{2}\omega) - \cot^{-1}\{\sin\omega \cos(l - \tfrac{1}{2}\omega)(\cos^2 l - \sin^2 \omega)^{-\tfrac{1}{2}}\},$$

where l is the latitude of the place and ω is the obliquity of the ecliptic.

The above remarks refer to the sun-dials in ordinary use. In 1892 General Oliver brought out in London a dial with a solid style, the section of the style being a certain curve whose form was determined empirically by the value of the equation of time as compared with the sun's declination[†]. The shadow of the style on the dial gives the local mean time, though of course in order to set the dial correctly at any place the latitude of the place must be known: the dial may be also set so as to give the mean time at any other locality whose longitude relative to the place of observation is known.

The *sun-ring* or *ring-dial* is another instrument for measuring solar time[‡]. One of the simplest type is figured in the accompanying diagram.

---

[*] 2 Kings, chap. XX, vv. 9–11.

[†] An account of this sun-dial with a diagram was given in *Knowledge*, July 1, 1892, pp. 133, 134.

[‡] See Ozanam, 1803 edition, vol. III, p. 317; 1840 edition, p. 526.

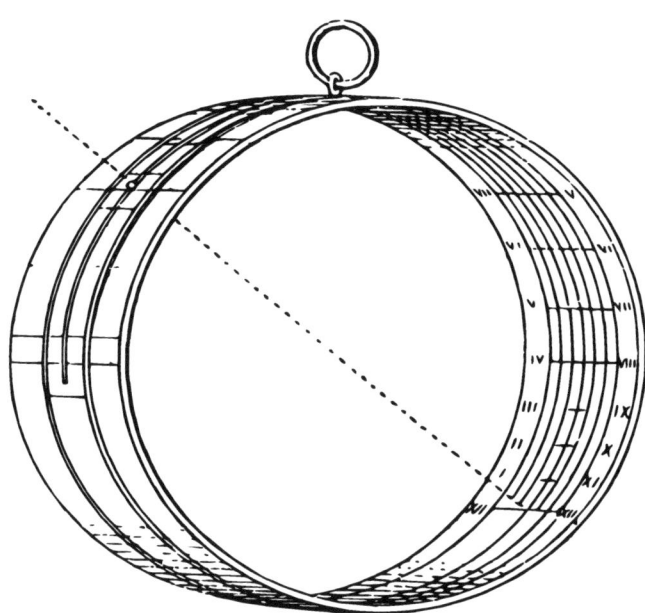

The sun-ring consists of a thin brass band, about a quarter of an inch wide, bent into the shape of a circle, which slides between two fixed circular rims—the radii of the circles being about one inch. At one point of the band there is a hole; and when the ring is suspended from a fixed point attached to the rims so that it hangs in a vertical plane containing the sun, the light from the sun shines through this hole and makes a bright speck on the opposite inner or concave surface of the ring. On this surface the hours are marked, and, if the ring is properly adjusted, the spot of light will fall on the hour which indicates the solar time. The adjustment for the time of year is made as follows. The rims between which the band can slide are marked on their outer or convex side with the names of the months, and the band containing the hole must be moved between the rims until the hole is opposite to that month for which the ring is being used.

For determining times near noon the instrument is reliable, but for other hours in the day it is accurate only if the time of year is properly chosen, usually near one of the equinoxes. This defect may be corrected by marking the hours on a curved brass band affixed to the concave surface of the rims. I possess two specimens of rings of this kind. These rings were distributed widely. Of my two specimens, one was bought in the Austrian Tyrol and the other in London. Astrolabes and sea-rings can be used as sun-rings.

Clepsydras or water-clocks, and *hour-glasses* or sand-clocks, afford other means of measuring time. The time occupied by a given amount of some liquid or sand in running through a given orifice under the same conditions is always the same, and by noting the level of the liquid which has run through the orifice, or which remains to run through it, a measure of time can be obtained.

The burning of graduated candles gives another way of measuring time, and we have accounts of those used by Alfred the Great for the purpose. Incense sticks were used by the Chinese in a similar way.

Modern *clocks* and *watches*[*] comprise a train of wheels turned by a weight, spring, or other motive power, and regulated by a pendulum, balance, fly-wheel, or other moving body whose motion is periodic and time of vibration constant. The direction of rotation of the hands of a clock was selected originally so as to make the hands move in the same direction as the shadow on a sun-dial whose face is horizontal—the dial being situated in our hemisphere.

The invention of clocks with wheels is attributed by tradition to Pacificus of Verona, circ. 850, and also to Gerbert, who is said to have made one at Magdeburg in 996: but there is reason to believe that these were sun-clocks. The earliest wheel-clock of which we have historical evidence was one sent by the Sultan of Egypt in 1232 to the Emperor Frederick II, though there seems to be no doubt that they had been made in Italy at least fifty years earlier.

The oldest clock in England of which we know anything was one erected in 1288 in or near Westminster Hall out of a fine imposed on a corrupt Lord Chief Justice. The bells, and possibly the clock, were staked by Henry VIII on a throw of dice and lost, but the site was marked by a sun-dial, destroyed about sixty years ago, and bearing the inscription *Discite justiciam moniti*. In 1292 a clock was erected in Canterbury Cathedral at a cost of £30. One erected at Glastonbury Abbey in 1325 is at present in the Kensington Museum and is in regular action. Another made in 1326 for St Alban's Abbey showed the astronomical phenomena, and seems to have been one of the earliest clocks that did so. One put up at Dover in 1348 is still in good working order.

---

[*] See *Clock and Watch Making* by Lord Grimthorpe, 7th edition, London, 1883.

The clocks at Peterborough and Exeter were of about the same date, and portions of them remain *in situ.* Most of these early clocks were regulated by horizontal balances: pendulums being then unknown. Of the elaborate clocks of a later date, that at Strasburg made by Dasypodius in 1571, and that at Lyons constructed by Lippeus in 1598, are especially famous: the former was restored in 1842, though in a manner which destroyed most of the ancient works.

In 1370, Vick constructed a clock for Charles V with a weight as motive power and a vibrating escapement—a great improvement on the rough time-keepers of an earlier date.

The earliest clock regulated by a pendulum seems to have been made in 1621 by a clockmaker named Harris, of Covent Garden, London, but the theory of such clocks is due to Huygens*. Galileo had discovered previously the isochronism of a pendulum, but did not apply it to the regulation of the motion of clocks. Hooke made such clocks, and possibly discovered independently this use of the pendulum: he invented or re-invented the anchor pallet.

A watch may be defined as a clock which will go in any position. Watches, though of a somewhat clumsy design, were made at Nuremberg by P. Hele early in the sixteenth century—the motive power being a ribbon of steel, wound round a spindle, and connected at one end with a train of wheels which it turned as it unwound—and possibly a few similar time-pieces had been made in the previous century. By the end of the sixteenth century they were not uncommon. At this time they were usually made in the form of fanciful ornaments such as skulls, or as large pendants, but about 1620 the flattened oval form was introduced, rendering them more convenient to carry in a pocket or about the person. In the seventeenth century their construction was greatly improved, notably by the introduction of the spring balance by Huygens in 1674, and independently by Hooke in 1675—both mathematicians having discovered that small vibrations of a coiled spring, of which one end is fixed, are practically isochronous. The fusee had been used by R. Zech of Prague in 1525, but was re-invented by Hooke.

Clocks and watches are usually moved and regulated in the manner indicated above. Other motive powers and other devices for regulating the motion may be met with occasionally. Of these I may mention a clock in the form of a cylinder, usually attached to another weight as

---

* *Horologium Oscillatorium*, Paris, 1673.

in Atwood's machine, which rolls down an inclined plane so slowly that it takes twelve hours to roll down, and the highest point of the face always marks the proper hour[*].

A water-clock made on a somewhat similar plan is described by Ozanam[†] as one of the sights of Paris at the beginning of the last century. It was formed of a hollow cylinder divided into various compartments each containing some mercury, so arranged that the cylinder descended with uniform velocity between two vertical pillars on which the hours were marked at equidistant intervals.

Other ingenious ways of concealing the motive power have been described in the columns of *La Nature*[‡]. Of such mysterious timepieces the following are not uncommon examples, and probably are known to most readers of this book. One kind of clock consists of a glass dial suspended by two thin wires; the hands however are of metal, and the works are concealed in them or in the pivot. Another kind is made of two sheets of glass in a frame containing a spring which gives to the hinder sheet a very slight oscillatory motion–imperceptible except on the closest scrutiny–and each oscillation moves the hands through the requisite angles. Some so-called perpetual motion timepieces were described above on pages 77–78. Lastly, I have seen in France a clock the hands of which were concealed at the back of the dial, and carried small magnets; pieces of steel in the shape of insects were placed on the dial, and, following the magnets, served to indicate the time.

The position of the sun relative to the points of the compass determines the solar time. Conversely, if we take the time given by a watch as being the solar time—and it will differ from it by only a few minutes at the most—and we observe the position of the sun, we can find the points of the compass[§]. To do this it is sufficient to point the hour-hand to the sun, and then the direction which bisects the angle between the hour and the figure XII will point due south. For instance, if it is four o'clock in the afternoon, it is sufficient to point the hour-hand (which is then at the figure IIII) to the sun, and the figure II on the watch will indicate the direction of south. Again, if it is eight

---

[*] Ozanam, 1803 edition, vol. II, p. 39; 1840 edition, p. 212; or *La Nature*, Jan. 23, 1892, pp. 123, 124.
[†] Ozanam, 1803 edition, vol. II, p. 68; 1840 edition, p. 225.
[‡] See especially the volumes issued in 1874, 1877, and 1878.
[§] The rule is given by W.H. Richards, *Military Topography*, London, 1883, p. 31, though it is not stated quite correctly. I do not know who first enunciated it.

o'clock in the morning, we must point the hour-hand (which is then at the figure VIII) to the sun, and the figure X on the watch gives the south point of the compass.

Between the hours of six in the morning and six in the evening the angle between the hour and XII which must be bisected is less than 180°, but at other times the angle to be bisected is greater than 180°; or perhaps it is simpler to say that at other times the rule gives the north point and not the south point.

The reason is as follows. At noon the sun is due south, and it makes one complete circuit round the points of the compass in 24 hours. The hour-hand of a watch also makes one complete circuit in 12 hours. Hence, if the watch is held in the plane of the ecliptic with its face upwards, and the figure XII on the dial is pointed to the south, both the hour-hand and the sun will be in that direction at noon. Both move round in the same direction, but the angular velocity of the hour-hand is twice as great as that of the sun. Hence the rule. The greatest error due to the neglect of the equation of time is less than 2°. Of course in practice most people, instead of holding the face of the watch in the ecliptic, would hold it horizontal, and in our latitude no serious error would be thus introduced.

In the southern hemisphere where at noon the sun is due north the rule requires modification. In such places the hour-hand of a watch (held face upwards in the plane of the ecliptic) and the sun move in opposite directions. Hence, if the watch is held so that the figure XII points to the sun, then the direction which bisects the angle between the hour of the day and the figure XII will point due north.

CHAPTER XIV.

MATTER AND ETHER THEORIES.

MATTER, like space and time, cannot be defined, but either the statement that matter is whatever occupies space or the statement that it is anything which can be seen, touched, or weighed, suggests its more important characteristics to anyone already familiar with it.

The means of measuring matter and some of its properties are treated in most text-books on mechanics, and I do not propose to discuss them. I confine the chapter to an account of some of the hypotheses by physicists as to the ultimate constitution of matter, but I exclude metaphysical conjectures, which from their nature are mere assertions incapable of proof and are not subject to mathematical analysis. The question is intimately associated with the explanation of the phenomena of attraction, light, chemistry, electricity, and other branches of physics.

I commence with a list of some of the more plausible of the hypotheses formerly proposed which accounted for the obvious properties of matter, and shall then discuss how far they explain or are consistent with other facts[*]. The interest of the list is largely historical, for within the last few years new views as to the constitution of matter

---

[*] I have based my account mainly on *Recent Advances in Physical Science*, by P.G. Tait, Edinburgh, 1876 (chaps, XII, XIII); and on the article *Atom* by J. Clerk Maxwell in the Encyclopaedia Britannica or his *Collected Works*, vol. II, pp. 445–484: see also W.M. Hicks's address, Report of the British Association (Ipswich meeting), 1895, vol. LXV, pp. 595–606. For the more recent speculations see J.J. Thomson, *Electricity and Matter*, Westminster, 1904, and J. Larmor, *Aether and Matter*, Cambridge, 1900.

have been propounded, but the details of these more recent hypotheses are so complicated and technical that only professional mathematicians can understand them. Accordingly I allude to them only briefly.

I. HYPOTHESIS OF CONTINUOUS MATTER. It may be supposed that matter is homogeneous and continuous, in which case there is no limit to the infinite divisibility of bodies. This view was held by Descartes*.

This conjecture is consistent with the facts deducible by untrained observation, but there are many other phenomena for which it does not account; moreover there seems to be no way of reconciling such a structure of matter either with the facts of chemical changes or with the results of spectrum analysis. At any rate the theory must be regarded as extremely improbable.

II. ATOMIC THEORIES. If matter is not continuous we must suppose that every body is composed of aggregates of molecules. If so, it seems probable that each such molecule is built up by the association of two or more atoms, that the number of kinds of atoms is finite, and that the atoms of any particular kind are alike. As to the nature of the atoms the following hypotheses have been made.

(i) *Popular Atomic Hypothesis.* The popular view is that every atom of any particular kind is a minute indivisible article possessing definite qualities, everlasting in its form and properties, and infinitely hard.

This statement is plausible, but the difficulties to which it leads appear to be insuperable. In fact we have reason to think that the atoms which form a molecule are composite systems in incessant vibration at a rate characteristic of the molecule, and it is most probable that they are elastic.

Newton seems to have hazarded a conjecture of this kind when he suggested† that the difficulties, connected with the fact that the velocity of sound was one-ninth greater than that required by theory, might be overcome if the particles of air were little rigid spheres whose distance from one another under normal conditions was nine times the

---

* Descartes, *Principia*, vol. II, pp. 18, 23.
† Newton, *Principia*, bk. II, prop. 50.

diameter of any one of them. This was ingenious, but obviously the view is untenable, because, if such a structure of air existed, the air could not be compressed beyond a certain limit, namely, about 1/1021st part of its original volume, which has been often exceeded. The true explanation of the difficulty noticed by Newton was given by Laplace.

(ii) *Boscovich's Hypothesis.* In 1759 Boscovich suggested* that the facts might be explained by supposing that an atom was an infinitely small indivisible mass which was a centre of force—the law of force being attractive for sensible distances, alternately attractive and repulsive for minute distances, and repulsive for infinitely small distances. In this theory all action between bodies is action at a distance.

He explained the apparent extension of bodies by saying that two parts are consecutive (or similarly that two bodies are in contact) when the nearest pair of atoms in them are so close to one another that the repulsion at any point between them is sufficiently great to prevent any other atom coming between them. It is essential to the theory that the atom shall have a mass but shall not have dimensions.

This hypothesis is not inconsistent with any known facts, but it has been described, perhaps not unjustly, as a mere mathematical fiction, and certainly it is opposed to the apparent indications of our senses. At any rate it is artificial, though it may be a prejudice to regard that as an argument against its adoption. To some extent this view was accepted by Faraday.

Lord Kelvin, better known as Sir William Thomson, has shown† that, if we assume the existence of gravitation, then each of the above hypotheses will account for cohesion.

(iii) *Hypothesis of an Elastic Solid Ether.* Some physicists have tried to explain the known phenomena by properties of the medium through which our impressions are derived. By postulating that all space is filled with a medium possessed of many of the characteristics of an elastic solid, it has been shown by Fresnel, Green, Cauchy, Neumann, MacCullagh, and others that a large number of the properties of light and electricity may be explained. In spite of the difficulties to which this hypothesis necessarily leads, and of its inherent improbability, it has been discussed by Stokes, Lamé, Boussinesq, Sarrau, Lorenz,

---

* *Philosophiae naturalis Theoria redacta ad unicam Legem Virium*, Vienna, 1759.
† *Proceedings of the Royal Society of Edinburgh*, April 21, 1862, vol. IV, pp. 604–606.

Lord Rayleigh, and Kirchhoff.

This hypothesis has been modified and rendered somewhat more plausible by von Helmholtz, Lommel, Ketteler[*], and Voigt, who based their researches on the assumption of a mutual reaction between the ether and the material molecules located in it: on this view the problems connected with refraction and dispersion have been simplified. Finally, Sir William Thomson, in his Baltimore Lectures, 1885, suggested a mechanical analogue to represent the relations between matter and this ether, by which a possible constitution of the ether can be realized. He also suggested later a form of *labile ether*, from whose properties most of the more familiar physical phenomena can be deduced, provided the arrangement can be considered stable; a labile ether is an elastic solid, and its properties in two dimensions may be compared with those of a soap-bubble film, in three dimensions.

It is, however, difficult to criticise any of these hypotheses as a theory of the constitution of matter until the arrangement of the atoms or their nature is more definitely expressed.

III. DYNAMICAL THEORIES. In recent years the suggestion has been made that the so-called atoms may be forms of motion (*ex. gr.* permanent eddies) in one elementary material known as the ether; on this view all the atoms are constituted of the same matter, but the physical conditions are different for the different kinds of atoms. It has been said that there is an initial difficulty in any such hypothesis, since the all-pervading elementary fluid must possess inertia, so that to explain matter we assume the existence of a fluid possessing one of the chief characteristics of matter. This is true as far as it goes, but it is not more unreasonable than to attribute all the fundamental properties of matter to the atoms themselves, as is done by many writers. The next paragraph contains a statement of one of the earliest attempts to formulate a dynamical atomic hypothesis.

(i) *The Vortex Ring Hypothesis.* This hypothesis assumes that each atom is a vortex ring in an incompressible frictionless homogeneous fluid.

Vortex rings—though, since friction is brought into play, of an imperfect character—can be produced in air by many smokers. Better

---

[*] *Theoretische Optik*, Braunschweig, 1885.

specimens can be formed by taking a cardboard box in one side of which a circular hole is cut, filling it with smoke, and hitting the opposite side sharply. The tendency of the particles forming a ring to maintain their annular connection may be illustrated by placing such a box on one side of a room in a direct line with the flame of a lighted candle on the other side. If properly aimed, the ring will travel across the room and put out the flame. If the box is filled only with air, so that the ring is not visible, the experiment is more effective.

In 1858 von Helmholtz[*] showed that a closed vortex filament in a perfect fluid is indestructible and retains certain characteristics always unaltered. In 1867 Sir William Thomson propounded[†] the idea that matter consists of vortex rings in a fluid which fills space. If the fluid is perfect we could neither create new vortex rings nor destroy those already created, and thus the permanence of the atoms is explained. Moreover the atoms would be flexible, compressible, and in incessant vibration at a definite fundamental rate. This rate is very rapid, and Sir William Thomson gave the number of vibrations per second of a sodium ring as probably being greater than 10^{14}.

By a development of this hypothesis Prof. J.J. Thomson[‡] showed, some years ago, that chemical combination may be explained. He supposed that a molecule of a compound is formed by the linking together of vortex filaments representing atoms of different elements: this arrangement may be compared with that of helices on an anchor ring. For stability not more than six filaments may be combined together, and their strengths must be equal. Another way of explaining chemical combination on the vortex atom hypothesis has been suggested by W.M. Hicks. It is known[§] that a spherical mass of fluid, whose interior possesses vortex motion, can move through liquid like a rigid sphere, and he has shown that one of these spherical vortices can swallow up another, thus forming a compound element.

(ii) *The Vortex Sponge Hypothesis.* Any vortex atom hypothesis labours under the difficulty of requiring that the density of the fluid

---

[*] *Crelle's Journal*, 1858, vol. LV, pp. 25–55; translated by Tait in the *Philosophical Magazine*, June, 1867, supplement, series 4, vol. XXXIII, pp. 485–512.
[†] *Proceedings of the Royal Society of Edinburgh*, Feb. 18, 1867, vol. VI, pp. 94–105.
[‡] *A Treatise on the Motion of Vortex Rings*, Cambridge, 1883.
[§] See a memoir by M.J.M. Hill in the *Philosophical Transactions of the Royal Society*, London, 1894, part i, pp. 213–246.

ether shall be comparable with that of ordinary matter. In order to obviate this and at the same time to enable it to transmit transversal radiations Sir William Thomson suggested what has been termed, not perhaps very happily, the vortex sponge hypothesis[*]: this rests on the assumption that laminar motion can be propagated through a turbulently moving inviscid liquid. The mathematical difficulties connected with such motion have prevented an adequate discussion of this hypothesis, and I therefore confine myself to merely mentioning it.

These hypotheses, of vortex motion in a fluid, account for the indestructibility of matter and for many of its properties. But in order to explain statical electrical attraction it would seem necessary to suppose that the ether is elastic; in other words, that an electric field must be a field of strain. If so, complete fluidity in the ether would be impossible, and hence the above theories are now regarded as untenable.

(iii) *The Ether-Squirts Hypothesis.* Prof. Karl Pearson[†] has suggested another dynamical theory in which an atom is conceived as a point at which ether is pouring into our space from space of four dimensions.

If an observer lived in two dimensional space filled with ether and confined by two parallel and adjacent surfaces, and if through a hole in one of these surfaces fresh ether were squirted into this space, the variations of pressure thereby produced might give the impression of a hard impenetrable body. Similarly an ether-squirt from space of four dimensions into our space might give us the impression of matter.

It seems necessary on this hypothesis to suppose that there are also ether-sinks, or atoms of negative mass; but as ether-squirts and ether-sinks would repel one another we may suppose that the latter have moved out of the universe known to our senses.

By defining the mass of an atom as the mean rate at which ether is squirting into our space at that point, we can deduce the Newtonian law of gravitation, and by assuming certain periodic variations in the rate of squirting we can deduce some of the phenomena of cohesion, of chemical action, and of electromagnetism and light. But of course the hypothesis rests on the assumption of the existence of a world beyond our senses.

---

[*] *Philosophical Magazine*, London, October, 1887, series 5, vol. XXIV, pp. 342–353.
[†] *American Journal of Mathematics*, 1891, vol. XIII, pp. 309–362.

(iv) *The Electron Hypothesis.* MacCullagh, in 1837 and 1839, proposed to account for optical phenomena on the assumption of an elastic ether possessing elasticity of the type required to enable it to resist rotation. This suggestion has been recently modified and extended by Dr J. Larmor*, and, as now enunciated, it accounts for many of the electrical and magnetic (as well as the optical) properties of matter.

The hypothesis is however very artificial. The assumed ether is a rotationally elastic incompressible fluid. In this fluid Larmor introduces monad electric elements or *electrons*, which are nuclei of radial rotational strain. He supposes that these electrons constitute the basis of matter. He further supposes that an electrical current consists of a procession of these electrons, and that a magnetic particle is one in which these entities are revolving in minute orbits. Dynamical considerations applied to such a system lead to an explanation of nearly all the more obvious phenomena. By further postulating that the orbital motion of electrons in the atom constitute it a fluid vortex it is possible to apply the hydrodynamical pulsatory theory of Bjerknes or Hicks and obtain an explanation of gravitation.

Thus on this view mass is explained as an electrical manifestation. Electricity in its turn is explained by the existence of electrons, that is, of nuclei of strain in the ether, which are supposed to be in incessant and rapid motion. Whilst, to render this possible, properties are attributed to the ether which are apparently inconsistent with our experience of the space it fills. Put thus, the hypothesis seems very artificial. Perhaps the utmost we can say for it is that, from some points of view, it may, so far as analysis goes, be an approximation to the true theory; in any case much work will have to be done before it can be considered established even as a working hypothesis.

Most of the above was written in 1891. Since then investigations on radio-activity have opened up new avenues of conjecture which tend to strengthen the electron theory as a working hypothesis. More than thirty years ago Clerk Maxwell had shown that light and electricity were closely connected phenomena. It was then believed that both were due to waves in the hypothetical ether, but it was supposed that the phenomena of matter on the one side and of light and electricity on the other were sharply distinguished one from the other. The differences,

---

* *Philosophical Transactions of the Royal Society*, London, 1894, pp. 719–822; 1895, pp. 695–743.

however, between matter and light tend to disappear as investigations proceed. In 1895 Röntgen established the existence of rays which could produce light, which had the same velocity as light, which were not affected by a magnet, and which could traverse wood and certain other opaque substances like glass. A year later Becquerel showed that uranium was constantly emitting rays which, though not affecting the eye as light, were capable of producing an image on a photographic plate. Like Röntgen rays they can go through thin sheets of metal; like heat rays they burn the skin; like electricity they generate ozone from oxygen. Passed into the air they enable it to conduct the electric current. Their speed has been measured and found to be rather more than half that of light and electricity. It was soon found that thorium possessed a similar property, but in 1903 Prof. Curie showed that radium possessed radio-activity to an extent previously unsuspected in any body, and in fact the rays were so powerful as to make the substance directly visible. Further experiments showed that numerous bodies are radio-active, but the effects are so much more marked in radium that it is convenient to use that substance for most experimental purposes.

Radium gives off no less than three kinds of rays besides a radio-active emanation. In these discharges there appears to be a gradual change from what had been supposed to be an elementary form of matter to another. This leads to the belief that of the known forms of matter some, perhaps even all, are not absolutely stable. On the other hand, it may be that only radio-active bodies are unstable, and that in their disintegration we are watching the final stage in the evolution of stable and constant forms of matter. It may, however, in any case turn out that some, or perhaps all, of the so-called elements may be capable of resolution into different combinations of electrons or electricity.

At an earlier date J.J. Thomson had concluded that the glow, seen when an electric current passes through a high vacuum tube, is due to a rush of minute particles across the tube. He calculated their weight, their velocity, and the charge of electricity transported by or represented by them, and found these to be constant. They were deflected like Becquerel rays. All space seems to contain them, and electricity, if not identical with them, is at least carried by them. This suggested that these minute particles might be electrons. If so, they might thus give the ultimate explanation of electricity as well as matter, and the atom of the chemist would be not an irreducible unit of matter, but a system comprising numerous such minute particles. These conclusions are consistent with those subsequently deduced from experiments with radium. In 1904 the hypothesis was carried one stage further. In that year J.J. Thomson investigated the conditions of stability of certain systems of revolving particles; and on the hypothesis that an atom of matter consists of a number of particles carrying negative charges of electricity revolving in orbits within a sphere of positive electrification he deduced many of the properties of the different chemical atoms corresponding to different possible stable systems of this kind. His scheme led to results agreeing closely with the results of Mendeléeff's periodic hypothesis. An interesting consequence of this view is that Franklin's description of electricity as subtle particles pervading all bodies, may turn out to be substantially correct. It is also remarkable that corpuscles somewhat analogous to those whose existence was suggested in Newton's corpuscular theory of light should be now supposed to exist in cathode and Becquerel rays.

(v) *The Bubble Hypothesis*[*]. The difficulty of conceiving the motion of matter through a solid elastic medium has been met in another way, namely, by suggesting that what we call matter is a deficiency of the ether, and that this region of deficiency can move through the ether in a manner somewhat analogous to that in which a bubble can move in a liquid. To express this in technical language we may suppose the ether to consist of an arrangement of minute uniform spherical grains piled together so closely that they cannot change their neighbours, although they can move relatively one to another. Places where the number of grains is less or greater than the number necessary to render the piling normal, move through the medium, as a wave moves through water, though the grains do not move with them. Places where the ether is in excess of the normal amount would repel one another and move away out of our ken, but places where it is below the normal amount would attract each other according to the law of gravity, and constitute particles of matter which would be indestructible. It is alleged that the theory accounts for the known phenomena of gravity, electricity, and light, provided the size of its grains is properly chosen. Reynolds has calculated that for this purpose their diameter should be rather more than 5×10^{-18} centimetres, and that the pressure in the medium would be about 10^4 tons per square centimetre. This theory is in itself more

---

[*] O. Reynolds, *Submechanics of the Universe*, Cambridge, 1903.

plausible than the electron hypothesis, but its consequences have not yet been fully worked out.

Returning from these novel hypotheses to the classical theories of matter, we may now proceed a step further. Before a hypothesis on the structure of matter can be ranked as a scientific theory we may reasonably expect it to afford some explanation of three facts. These are (*a*) the Newtonian law of attraction; (*b*) the fact that there are only a finite number of ultimate kinds of matter—such as oxygen, iron, etc.—which can be arranged in a series such that the properties of the successive members are connected by a regular law; and (*c*) the main results of spectrum analysis.

In regard to the first point (*a*), we can say only that none of the above theories are inconsistent with the known laws of attraction; and as far as the ether-squirts, the electron, and the bubble hypotheses are concerned, they have been elaborated into a form from which the gravitational law of attraction can be deduced. But we may still say that as to the cause of gravity—or indeed of force—we know nothing.

Newton, in his Letters to Bentley, while declaring his ignorance of the cause of gravity, refused to admit the possibility of force acting at a finite distance through a vacuum. "You sometimes speak of gravity," said he[*], "as essential and inherent to matter: pray do not ascribe that notion to me, for the cause of gravity is what I do not pretend to know." And in another place he wrote[†], "'Tis inconceivable, that inanimate brute matter should (without the mediation of something else which is not material) operate upon and affect other matter without mutual contact; as it must if gravitation in the sense of Epicurus, be essential and inherent in it... That gravity should be innate, inherent, and essential to matter, so that one body may act upon another at a distance thro' a vacuum, without the mediation of anything else, by and through which their action and force may be conveyed from one to another, is to me so great an absurdity, that I believe no man who has in philosophical matters a competent faculty of thinking can ever fall into it. Gravity must be caused by an agent acting constantly according to

---

[*] Letter dated Jan. 17, 1693. I quote from the original, which is in the Library of Trinity College, Cambridge; it is printed in the *Letters to Bentley*, London, 1756, p. 20.
[†] Letter dated Feb. 25, 1693; *ibid.*, pp. 25, 26.

certain laws, but whether this agent be material or immaterial, I have left to the consideration of my readers."

I have already alluded to conjectural explanations of gravity dependent on the ether-squirts, the electron, and the bubble hypotheses. Of other conjectures as to the cause of gravity, three, which do not involve the idea of force acting at a distance, may be here mentioned:

(1) The first of these conjectures was propounded by Newton in the *Queries* at the end of his *Opticks*, where he suggested as a possible explanation the existence of a stress in the ether surrounding a particle of matter[*].

This has been elaborated on a statical basis by Maxwell, who showed[†] that the stress would have to be at least 3000 times greater than that which the strongest steel would support. Sir William Thomson (Lord Kelvin) has suggested[‡] a dynamical way of producing the stress by supposing that space is filled with an incompressible fluid, constantly being annihilated by each atom of matter at a rate proportional to its mass, a constant supply being kept up at an infinite distance. It is true that this avoids Maxwell's difficulty, but we have no right to introduce such sinks and sources of fluid unless we have other grounds for believing in their existence. The conclusion is that Newton's conjecture is very improbable unless we adopt the ether-squirts theory: on that hypothesis it is a plausible explanation.

I should add that Maclaurin implies[§] that though the above explanation was Newton's early opinion, yet his final view was that he could not devise any tenable hypothesis about the cause of gravitation.

(2) In 1782 Le Sage of Geneva suggested[‖] that gravity was caused by the bombardment of streams of ultramundane corpuscles. These corpuscles are supposed to come in all directions from space and to be so small that inter-collisions are rare.

---

[*] Quoted by S.P. Rigaud in his *Essay* on the *Principia*, Oxford, 1838, appendix, pp. 68–70. On other guesses by Newton see Rigaud, text, pp. 61–62, and references there given.

[†] Article *Attraction*, in *Encyclopaedia Britannica*, or *Collected Works*, vol. II, p. 489.

[‡] *Proceedings of the Royal Society of Edinburgh*, Feb. 7, 1870, vol. VII, pp. 60–63.

[§] *An Account of Sir Isaac Newton's Philosophical Discoveries*, London, 1748, p. 111.

[‖] *Mémoires de l'Académie des Sciences* for 1782, Berlin, 1784, pp. 404–432: see also the first two books of his *Traité de Physique*, Geneva, 1818.

A body by itself in space would receive on an average as many blows on one side as on another, and therefore would have no tendency to move. But, if there are two bodies, each will screen the other from some of the bombarding corpuscles. Thus each body will receive more blows on the side remote from the other body than on the side turned towards it. Hence the two bodies will be impelled each towards the other.

In order to make this force between two particles vary directly as the product of their masses and inversely as the square of the distance between them, Le Sage showed that it was sufficient to suppose that the mass of a body was proportional to the area of a section at right angles to the direction in which it was attracted. This requires that the constitution of a body shall be molecular, and that the distances between consecutive molecules shall be very large compared with the sizes of the molecules. On the vortex hypothesis we may suppose that the ultramundane corpuscles are vortex rings.

This is ingenious, and it is possible that if the corpuscles were perfectly elastic the theory might be tenable[*]. But the results of Maxwell's numerical calculation show, first, that the particles must be imperfectly elastic; second, that merely to produce the effect of the attraction of the earth on a mass of one pound would require that Le Sage's corpuscles should expend energy at the rate of at least billions[†] of foot-pounds per second; and third, that it is probable that the effect of such a bombardment would be to raise the temperature of all bodies beyond a point consistent with our experience. Finally, it seems probable that the distance between consecutive molecules would have to be considerably greater than is compatible with the results given below.

Tait summed up the objections to these two hypotheses by saying[‡], "One common defect of these attempts is... that they all demand some prime mover, working beyond the limits of the visible universe or inside each atom: creating or annihilating matter, giving additional speed to spent corpuscles, or in some other way supplying the exhaustion suffered in the production of gravitation. Another defect is that they all make gravitation a mere difference-effect, as it were; thereby implying the presence of stores of energy absolutely gigantic in com-

---

[*] See a paper by Sir William Thomson (Lord Kelvin) in the *Proceedings of the Royal Society of Edinburgh*, Dec. 18, 1871, vol. VII, pp. 577–589.

[†] I use billion with the English (and not the French) meaning, that is, a billion $= 10^{12}$.

[‡] *Properties of Matter*, London, 1885, art. 164.

parison with anything hitherto observed, or even suspected to exist, in the universe; and therefore demanding the most delicate adjustments, not merely to maintain the conservation of energy which we observe, but to prevent the whole solar and stellar systems from being instantaneously scattered in fragments through space. In fact, the cause of gravitation remains undiscovered."

(3) There is another conjecture on the cause of gravity which I may mention[*]. It is possible that the attraction of one particle on another might be explained if both of them rested on a homogeneous elastic body capable of transmitting energy. This is the case if our three-dimensional universe rests in the direction of a fourth dimension on a four-dimensional homogeneous elastic body (which we may call the ether) whose thickness in the fourth dimension is small and constant.

The results of spectrum analysis lead us to suppose that every molecule of matter in our universe is in constant vibration. On the above hypothesis these vibrations would cause a disturbance in the supporting space, *i.e.* in the ether. This disturbance would spread out uniformly in all directions; the intensity diminishing as the square of the distance from the centre of vibration, but the rate of vibration remaining unaltered. The transmission of light and radiant heat may be explained by such vibrations transversal to the direction of propagation. It is possible that gravity may be caused by vibrations in the supporting space which are wholly longitudinal or are compounded of vibrations which are partly longitudinal and partly transversal in any of the three directions at right angles to the direction of propagation. If we define the mass of a molecule as proportional to the intensity of these vibrations caused by it, then at any other point in space the intensity of the vibration there would vary as the mass of the molecule and inversely as the square of the distance from the molecule; hence, if we may assume that such vibrations of the medium spreading out from any centre would draw to that centre a particle of unit mass at any other point with a force proportional to the intensity of the vibration there, then the Newtonian law of attraction would follow. This conjecture is consistent either with Boscovich's hypothesis or with the vortex theory. It would be interesting if the results of a branch of pure mathematics so abstract as the theory of hyperspace should be found

---

[*] See an article by myself in the *Messenger of Mathematics*, Cambridge, 1891, vol. XXI, pp. 20–24.

to be closely connected with one of the most fundamental problems of material science.

I should sum up the effect of this discussion on gravity on the relative probabilities of the hypotheses as to the constitution of matter enumerated above, by saying that it does not enable us to discriminate between them.

The fact that the number of kinds of matter (chemical elements) is finite and the consequences of spectrum analysis are closely related and may be treated together. The results of spectrum analysis show that every molecule of any species of matter, such as hydrogen, vibrates with (so far as we can tell) exactly equal sets of periods of vibration. This then is one of the characteristics of the particular kind of matter, and it is probable that any explanation of why the molecules of each kind have a definite set of periods of vibration will account also for the fact that the number of kinds of matter is finite.

Various attempts to explain why the molecules of matter are capable only of certain definite periods of vibration have been made, and it may be interesting if I give them briefly.

(1) To begin with, I may note the conjecture that it depends on properties of time. This, however, is impossible, for the continuity of certain spectra proves that in these cases there is nothing which prevents the period of vibration from taking any one of millions of different values: thus no explanation dependent on the nature of time is permissible.

(2) It has been suggested that there may have been a sorting agency, and only selected specimens of the infinite number of species formed originally have got into our universe. The objection to this is that no explanation is offered as to what has become of the excluded molecules.

(3) The finite number of species might be explained by supposing a physical connection to exist between all the molecules in the universe, just as two clocks whose rates are nearly the same tend to go at the same rate if their cases are connected.

Maxwell's objection to this is that we have no other reason for supposing that such a connection exists, but if we are living in a space of four dimensions as suggested above in chapter XII, this connection does exist, for all the molecules rest on one and the same body. This body is capable of transmitting vibrations, hence, no matter how the molecules were set vibrating originally, they would fall into certain groups, and all the members of each group would vibrate at the same rate. It was the possibility of obtaining thus a physical connection between the various particles in our universe that first suggested to me the idea of a supporting medium in a fourth dimension.

(4) If we accept Boscovich's hypothesis or that of an elastic solid ether, and if we may lay it down as axiomatic that the mass of every sub-atom is the same, we may conceive that the number of ways of combining the sub-atoms into a permanent system is limited, and that the period of vibration depends on the form in which the sub-atoms are combined into an atom. This view is not inconsistent with any known facts. I may add that it is probable that the chemical atom is the essential vibrating system, for the sodium spectrum, to take one instance, is the same as that of all its compounds.

(5) In the same way we may suppose that the vortex rings are formed so that they can have only a definite number of stable forms produced by interlinking or kinking.

(6) Similarly we may modify the popular hypothesis by treating the atoms as indivisible aggregates of sub-atoms which are in all respects equal and similar, and can be combined in only a limited number of forms which are permanent. But most of the old difficulties connected with the atoms arise again in connection with the sub-atoms.

(7) I am not aware that Maxwell discussed any other hypotheses in connection with this point, but it has been suggested recently that, if the various forms of matter were evolved originally out of some one primitive material, then there may have been periodic disturbances in this matter when the atoms were being formed, such that they were produced only at some definite phase in the period[*].

Thus, if the disturbance is represented by the swinging of a pendulum in a resisting medium, it might be supposed that the atoms were formed at the points of maximum amplitude, and we should expect that the atoms successively thrown off would form a series having the properties of its successive members connected by a regular periodic law. This conjecture, when worked out in some detail, led to the conclusion that some elements which ought to have appeared in the series were

---

[*] See Crookes on Mendeléeff's periodic law, *Nature*, Sept. 2, 1886, vol. XXXIV, pp. 423–432.

missing, but it was possible to predict their properties and to suggest the substances with which they were most likely to be found in combination. Guided by these theoretical conclusions a careful chemical analysis revealed the fact that such elements did exist.

That this hypothesis has led to new discoveries is something in its favour, but I do not wish to be understood to say that it is a theory which leads to results that have been verified subsequently. I should say rather that we have obtained an analogy which is sufficiently like the truth to suggest new discoveries. Such analogies are often the precursors of laws, so that it is not unreasonable to hope that ere long our knowledge of this border-land of chemistry and physics may be more definite, and thus that molecular physics may be brought within the domain of mathematics. It is however very remarkable that J.J. Thomson's conclusions on the stability of the orbital systems he devised should agree so closely with Mendeléeff's periodic law.

On the whole Maxwell thought that the phenomena point to a common origin of all molecules of the same kind, that this was an event not belonging to that order of nature under which we live, but must have originated when or before the existing order was established, and that so long as the present order exists it is immutable.

This is equivalent to saying that we have arrived at a point beyond which our limited experience does not enable us to carry the explanation.

That we should be able to form an approximate idea of the size of the molecules of matter is a testimony to the extraordinary advance which mathematical physics has made recently.

Sir William Thomson, now Lord Kelvin, whose ingenuity seems to know no limits—has suggested[*] four distinct methods of attacking the problem. They lead to results which are not very different.

The first of these rests on an assertion of Cauchy that the phenomena of prismatic colours show that the distance between consecutive molecules of matter is comparable with the wave-lengths of light. Taking the most unfavourable case this would seem to indicate that in a transparent homogeneous solid or liquid medium there are not more

---

[*] See *Nature*, March 31, 1870, vol. I, pp. 531–553; and Tait's *Recent Advances*, pp. 303–318. The fourth method had been proposed by Loschmidt in 1863.

than 64×10^{24} molecules in a cubic inch, that is, that the distance between consecutive molecules is greater than $1/(4 \times 10^8)$th of an inch.

The second method is founded on the amount of work required to draw out a film of liquid, such as a soap-bubble, to a given thickness. This can be calculated from experiments in a capillary tube, and it is found that, if a soap-bubble could be drawn out to a thickness of $1/10^8$th of an inch there would be but a few molecules in its thickness. This method is not quantitative.

Thirdly, Sir William Thomson proved that the contact phenomena of electricity require that in an alloy of brass the distance between two molecules, one of zinc and one of copper, shall be greater than $1/(7 \times 10^8)$th of an inch; hence the number of molecules in a cubic inch of zinc or copper is not greater than 35×10^{25}.

Lastly, the kinetic theory of gases leads to the conclusion that certain phenomena of temperature and viscosity depend, *inter alia*, on inter-molecular collisions, and so on the sizes and velocities of the molecules, while the average velocity with which the molecules move increases with the temperature. This leads to the conclusion that the distance between two consecutive molecules of a gas at normal pressure and temperature is greater than $1/(6 \times 10^6)$th of an inch, and is less than $1/10^7$th of an inch; while the actual size of the molecule is a trifle greater than $1/(3 \times 10^{20})$th of a cubic inch; and the number of molecules in a cubic inch is about 3×10^{20}.

Thus it would seem that a cubic inch of gas at ordinary pressure and temperature contains about 3×10^{20} molecules, all similar and equal, and each molecule has a volume of about $1/(3 \times 10^{25})$th of a cubic inch; while a cubic inch of the simplest solid or liquid contains rather less than 10^{27} molecules, and perhaps each molecule has a volume of about $1/(3 \times 10^{26})$th of a cubic inch. For instance, if a pea or a drop of water whose radius is 1/16th inch was magnified to the size of the earth, then there would be about thirty molecules in every cubic foot of it, and probably the size of a molecule would be about the same as that of a fives-ball. The average size of the minute drops of water in a very light cloud can be calculated from the coloured rings produced when the sun or moon shines through it. The radius of a drop is about 1/30000th of an inch. Such a drop therefore would contain about 2×10^{13} separate molecules. In gases and vapours, the number of atoms required to make up one of these molecules can be estimated, but in liquids the number is not as yet known.

Loschmidt asserted that a cube whose side is 1/4000th of a millimetre is the smallest object which can be made visible at the present time. Such a cube of oxygen or nitrogen would contain from 60 to 100 millions of molecules of the gas. Also on an average about 50 elementary molecules of the so-called elements are required to constitute one molecule of organic matter. At least half of every living organism consists of water, and we may for the moment suppose that the remainder consists of organic matter. Hence the smallest living being which is visible under the microscope contains from 30 to 50 millions of elementary molecules which are combined in the form of water, and from 30 to 50 millions of elementary molecules which are combined so as to make not more than one million organic molecules.

Hence a very simple organism might be built up out of as few as a million similar organic molecules. Maxwell did not consider that this was sufficient to justify the current conclusions of physiologists, and said that they must not suppose that structural details of infinitely small dimensions can furnish by themselves an explanation of the variety known to exist in the properties and functions of the most minute organisms; but physiologists have replied that whether their conjectures be right or wrong Maxwell's argument is vitiated by his non-consideration of differences due to the physical (as opposed to the chemical) structure of the organism and the consequent motions of the component parts.

INDEX

Abbot, W., 177
Abbott, E.A., 277
Achilles and the Tortoise, 67
Acts or Disputations, chap. VII
Agrippa, Cornelius, 122
Ahrens, 97, 103, 123
Airy, Sir Geo., 87, 194
Aix, Labyrinth at, 152
Albohazen on Astrology, 238
Alcuin, 2, 55
Alfred the Great, 301
Alkarisimi on π, 217
Alkborough, Labyrinth at, 152
Anallagmatic Squares, 51
Anaxagoras, 215
Anderson, T., 272
Angular Motion, 69
Anstice, 103, 105–106
Antipho, 215
Apollonius, 206, 208, 216
Archimedes, 72, 212, 215, 219
Archytas on Delian Problem, 207
Argyle, 258
ARITHMETIC, HIGHER, 29–34
Arithmetical Fallacies, 20–24
ARITHMETICAL PUZZLES, 4–26
— RECREATIONS, chap. I
Arya Bhata on π, 216
Asenby, Labyrinth at, 152
Astrological Planets, 122, 240, 291

ASTROLOGY, chap. X
ATOMIC THEORIES, chap. XIV
Atoms, Size of, 320
Attraction, Law of, 314–318
Augustine on Astrology, 244
Augustus, 269, 292, 293
Ayrton on Magic Mirrors, 87

Babbage, C., 192
Bachet's *Problèmes*, 2–20, 27–29, 55, 113, 115, 122, 125, 132
Bacon, Francis, 269
Bailey, J.E., 251
Ball, 171, 173, 205, 208, 224, 282
Bardesan on Fate, 237
Barrow, I., 172
Baudhayana on π, 216
Becquerel Rays, 312, 313
Bedwell, T., 172
Beltrami on Space, 277
Benham on Spectrum Top, 87
Bentley, Newton to, 314
Bentley, R., 176
Bernoulli, John, 22
Berosus, 298
Berri, de, 267
Bertrand, J.L.F., 21
Bertrand, L. (of Geneva), 159
Besant on Hauksbee's Law, 82
Bhaskara on π, 217
Bickmore, C.E., 230, 231, 236

Biering on Delian Problem, 205
Billingsley, H., 172
Bills on Kirkman's Problem, 109
BINARY POWERS, Fermat on, 31–32
Birch, J.G., 233
Birds, Flight of, 85–86
Bjerknes, 311
Blackburn, J., 185
Blundeville, T., 172
Board, Mathematical, 198
Boat-racing with a rope, 81–82
Bolyai, J., 285
Bonnycastle, J., 189
Bordered Magic Squares, 135–136
Boscovich on Matter, 307, 317, 319
Boughton Green, Labyrinth at, 152
Bouniakowski, V., on shuffling, 110
Bourget, M.J., on shuffling, 110
Bourlet, 18, 19
Boussinesq on Ether, 307
Brackets, in Tripos, 182, 188, 197, 198
Brahmagupta on π, 217
Breton on Mosaics, 152
Brewster, Sir David, 249
Briggs, H., 172
Bristed, C.A., 192
Bromton, 151
Brouncker on π, 220
Brown, J. (Saint), 190
Bryan on Bird Flight, 86
Bryso, 215
Bubble Theory of Matter, 313
Buckley, W., 172
Burnside, Kirkman's Problem, 103
Butcher on the Calendar, 297

Caesar, Julius, 244, 269, 292

CALENDAR, the Civil, 292–294
— the Ecclesiastical, 294–297
— the Gregorian, 294–296
— the Julian, 292–293
Calendars, University, 186
CAMBRIDGE, MATHEMATICS, chap. VII
— STUDIES AT, chap. VII
Cantor on π, 215
Cardan, 2, 93, 95, 237, 247–249, 275
Cards, Problems with, 14–15, 26, 55, 109–120
Carpmael, Kirkman's Problem, 103
Cartwright, W., 271
Cauchy, 48, 307, 320
Cayley, 44, 46, 103, 119, 154, 220
Cellini, 249
Cells of a Chess-board, 158
Centrifugal Force, 71–72
Ceulen, van, on π, 218, 219
Challis MSS, 183
Charles I, 269–271, 272, 297
Charles V of Germany, 302
Chartres, Labyrinth at, 152
Chartres, R., 21, 41, 223
Chasles on Trisection of Angle, 211
Chaucer on the Sun-cylinder, 298
Cheke, Sir John, 247
CHESS-BOARD, GAMES ON, 60–64, 103, 158–169
— knights' moves on, 63, 158–169
— problems, 25, 60–64, 103, 158–169
Chilcombe, Labyrinth at, 152
Chinese on π, 217
CHINESE RINGS, 93–97
Ciccolini on Chess, 164
Cicero on Astrology, 244

INDEX.

CIPHERS, chap. XI
— Definition of, 252
— Four types of, 259–267
— Historical, 269–276
— Requisites for good, 268–269
CIRCLE, QUADRATURE OF, 212–223
Cissoid, the, 206, 208, 213
Clairaut on Trisection of Angle, 211
Clarke, S., 172
Classical Tripos, 196
Claus, 91
Clausen on π, 221
Clavius on Calendar, 293, 294, 297
Clepsydras, 301
Clerk Maxwell, *see* Maxwell
Clerke, G., 172
Clifford, 70
Clocks, 77, 301–303
Cnossus, Coins of, 151
Coat and Waistcoat Trick, 64
Coccoz, 37, 137
Code-Book Ciphers, 274
Cole, F.N., 225, 227, 230, 232, 234
Colebrooke on Indian Algebra, 217
Collini on Chess, 164
Collins, Letter from J. Gregory, 220
COLOUR-CUBE PROBLEM, 51–52
COLOURING MAPS, 44–46
Columbus, 297
Columbus's Egg Puzzle, 74
Comberton, Labyrinth at, 152
Compasses, Watches as, 303–304
Competition, in Tripos, 188, 192
Composite Magic Squares, 134
Conchoid, the, 206, 210, 213
Cones moving uphill, 75
Conradus, D.A., 261
Continuity of Matter, 306
Contour-lines, 47
Cotes, R., 172

Counters, Games with, 48–50, 58–63
Craig, J., 172
Crassus, 244
Cretan Labyrinth, 151, 152
Cricket-Ball, Spin on, 85
Crookes on Mendeléeff's Laws, 319
Cross-fours, 51
CRYPTOGRAPHS, Definition of, 251
— Three types of, 253–259
CRYPTOGRAPHY, chap. XI
CUBE, DUPLICATION OF, 205–209
Cubes, Coloured, 51–52
Cubes, Skeleton, 26
Cudworth on Sharp, 220
Cumberland, R., 176–177
Cumulative Vote, 26
Cunningham, A.J.C., 32, 225, 228, 229, 230, 231, 234, 235
Cureton on Syriac Astrology, 237
Curie on radio-activity, 312
Curiosa Physica, 86–87
Cursor, 298
Cusa on π, 217
Cusps, Astrological, 238
Cut on a Tennis-ball, 83–85
Cutting Cards, Problems on, 15
Cylinders, Sun-, 298

Dacres, A., 172
Daedalus, Labyrinth of, 151
D'Alembert, 21, 24
Darwin, G.H., 42
Dase on π, 221
Dasypodius, 302
Day, Definition of, 289
— Commencement of, 291
— Sidereal and Solar, 290
Days of Week from date, 297
Days of Week, Names of, 291–292
Dealtry, W., 189, 190, 191

De Berri, 267
Decimation, 19–20
Dee, J., 172
De Fonteney on Ferry Problem, 55
De Fouquières, 49
De Haan on π, 215, 218
De Lagny on π, 221
De la Hire on Magic Squares, 122, 123, 126–127, 128, 132–134
De la Loubère on Magic Squares, 124–125, 137
Delambre on Calendar, 296, 297
Delannoy, 55, 60
De la Pryme, 175
DELIAN PROBLEM, 205–209
De Longchamps, G., 235
De Moivre, on Knight's Move, 159
De Montmort, 1, 159
De Morgan, A., 44, 67, 170, 173, 189, 190, 212, 214, 215, 223, 246, 294
Denary scale of notation, 10, 11
De Parville on Tower of Hanoï, 92
De Polignac on Knight's Move, 167
De Rohan, 272
De St Laurent, 110
Descartes, 209, 211, 220, 306
Des Ourmes on Magic Squares, 122
Diabolic Squares, 136–137
Dials, Sun-, 297–299
Dickson, L.E., 118
Diego Palomino, 19
Digby, Lord, 271
Digges, T., 172
Diocles on Delian Problem, 208
Diodorus on Lake Moeris, 150
Dircks, H., 272
Dirichlet on Fermat's Theorem, 33
Dissection, Proofs by, 42–44
Dixon, A.C., 46, 103
DODECAHEDRON GAME, 155–158

Dominical Letter, 296
Dominoes, 48, 141
D'Ons-en-bray, Magic Squares, 122
Doubly Magic Squares, 137
Douglas, S., 192, 194
Drach on Magic Squares, 123
Drayton, 151
DUPLICATION OF CUBE, 205–209
Durations, see Time
Dürer, A., 122
DYNAMICAL GAMES, 52–64

Earnshaw, S., 87, 197
Easter, Date of, 294–297
Edward VI, 240, 247–249, 260
— Horoscope of, 248
EIGHT QUEENS PROBLEM, 97–103
Eisenlohr on Ahmes, 215
Eisenstein, 32
Electrons, 311–313
Elliptic Geometry, 285–286
Eneström on π, 214
Epicurus on Gravitation, 314
Equilibrium, Puzzles on, 72–74
Eratosthenes, 205
Escott, E.B., 236
Ether Theories, 307–310
Ether-Squirts, 310
Etten, van, 2, 10
Euclid, 30, 36, 215
Euclid's Axioms &c., 285
— Parallel Postulate, 285
Euclidean Geometry, 284–287
Euclidean Space, 284–287
Euclid I. 32, 42
Euclid I. 47, 42
Euler, 29, 30, 32, 48, 122, 123, 136, 140, 159–163, 214, 220, 221, 225, 227, 228
EULER'S UNICURSAL PROBLEM, 143–149

INDEX.

Examination, Printed, 184, 196
Exploration Problems, 26

FALLACIES, ARITHMETICAL, 20–24
— GEOMETRICAL, 35–41
— MECHANICAL, 67–69, 75–78
Faraday on Matter, 307
Fauquembergue, E., 230
Fenn, J., 177
Fermat on Binary Powers, 31–32
Fermat, P., 29, 31–34, 123, 225, 226, 227, 228, 235
FERMAT'S LAST THEOREM, 32–34
FERRY-BOAT PROBLEMS, 55–57
FIFTEEN PUZZLE, 88–91
FIFTEEN SCHOOL-GIRLS, 103–109
Figulus on Astrology, 244
Firmicus on Astrology, 238
Fitzpatrick, J., 59
Flamsteed on Astrology, 246
Flamsteed, J., 172
Flat-land, 278–283
Fluid Motion, 82–86
Fluxions, 180, 183, 187, 189, 190, 192, 193
Fonteney on Ferry Problem, 55
Force, Definition of, 70
Foster, S., 172
Fouquières on Ancient Games, 49
Four-Colour Theorem, 44–46
Fox on π, 223
Frankenstein on Magic Pencils, 138
Franklin, B., 313
Frederick II of Germany, 301
Frénicle, Magic Squares, 122, 135
Frere, J., 178
Fresnel on Ether, 307
Frolow on Magic Squares, 136
Frost, A.H., 103, 104–105, 123, 136

Galileo on Pendulum, 302
Galton, 21
GAMES, Dynamical, 52–64
— Statical, 48–52
— with Counters, 58–63
Gases, Theory of, 321
Gauss, 29, 34, 231, 277, 295
GEODESIC PROBLEMS, 57–58
GEOGRAPHY, PHYSICAL, 46–48
GEOMETRICAL FALLACIES, 35–41
GEOMETRICAL RECREATIONS, chap. II
GEOMETRY, NON-EUCLIDEAN, chap. XII
George I of England, 175
Gerbert, 216, 301
GERGONNE'S PROBLEM, 115–118
Gill, Kirkman's Problem, 106–107
Glaisher, J.W.L., 98, 99, 171, 215
Glamorgan, Earl of, 272
Gnomons, 297
Goldbach's Theorem, 31
Golden Number, 296
Gooch, W., 184
GRAVITY, Hypotheses on, 314–318
Green on Ether, 307
Greenwich, Labyrinth at, 153
Gregorian Calendar, 293, 294
Gregory XIII, 293–294
Gregory, Jas., 213, 219
Gregory of St Vincent, 209
Gregory's Series, 220
Grienberger on π, 219
Grille, The, 256
Grimthorpe on Clocks, 301
Gros on Chinese Rings, 95, 96
Guarini's Problem, 63
Gun, Report of, 87
Gunning, H., 184, 185
Günther, S., 97, 99, 123
Guthrie on colouring maps, 44

Haan, de, on π, 215, 218
Halley on π, 221
Halsted on Hyper-space, 277
Hamilton, Archbishop, 247
Hamilton, Sir Wm., 155–158
HAMILTONIAN GAME, 155–158
Hampton Court, Maze at, 149, 153
HANOÏ, TOWER OF, 91–93
Harris on pendulum clock, 302
Harvey, J., 172
Harvey, R., 172
HAUKSBEE'S LAW, 82–85
Hayward, J., 260
Heawood on Colouring Maps, 46
Hegesippus on Decimation, 19
Hele, P., 302
Helmholtz, 78, 277, 309
Helmholz, 308
Henrietta Maria, 270
Henry on Unicursal Problems, 143
Henry VIII of England, 248, 301
Hermary, 158
Hero of Alexandria on π, 206, 216
Herodotus on Lake Moeris, 150
Herschel, Sir John, 192, 292
Hezekiah, 299
Hicks on Matter, 305, 309, 311
Hiero of Syracuse, 72
HIGHER ARITHMETIC, 29–34
Hill, C., 273
Hill, M.J.M., 309
Hill, T., 172
HILLS AND DALES, 46–48
Hinton on Space, 277, 280, 282
Hipparchus on hours of day, 291
Hippias, 215
Hippocrates of Chios, 206, 215
Hodson, W., 183
Holditch on Magic Squares, 123
Homaloidal Geometry, 286
Honorary Optimes, 174, 176, 196

Hood, T., 172
Hooke on Timepieces, 302
Horary Astrology, 238
Hornbuckle, T.W., 190, 191
Horner on Magic Squares, 123
— Rules to cast, 239
HOROSCOPES, chap. X
— Example of, 248
— Rules to cast, 238
— Rules to read, 240–243
Horrox, J., 172
Hour-glasses, 301
Hours, definition of, 290, 291
Houses, Astrological, 238
Huddling, 172, 173
Hudson, C.T., on cards, 118
Hudson, W.H.H., on cards, 111
Hustler, J.D., 190
Hutton, C., 3, 221
Huygens, 209, 212, 213, 219, 220, 302
Hyper-magic Squares, 136–137
HYPER-SPACE, chap. XII
Hyperbolic Geometry, 285–286

ICOSIAN GAME, 155–158
Ideler on the Calendar, 297
Inertia, 70, 71
Inwards on the Cretan Maze, 151
Isaiah, 299

Jacob, E., 190, 191, 192
Jacobi, 231, 235
Jaenisch, 159, 164, 167, 168
James II of England, 258
Japanese Magic Mirrors, 87
Jebb, J., 177, 179–181
Johnson on Fifteen Puzzle, 88
Jones on π, 221
Jones on π, 214
Josephus on Decimation, 19
Julian Calendar, 292–293

Julian's Bowers, 152
Julius Caesar, 244, 269
Junior Optimes, 174, 181
Jurin, J., 172

Kelvin, 307, 308, 309, 315, 316, 320
Kempe on Colouring Maps, 45
Ketteler on Ether, 308
Kinetic Theory of Gases, 321
Kirchhoff on Ether, 308
Kirkman, T.P., 103, 108
KIRKMAN'S PROBLEM, 103–109
Klein, 204
KNIGHT'S PATH PROBLEM, 158–169
Knyghton, 151
Königsberg Problem, 143–149
Kummer on Fermat's Theorem, 33

Labile Ether, 308
Labosne on Magic Squares, 132
Labyrinths, 149–154
Lacroix, 212
Lagny on π, 221
Lagrange, 192, 228
Lagrange's Theorem, 31
La Hire, 122, 123, 126–127, 128, 132–134
Laisant, C.A., 11, 236
La Loubère, 124–125
Lambert on π, 212
Lambert on π, 212
Lamé, 33, 307
Landry, 225, 227, 229, 230, 232
Langley on Bird Flight, 86
Laplace, 192
Laplace on velocity of sound, 307
Laquière on Knight's Path, 167
Larmor on Electrons, 305, 311
Latruaumont, 272
Laughton, R., 172
Lawrence, F.W., 236

Lax, W., 184, 185
Lea on Kirkman's Problem, 109
Leap-year, 292–293
Lebesgue on Fermat's Theorem, 33
Legendre, 33, 160, 168, 192, 212, 228, 231
Leibnitz on Games, 1
Lejeune Dirichlet on Fermat, 33
Le Lasseur, 225, 227, 228, 229, 230
Leonardo of Pisa on π, 217
Le Sage on Gravity, 315, 316
Leslie, J., 206, 210
Leurechon, 2
Lilius on the Calendar, 293
Lilly on Astrology, 246
Linde on Knight's Path, 158
Lindemann on π, 212
Line-land, 279
Lines of Slope, 47
Lippeus, 302
Listing's *Topologie*, 65, 145
Liveing on the Spectrum Top, 87
Lobatschewsky, N.I., 277, 285
Locke, J., 182, 184
Lommel on Ether, 308
London and Wise, 153
Lorentz on Ether, 307
Loschmidt on Molecules, 320, 322
Loubère, de la, 124–125
Louis XI of France, 245
Louis XIV of France, 124
Loyd, S., 17
Lucas di Burgo, 2
Lucas, E., 50, 55, 60, 61, 63, 91, 95, 103, 107, 143, 155, 229, 230
Lucca, Labyrinth at, 152
Lydgate on the Sun-cylinder, 298

MacCullagh on Ether, 307
Machin's series for π, 221, 222
Maclaurin on Newton, 315
MacMahon, 28–29, 51
Magic Bottles, 73
Magic Mirrors, 87
MAGIC PENCILS, 137–140
MAGIC SQUARES, chap. V
Magic Square Puzzles, 140–142
Magnus on Hauksbee's Law, 83
MAP COLOUR THEOREM, 44–46
Marie Antoinette, 273
MATHEMATICS, CAMBRIDGE, chap. VII
Mathews, G.B., 234
Matter, Constitution of, chap. XIV
— Hypotheses on, 306–314
— Kinds of, limited, 318–320
— Size of Molecules, 320–322
Maxim on Bird Flight, 86
Maxwell's Demon, 87
Maxwell, J. Clerk, 46, 87, 305, 311, 315, 316, 318, 319, 322
Mazes, 149–154
Mean Time, 290, 291
MECHANICAL RECREATIONS, chap. III
Medieval Problems, 16–20
Menaechmus, 207
Mendeléeff, 313, 319, 320
Mersenne on Primes, 29
MERSENNE'S NUMBERS, chap. IX, 30, 224, 225
Mesolabum, 207
Metius on π, 218
Meton, 298
Méziriac, *see* Bachet
Milner, I., 182, 190
Minding on Knight's Path, 168
Minos, 149, 205
Minotaur, 151

Minutes, def. of, 290, 291
Mirrors, Magic, 87
MODELS, 78
Moderators, chap. VII
Mohammed's sign-manual, 148
Moivre, A. De, 159
MOLECULES, SIZE OF, 320–322
Money, Question on, 8–9
Monge on Shuffling Cards, 109–111
Months, 292
Montmort, De, 1, 159
Montucla, 3, 50, 72, 73, 122, 123, 132, 133, 159, 212, 213
Moon, R., 123, 167, 168
Morgan, A. De, *see* De Morgan
Morland, S., 172
Morley on Cardan, 247
Mosaic Pavements, 50, 152
Moschopulus, 122, 126
Motion in Fluids, 82–86
Motion, Laws of, 66, 70–75
— Paradoxes on, 67–69
— Perpetual, 75–78
MOUSETRAP, GAME OF, 119–120
Müller (Regiomontanus), 217
Mullinger, J.B., 202
Mydorge, 2

Nasik Squares, 136–137
Natal Astrology, 238
Nauck, 97
Neumann on Ether, 307
Newton, 75, 172, 189, 190, 191, 209, 211, 213, 306, 313, 314, 315
Newtonian Laws of Motion, 66–75
Nicene Council on Easter, 294
Niceron, 275
Nicomedes, 206
Nigidius on Astrology, 244
NON-EUCLIDEAN GEOMETRY, 284–287

INDEX. 331

Nonez on Sun-dials, 298
Notation, Denary scale of, 10
Noughts and Crosses, 48
Numa on the Year, 292
NUMBERS, PERFECT, 225
— PUZZLES WITH, 4–20
— THEORY OF, 29–34

Oliver on Sun-dials, 299
Ons-en-bray on Magic Squares, 122
Oppert on π, 215
Optimes, chap. VII, 174, 177, 196
Oram on Eight Queens, 101
Oughtred, W., 172
Oughtred's *Recreations*, 2, 10, 13, 16, 18, 73, 74
Ourmes on Magic Squares, 122
Ovid, 150
Ozanam's *Récréations*, 2, 10, 16, 20, 43, 50, 55, 64, 72, 73, 74, 75, 77, 79, 93, 122, 123, 132, 140, 159, 298, 299, 303
Ozanam, A.F., on Labyrinths, 152

π, 212–223, *see* table of contents
Pacificus on Clocks, 301
Pacioli di Burgo, 2
Pairs of Cards Trick, 113–115
Paley, W., 178, 184, 186, 190, 195
Palomino, 19
Pappus, 206, 210, 211
Parabolic Geometry, 286
PARADROMIC RINGS, 64–65
Parallels, Theory of, 285
Parmentier on Knight's Path, 159
Parry on Sound, 87
Parville on Tower of Hanoï, 92
PAWNS, GAMES WITH, 58–63
Paynell, N., 172
Pearson on Ether-Squirts, 310
Pein on Ten Queens, 101
PENCILS, MAGIC, 137–140

Pepys, S., 271–272
PERFECT NUMBERS, 30, 225
Permutation Problems, 26
PERPETUAL MOTION, 75–78
Perrin, 11
Perry on Magic Mirrors, 87
Peterson on maps, 46
Philo, 206
Philoponus on Delian Problem, 205
PHYSICAL GEOGRAPHY, 46–48
Pierce on Kirkman's Problem, 103
PILE PROBLEMS, 115–119
Pirie on π, 220
Pitatus on the Calendar, 293
Pittenger, 71
Plana, G.A.A., 225, 227, 229, 232
Planets (astrological), 122, 240, 291
— Signification of, 240–242
Plato on Delian Problem, 205, 206, 207
Pliny, 150, 244
Poe, E.A., 261, 264, 275
Poignard, Magic Squares, 122, 123, 127
Poincaré, H., 286
Poitiers, Labyrinth at, 152
Polignac on Knight's Path, 167
Poll Examinations, 196
Poll-Men, 174
Pollock, F., 189–191
Pompey, 244
Porta, G., 275
Power, Kirkman's Problem, 103, 105
Pratt on Knight's Path, 164
Pretender, The Young, 258
PRIMES, 29
Probabilities and π, 222
Probabilities, Fallacies in, 24, 42
Problem Papers, 183, 184, 186
Ptolemy, 216, 237, 238, 291

Purbach on π, 217
PUZZLES, Arithmetical, 4–29
— Geometrical, 48–65
— Mechanical, 67–75
Pythagorean Symbol, 148

QUADRATURE OF CIRCLE, 212–223
QUEENS PROBLEM, EIGHT, 97–103
Queens, Problems with, 97–103

Racquet-ball, Cut on, 83–85
Railway Puzzles (shunting), 53–54
Ramification, 154, 155
Raphael on Astrology, chap. X
Ravenna, Labyrinth at, 152
Rayleigh, 84, 86, 308
Record, R., 172
Regiomontanus on π, 217
Reimer on Delian Problem, 205
Reiss, 26, 63
Relative Motion, 69
Reneu, W., 175
Renton, 42
Reynolds, O., 313
Rhind Papyrus, 215
Riccioli on the Calendar, 293
Rich, J., 271
Richard, J., 69, 285
Richards on use of compass, 303
Richter on π, 222
Riemann, G.F.B., 277, 285, 287
Rigaud, S.P., 315
Ring-Dial, 299–301
Rockliff Marshes, Labyrinth at, 152
Rodet on Arya-Bhata, 216
Rodwell on Hyper-space, 277
Roget, P.M., 163, 164–167
Romanus on π, 218
Rome, Labyrinth at, 152
Röntgen Rays, 312
Rooke, L., 172
Rosamund's Bower, 151

Rosen on Arab values of π, 217
Rothschild, F., 245
Routes on a Chess-board, 25, 158
Row, Counters in a, 48–50, 58–61
Rudio on π, 212
Russell, B.A.W., 68
Rutherford on π, 221

Saccheri, J., 277, 285
Saffron Walden, Labyrinth at, 152
SAILING, Theory of, 79–82
Sand-clocks, 301
Sarrau on Ether, 307
Saunderson, N., 172
Sauveur, Magic Squares, 122, 123, 136
Scale of Notation, Denary, 10
— Puzzles dependent on, 10–12
SCHOOL-GIRLS, FIFTEEN, 103–109
Schubert on π, 212
Schumacher, 221
Scott, Sir Walter, 245
Scytale, The, 258
Seconds, def. of, 290, 291
SECRET COMMUNICATIONS, chap. XI
Seelhoff, 225, 227, 230, 232
Selander on π, 214
SENATE-HOUSE EXAMINATION, chap. VII
Seneca on Astrology, 244
Senior Optimes, 174, 177, 178
Seventy-seven Puzzle, 44
Shanks on π, 221
Sharp on π, 220
Shelton, T., 271, 272
Sherwin's Tables, 220
SHUFFLING CARDS, 109–111
SHUNTING PROBLEMS, 53–54
Sidereal Time, 290
Simpson's Euclid, 189

INDEX. 333

Sixty-five Puzzle, 42
Skeleton Cubes, 26
Smith, A., on π, 223
— Hen., on Numbers, 33
— R., 172, 178
— R.C., see Raphael
Snell on π, 219, 220
Solar Time, 290
Sosigenes on Calendar, 292, 293
Sound, Problem in, 87
— Velocity of, 306
Southey on Astrology, 249
Southwark, Labyrinth at, 153
SPACE, Properties of, chap. XII
Spectrum Analysis, 306
Spectrum Top, 87
Spin on Cricket-ball, 85
Spirits, Raising, 249
Sporus on Delian Problem, 208
Sprague on Eleven Queens, 101
SQUARING THE CIRCLE, 212–223
Stability of Equilibrium, 72–74
STATICAL GAMES, 48–52
Steen on the Mousetrap, 119
St Laurent on cards, 110
Stöffler on the Calendar, 293
Stokes on Ether, 307
St Omer, Labyrinth at, 152
Story on the Fifteen Puzzle, 88
Strabo on Lake Moeris, 150
Stringham on Hyper-space, 284
Sturm, A., 205
St Vincent, Gregory of, 209
Styles, 297
Suetonius, 269
Sun-cylinders, 298
Sun-dials, 297–299
Sun-rings, 299–301
Sun, the Mean, 291
Svastika, 152
Swift, 67

Sylvester, 49, 51, 108, 109

Tacitus on Astrology, 244
Tait, 46, 59, 60, 145, 148, 305, 309, 316, 320
Tanner on Shuffling Cards, 110
Tarry, 57, 140, 150
Tartaglia, 2, 16, 20, 27, 55
Tate, 270
Tavel, G.F., 191
Taylor, B., 172
Taylor, Ch., on Trisection, 211
Tennis-ball, Cut on, 83–85
Tesselation, 50–51
THEORY OF NUMBERS, chap. IX
Thibaut on Baudhayana, 216
Thompson on Magic Squares, 123
Thomson, J.J., 305, 309, 312, 313, 320
Thomson, Sir Wm., see Kelvin
Thrasyllus on Astrology, 245
THREE-IN-A-ROW, 48–50
THREE-PILE PROBLEM, 115–119
Three-Things Problem, 18–19
Tiberius on Astrology, 245
TIME, chap. XIII
— Equation of, 291
— Measurement of, 288–291
— Units of, 288–292
Tissandier, 70
Todhunter, J., 199
Tonstall, C., 172
TOWER OF HANOÏ, 91–93
Trastevere, Labyrinth at, 152
TREES, GEOMETRICAL, 154–155
Treize, Game of, 120
TRICKS WITH NUMBERS, 4–25
TRIPOS, MATHEMATICAL, chap. VII
Tripos, Origin of term, 201–203
TRISECTION OF ANGLE, 210–212
Tritheim, J., 275

Trollope on Mazes, 151
Troy-towns, 152
Turton, W.H., 39, 44, 102

Uhlemann on Astrology, 237
UNICURSAL PROBLEMS, chap. VI

Van Ceulen on π, 218, 219
Vandermonde, 63, 159, 163
Van Etten, 2, 10
Vase Problem, 16
Vega on π, 221
Vick on Clocks, 302
Vieta, 209, 218
Vince, S., 189, 190, 191
Violle, Magic Squares, 122, 136
Virgil, 150
Voigt on Ether, 308
Volpicelli on Knight's Path, 158
Von Helmholz, 78, 277, 308, 309
Vortex rings, 308, 309
— Spheres, 309
— Sponges, 309, 310
Voting, Question on, 26

Walecki, 107
Walker, G.T., 21
Wallis, J., 93, 95, 172, 220, 270
Ward, S., 172
Waring, E., 177, 185, 187, 190, 191
Warnsdorff, Knight's Path, 164
Watch Problem, 13
Watches, 77, 291, 301, 302
— as Compasses, 303–304
Water-clocks, 301, 303
Waterloo, Battle of, 297
Watersheds and Watercourses, 48
Watson, R., 178, 186
Waves, Superposition of, 87

Weber-Wellstein, 230
Week, Days of, from date, 297
Week, Names of Days, 291–292
WEIGHTS PROBLEM, THE, 27–29
Western on Binary Powers, 32
Wheatstone on Ciphers, 269, 271
Whewell, W., 171, 190, 191, 192, 193, 237, 245
Whist, Number of Hands at, 26
Whiston, W., 172
Wiedemann on Lake Moeris, 150
Wilkins on Ciphers, 253, 254, 258, 266, 275
William III of England, 153
Willis on Hauksbee's law, 82
Wilson on Ptolemy, 238
Wilson's Theorem, 191
Wing, Labyrinth at, 152
Wood, J., 185, 189, 190
Woodhouse, R., 191, 192, 193
Woolhouse, Kirkman's Problem, 103
Worcester, Marquis of, 272
Wordsworth, C., 171, 177, 184, 186, 203
Work, 72–75
Wranglers, chap. VII, 174
Wright, E., 172
Wright, J.M.F., 194

Year, Civil, 292–294
Year, Mohammedan, 294

Zach on π, 221
Zech, R., 302
Zeller, 297
Zeno on Motion, 67–68
Zodiac, Signs in Astrology, 239, 242

A SHORT ACCOUNT OF THE
HISTORY OF MATHEMATICS

By W.W. ROUSE BALL.

[*Third Edition. Pp.* xxiv + 527. *Price* 10*s. net.*]

MACMILLAN AND CO. Ltd., LONDON AND NEW YORK.

This book gives an account of the lives and discoveries of those mathematicians to whom the development of the subject is mainly due. The use of technicalities has been avoided and the work is intelligible to any one acquainted with the elements of mathematics.

The author commences with an account of the origin and progress of Greek mathematics, from which the Alexandrian, the Indian, and the Arab schools may be said to have arisen. Next the mathematics of medieval Europe and the renaissance are described. The latter part of the book is devoted to the history of modern mathematics (beginning with the invention of analytical geometry and the infinitesimal calculus), the account of which is brought down to the present time.

This excellent summary of the history of mathematics supplies a want which has long been felt in this country. The extremely difficult question, how far such a work should be technical, has been solved with great tact.... The work contains many valuable hints, and is thoroughly readable. The biographies, which include those of most of the men who played important parts in the development of culture, are full and general enough to interest the ordinary reader as well as the specialist. Its value to the latter is much increased by the numerous references to authorities, a good table of contents, and a full and accurate index.—*The Saturday Review*.

Mr. Ball's book should meet with a hearty welcome, for though we possess other histories of special branches of mathematics, this is the first serious attempt that has been made in the English language to give a systematic account of the origin and development of the science as a whole. It is written too in an attractive style. Technicalities are not too numerous or obtrusive, and the work is interspersed with biographical sketches and anecdotes likely to interest the general reader. Thus the tyro and the advanced mathematician alike may read it with pleasure and profit.—*The Athenæum.*

A wealth of authorities, often far from accordant with each other, renders a work such as this extremely formidable; and students of mathematics have reason to be grateful for the vast amount of information which has been condensed into this short account.... In a survey of so wide extent it is of course impossible to give anything but a bare sketch of the various lines of research, and this circumstance tends to render a narrative scrappy. It says much for Mr. Ball's descriptive skill that his history reads more like a continuous story than a series of merely consecutive summaries.—*The Academy.*

We can heartily recommend to our mathematical readers, and to others also, Mr. Ball's *History of Mathematics*. The history of what might be supposed a dry subject is told in the pleasantest and most readable style, and at the same time there is evidence of the most careful research.—*The Observatory.*

All the salient points of mathematical history are given, and many of the results of recent antiquarian research; but it must not be imagined that the book is at all dry. On the contrary the biographical sketches frequently contain amusing anecdotes, and many of the theorems mentioned are very clearly explained so as to bring them within the grasp of those who are only acquainted with elementary mathematics.—*Nature.*

Le style de M. Ball est clair et élégant, de nombreux aperçus rendent facile de suivre le fil de son exposition et de fréquentes citations permettent à celui qui le désire d'approfondir les recherches que l'auteur n'a pu qu'effleurer.... Cet ouvrage pourra devenir très utile comme manuel d'histoire des mathématiques pour les étudiants, et il ne sera pas déplacé dans les bibliotheques des savants.—*Bibliotheca Mathematica.*

The author modestly describes his work as a compilation, but it is thoroughly well digested, a due proportion is observed between the various parts, and when occasion demands he does not hesitate to give an independent judgment on a disputed point. His verdicts in such instances appear to us to be generally sound and reasonable.... To many readers who have not the courage or the opportunity to tackle the ponderous volumes of Montucla or the (mostly) ponderous treatises of German writers on special periods, it may be somewhat of a surprise to find what a wealth of human interest attaches to the history of so "dry" a subject as mathematics. We are brought into contact with many remarkable men, some of whom have played a great part in other fields, as the names of Gerbert, Wren, Leibnitz, Descartes, Pascal, D'Alembert, Carnot, among others may testify, and with at least one thorough blackguard (Cardan); and Mr. Ball's pages abound with quaint and amusing touches characteristic of the authors under consideration, or of the times in which they lived.—*Manchester Guardian.*

There can be no doubt that the author has done his work in a very excellent way.... There is no one interested in almost any part of mathematical science who will not welcome such an exposition as the present, at once popularly written and exact, embracing the entire subject.... Mr. Ball's work is destined to become a standard one on the subject.—*The Glasgow Herald.*

A most interesting book, not only for those who are mathematicians, but for the much larger circle of those who care to trace the course of general scientific progress. It is written in such a way that those who have only an elementary acquaintance with the subject can find on almost every page something of general interest.—*The Oxford Magazine.*

A PRIMER OF THE
HISTORY OF MATHEMATICS

By W.W. ROUSE BALL.

[*Second Edition.   Pp.* iv + 148.   *Price 2s. net.*]

MACMILLAN AND CO. Ltd., LONDON AND NEW YORK.

This book contains a sketch in popular language of the history of mathematics; it includes some notice of the lives and surroundings of those to whom the development of the subject is mainly due as well as of their discoveries.

This Primer is written in the agreeable style with which the author has made us acquainted in his previous essays; and we are sure that all readers of it will be ready to say that Mr. Ball has succeeded in the hope he has formed, that "it may not be uninteresting" even to those who are unacquainted with the leading facts. It is just the book to give an intelligent young student, and should allure him on to the perusal of Mr. Ball's "Short Account." The present work is not a mere *réchauffé* of that, though naturally most of what is here given will be found in equivalent form in the larger work.... The choice of material appears to us to be such as should lend interest to the study of mathematics and increase its educational value, which has been the author's aim. The book goes well into the pocket, and is excellently printed.—*The Academy.*

We have here a new instance of Mr. Rouse Ball's skill in giving in a small space an intelligible account of a large subject. In 137 pages we have a sketch of the progress of mathematics from the earliest records up to the middle of this century, and yet it is interesting to read and by no means a mere catalogue.—*The Manchester Guardian.*

It is not often that a reviewer of mathematical works can confess that he has read one of them through from cover to cover without abatement of interest or fatigue. But that is true of Mr. Rouse Ball's wonderfully entertaining little "History of Mathematics," which we heartily recommend to even the quite rudimentary mathematician. The capable mathematical master will not fail to find a dozen interesting facts therein to season his teaching.—*The Saturday Review.*

A fascinating little volume, which should be in the hands of all who do not possess the more elaborate *History of Mathematics* by the same author.—*The Mathematical Gazette.*

This excellent sketch should be in the hands of every student, whether he is studying mathematics or no. In most cases there is an unfortunate lack of knowledge upon this subject, and we welcome anything that will help to supply the deficiency. The primer is written in a concise, lucid and easy manner, and gives the reader a general idea of the progress of mathematics that is both interesting and instructive.—*The Cambridge Review.*

Mr. Ball has not been deterred by the existence and success of his larger "History of Mathematics" from publishing a simple compendium in about a quarter of the space.... Of course, what he now gives is a bare outline of the subject, but it is ample for all except the most advanced proficients. There is no question that, as the author says, a knowledge of the history of a science lends interest to its study, and often increases its educational value. We can imagine no better cathartic for any mathematical student who has made some way with the calculus than a careful perusal of this little book.—*The Educational Times.*

The author has done good service to mathematicians by engaging in work in this special field.... The Primer gives, in a brief compass, the history of the advance of this branch of science when under Greek influence, during the Middle Ages, and at the Renaissance, and then goes on to deal with the introduction of modern analysis and its recent developments. It refers to the life and work of the leaders of mathematical thought, adds a new and enlarged value to well-known problems by treating of their inception and history, and lights up with a warm and personal interest a science which some of its detractors have dared to call dull and cold.—*The Educational Review.*

It is not too much to say that this little work should be in the possession of every mathematical teacher.... The Primer gives in a small compass the leading events in the development of mathematics.... At the same time, it is no dry chronicle of facts and theorems. The biographical sketches of the great workers, if short, are pithy, and often amusing. Well-known propositions will attain a new interest for the pupil as he traces their history long before the time of Euclid.—*The Journal of Education.*

This is a work which all who apprehend the value of "mathematics" should read and study..., and those who wish to learn how to think will find advantage in reading it.—*The English Mechanic.*

The subject, so far as our own language is concerned, is almost Mr. Ball's own, and those who have no leisure to read his former work will find in this Primer a highly-readable and instructive chapter in the history of education. The condensation has been skilfully done, the reader's interest being sustained by the introduction of a good deal of far from tedious detail.—*The Glasgow Herald.*

Mr. W.W. Rouse Ball is well known as the author of a very clever history of mathematics, besides useful works on kindred subjects. His latest production is *A Primer of the History of Mathematics*, a book of one hundred and forty pages, giving in non-technical language a full, concise, and readable narrative of the development of the science from the days of the Ionian Greeks until the present time. Anyone with a leaning towards algebraic or geometrical studies will be intensely interested in this account of progress from primitive usages, step by step, to our present elaborate systems. The lives of the men who by their research and discovery helped along the good work are described briefly, but graphically.... The Primer should become a standard text-book.—*The Literary World.*

This is a capital little sketch of a subject on which Mr. Ball is an acknowledged authority, and of which too little is generally known. Mr. Ball, moreover, writes easily and well, and has the art of saying what he has to say in an interesting style.—*The School Guardian.*

MATHEMATICAL RECREATIONS AND ESSAYS

By W.W. ROUSE BALL.

[*Fourth Edition.*   Pp. xvi + 402.   *Price 7s. net.*]

MACMILLAN AND CO. Ltd., LONDON AND NEW YORK.

This work is divided into two parts; the first is on mathematical recreations and puzzles, the second includes some miscellaneous essays and an account of some problems of historical interest. In both parts questions which involve advanced mathematics are excluded.

The mathematical recreations include numerous elementary questions and paradoxes, as well as problems such as the proposition that to colour a map not more than four colours are necessary, the explanation of the effect of a cut on a tennis ball, the fifteen puzzle, the eight queens problem, the fifteen school-girls, the construction of magic squares, the theory and history of mazes, and the knight's path on a chess-board.

The second part commences with sketches of the history of the Mathematical Tripos at Cambridge, of the three famous classical problems in geometry (namely, the duplication of the cube, the trisection of an angle, and the quadrature of the circle) and of Mersenne's Numbers. These are followed by essays on Astrology and Ciphers. The last three chapters are devoted to an account of the hypotheses as to the nature of Space and Mass, and the means of measuring Time.

Mr. Ball has already attained a position in the front rank of writers on subjects connected with the history of mathematics, and this brochure will add another to his successes in this field. In it he has collected a mass of information bearing upon matters of more general interest, written in a style which is eminently readable, and at the same time exact. He has done his work so thoroughly that he has left few ears for other gleaners. The nature of the work is completely indicated to the mathematical student by its title. Does he want to revive his acquaintance with the *Problèmes Plaisans et Délectables* of Bachet, or the *Récréations Mathématiques et Physiques* of Ozanam? Let him take Mr. Ball for his companion, and he will have the cream of these works put before him with a wealth of illustration quite delightful. Or, coming to more recent times, he will have full and accurate discussion of 'the fifteen puzzle,' 'Chinese rings,' 'the fifteen school-girls problem' *et id genus omne.* Sufficient space is devoted to accounts of magic squares and unicursal problems (such as mazes, the knight's path, and geometrical trees). These, and many other problems of equal interest, come under the head of 'Recreations.' The problems and speculations include an account of the Three Classical Problems; there is also a brief sketch of Astrology; and interesting outlines of the present state of our knowledge of hyper-space and of the constitution of matter. This enumeration badly indicates the matter handled, but it sufficiently states what the reader may expect to find. Moreover for the use of readers who may wish to pursue the several heads further, Mr. Ball gives detailed references to the sources from whence he has derived his information. These *Mathematical Recreations* we can commend as suited for mathematicians and equally for others who wish to while away an occasional hour.—*The Academy.*

The idea of writing some such account as that before us must have been present to Mr. Ball's mind when he was collecting the material which he has so skilfully worked up into his *History of Mathematics.* We think this because ... many bits of ore which would not suit the earlier work find a fitting niche in this. Howsoever the case may be, we are sure that non-mathematical, as well as mathematical, readers will derive amusement, and, we venture to think, profit withal, from a perusal of it. The author has gone very exhaustively over the ground, and has left us little opportunity of adding to or correcting what he has thus reproduced from his notebooks. The work before us is divided into two parts: mathematical recreations and mathematical problems and speculations. All these matters are treated lucidly, and with sufficient detail for the ordinary reader, and for others there is ample store of references.... Our analysis shows how great an extent of ground is covered, and the account is fully pervaded by the attractive charm Mr. Ball knows so well how to infuse into what many persons would look upon as a dry subject.—*Nature.*

A fit sequel to its author's valuable and interesting works on the history of mathematics. There is a fascination about this volume which results from a happy combination of puzzle and paradox. There is both milk for babes and strong meat for grown men.... A great deal of the information is hardly accessible in any English books; and Mr. Ball would deserve the gratitude of mathematicians for having merely collected the facts. But he has presented them with such lucidity and vivacity of style that there is not a dull page in the book; and he has added minute and full bibliographical references which greatly enhance the value of his work.—*The Cambridge Review.*

Mathematicians with a turn for the paradoxes and puzzles connected with number, space, and time, in which their science abounds, will delight in *Mathematical Recreations and Problems of Past and Present Times.*—*The Times.*

Mathematicians have their recreations; and Mr. Ball sets forth the humours of mathematics in a book of deepest interest to the clerical reader, and of no little attractiveness to the layman. The notes attest an enormous amount of research.—*The National Observer.*

Mr. Ball, to whom we are already indebted for two excellent Histories of Mathematics, has just produced a book which will be thoroughly appreciated by those who enjoy the setting of the wits to work.... He has collected a vast amount of information about mathematical quips, tricks, cranks, and puzzles—old and new; and it will be strange if even the most learned do not find something fresh in the assortment.—*The Observatory.*

Mr. Rouse Ball has the true gift of story-telling, and he writes so pleasantly that though we enjoy the fulness of his knowledge we are tempted to forget the considerable amount of labour involved in the preparation of his book. He gives us the history and the mathematics of many problems ... and where the limits of his work prevent him from dealing fully with the points raised, like a true worker he gives us ample references to original memoirs.... The book is warmly to be recommended, and should find a place on the shelves of every one interested in mathematics and on those of every public library.—*The Manchester Guardian.*

A work which will interest all who delight in mathematics and mental exercises generally. The student will often take it up, as it contains many problems which puzzle even clever people.—*The English Mechanic and World of Science.*

This is a book which the general reader should find as interesting as the mathematician. At all events, an intelligent enjoyment of its contents presupposes no more knowledge of mathematics than is now-a-days possessed by almost everybody.—*The Athenæum.*

An exceedingly interesting work which, while appealing more directly to those who are somewhat mathematically inclined, it is at the same time calculated to interest the general reader.... Mr. Ball writes in a highly interesting manner on a fascinating subject, the result being a work which is in every respect excellent.—*The Mechanical World.*

É um livro muito interessante, consagrado a recreios mathematicos, alguns dos quaes são muito bellos, e a problemas interessantes da mesma sciencia, que não exige para ser lido grandes conhecimentos mathematicos e que tem em gráo elevado a qualidade de instruir, deleitando ao mesmo tempo.—*Journal de sciencias mathematicas, Coimbra.*

The work is a very judicious and suggestive compilation, not meant mainly for mathematicians, yet made doubly valuable to them by copious references. The style in the main is so compact and clear that what is central in a long argument or process is admirably presented in a few words. One great merit of this, or any other really good book on such a subject, is its suggestiveness; and in running through its pages, one is pretty sure to think of additional problems on the same general lines.—*Bulletin of the New York Mathematical Society.*

A book which deserves to be widely known by those who are fond of solving puzzles ... and will be found to contain an admirable classified collection of ingenious questions capable of mathematical analysis. As the author is himself a skilful mathematician, and is careful to add an analysis of most of the propositions, it may easily be believed that there is food for study as well as amusement in his pages.... Is in every way worthy of praise.—*The School Guardian.*

Once more the author of a *Short History of Mathematics* and a *History of the Study of Mathematics at Cambridge* gives evidence of the width of his reading and of his skill in compilation. From the elementary arithmetical puzzles which were known in the sixteenth and seventeenth centuries to those modern ones the mathematical discussion of which has taxed the energies of the ablest investigator, very few questions have been left unrepresented. The sources of the author's information are indicated with great fulness.... The book is a welcome addition to English mathematical literature.—*The Oxford Magazine.*

A HISTORY OF THE STUDY OF
MATHEMATICS AT CAMBRIDGE

By W.W. ROUSE BALL.

[*Pp.* xvi + 264.   *Price* 6*s.*]

THE UNIVERSITY PRESS, CAMBRIDGE.

THIS work contains an account of the development of the study of mathematics in the university of Cambridge from the twelfth century to the middle of the nineteenth century, and a description of the means by which proficiency in that study was tested at various times.

The first part of the book is devoted to a brief account of the more eminent of the Cambridge mathematicians, the subject matter of their works, and their methods of exposition. The second part treats of the manner in which mathematics was taught, and of the exercises and examinations required of students in past times. A sketch is given of the origin and history of the Mathematical Tripos; this includes the substance of the earlier parts of the author's work on that subject, Cambridge, 1880. To explain the relation of mathematics to other departments of study an outline of the general history of the university and the organization of education therein is added.

The present volume is very pleasant reading, and though much of it necessarily appeals only to mathematicians, there are parts—*e.g.* the chapters on Newton, on the growth of the tripos, and on the history of the university—which are full of interest for a general reader.... The book is well written, the style is crisp and clear, and there is a humorous appreciation of some of the curious old regulations which have been superseded by time and change of custom. Though it seems light, it must represent an extensive study and investigation on the part of the author, the essential results of which are skilfully given. We can most thoroughly commend Mr. Ball's volume to all readers who are interested in mathematics or in the growth and the position of the Cambridge school of mathematicians.—*The Manchester Guardian.*

Voici un livre dont la lecture inspire tout d'abord le regret que des travaux analogues n'aient pas été faits pour toutes les Écoles célèbres, et avec autant de soin et de clarté.... Toutes les parties du livre nous out vivement intéressé.—*Bulletin des sciences mathématiques.*

A book of pleasant and useful reading for both historians and mathematicians. Mr. Ball's previous researches into this kind of history have already established his reputation, and the book is worthy of the reputation of its author. It is more than a detailed account of the rise and progress of mathematics, for it involves a very exact history of the University of Cambridge from its foundation.—*The Educational Times.*

Mr. Ball is far from confining his narrative to the particular science of which he is himself an acknowledged master, and his account of the study of mathematics becomes a series of biographical portraits of eminent professors and a record not only of the intellectual life of the *élite* but of the manners, habits, and discussions of the great body of Cambridge men from the sixteenth century to our own.... He has shown how the University has justified its liberal reputation, and how amply prepared it was for the larger freedom which it now enjoys.—*The Daily News.*

Mr. Ball has not only given us a detailed account of the rise and progress of the science with which the name of Cambridge is generally associated but has also written a brief but reliable and interesting history of the university itself from its foundation down to recent times.... The book is pleasant reading alike for the mathematician and the student of history.—*St. James's Gazette.*

A very handy and valuable book containing, as it does, a vast deal of interesting information which could not without inconceivable trouble be found elsewhere.... It is very far from forming merely a mathematical biographical dictionary, the growth of mathematical science being skilfully traced in connection with the successive names. There are probably very few people who will be able thoroughly to appreciate the author's laborious researches in all sorts of memoirs and transactions of learned societies in order to unearth the material which he has so agreeably condensed.... Along with this there is much new matter which, while of great interest to mathematicians, and more especially to men brought up at Cambridge, will be found to throw a good deal of new and important light on the history of education in general.—*The Glasgow Herald.*

Exceedingly interesting to all who care for mathematics.... After giving an account of the chief Cambridge Mathematicians and their works in chronological order, Mr. Rouse Ball goes on to deal with the history of tuition and examinations in the University ... and recounts the steps by which the word "tripos" changed its meaning "from a thing of wood to a man, from a man to a speech, from a speech to two sets of verses, from verses to a sheet of coarse foolscap paper, from a paper to a list of names, and from a list of names to a system of examination."—Never did word undergo so many alterations.—*The Literary World.*

In giving an account of the development of the study of mathematics in the University of Cambridge, and the means by which mathematical proficiency was tested in successive generations, Mr. Ball has taken the novel plan of devoting the first half of his book to ... the more eminent Cambridge mathematicians, and of reserving to the second part an account of how at various times the subject was taught, and how the result of its study was tested.... Very interesting information is given about the work of the students during the different periods, with specimens of problem-papers as far back as 1802. The book is very enjoyable, and gives a capital and accurate digest of many excellent authorities which are not within the reach of the ordinary reader.—*The Scots Observer.*

AN ESSAY ON
THE GENESIS, CONTENTS, AND HISTORY OF
NEWTON'S "PRINCIPIA"

By W.W. ROUSE BALL.

[*Pp.* x + 175.   *Price* 6*s. net.*]

MACMILLAN AND CO. Ltd., LONDON AND NEW YORK.

This work contains an account of the successive discoveries of Newton on gravitation, the methods he used, and the history of his researches.

It commences with a review of the extant authorities dealing with the subject. In the next two chapters the investigations made in 1666 and 1679 are discussed, some of the documents dealing therewith being here printed for the first time. The fourth chapter is devoted to the investigations made in 1684: these are illustrated by Newton's professorial lectures (of which the original manuscript is extant) of that autumn, and are summed up in the almost unknown memoir of February, 1685, which is here reproduced from Newton's holograph copy. In the two following chapters the details of the preparation from 1685 to 1687 of the *Principia* are described, and an analysis of the work is given. The seventh chapter comprises an account of the researches of Newton on gravitation subsequent to the publication of the first edition of the *Principia*, and a sketch of the history of that work.

In the last chapter, the extant letters of 1678–1679 between Hooke and Newton, and of those of 1686–1687 between Halley and Newton, are reprinted, and there are also notes on the extant correspondence concerning the production of the second and third editions of the *Principia*.

For the essay which we have before us, Mr Ball should receive the thanks of all those to whom the name of Newton recalls the memory of a great man. The

Principia, besides being a lasting monument of Newton's life, is also to-day the classic of our mathematical writings, and will be so for some time to come.... The value of the present work is also enhanced by the fact that, besides containing a few as yet unpublished letters, there are collected in its pages quotations from all documents, thus forming a complete summary of everything that is known on the subject.... The author is so well known a writer on anything connected with the history of mathematics, that we need make no mention of the thoroughness of the essay, while it would be superfluous for us to add that from beginning to end it is pleasantly written and delightful to read. Those well acquainted with the *Principia* will find much that will interest them, while those not so fully enlightened will learn much by reading through the account of the origin and history of Newton's greatest work.—*Nature.*

An Essay on Newton's Principia will suggest to many something solely mathematical, and therefore wholly uninteresting. No inference could be more erroneous. The book certainly deals largely in scientific technicalities which will interest experts only; but it also contains much historical information which might attract many who, from laziness or inability, would be very willing to take all its mathematics for granted. Mr. Ball carefully examines the evidence bearing on the development of Newton's great discovery, and supplies the reader with abundant quotations from contemporary authorities. Not the least interesting portion of the book is the appendix, or rather appendices, containing copies of the original documents (mostly letters) to which Mr. Ball refers in his historical criticisms. Several of these bear upon the irritating and unfounded claims of Hooke.—*The Athenæum.*

La savante monographie de M. Ball est rédigée avec beaucoup de soin, et à plusieurs égards elle peut servir de modèle pour des écrits de la même nature.—*Bibliotheca Mathematica.*

Newton's *Principia* has world-wide fame as a classic of mathematical science. But those who know thoroughly the contents and the history of the book are a select company. It was at one time the purpose of Mr. Ball to prepare a new critical edition of the work, accompanied by a prefatory history and notes, and by an analytical commentary. Mathematicians will regret to hear that there is no prospect in the immediate future of seeing this important book carried to completion by so competent a hand. They will at the same time welcome Mr. Ball's *Essay on the Principia* for the elucidations which it gives of the process by which Newton's great work originated and took form, and also as an earnest of the completed plan.—*The Scotsman.*

In this essay Mr. Ball presents us with an account highly interesting to mathematicians and natural philosophers of the origin and history of that remarkable product of a great genius *Philosophiae Naturalis Principia Mathematica*, 'The Mathematical Principles of Natural Philosophy,' better known by the short term *Principia*.... Mr. Ball's essay is one of extreme interest to students of physical science, and it is sure to be widely read and greatly appreciated.—*The Glasgow Herald.*

To his well-known and scholarly treatises on the *History of Mathematics*

Mr. W.W. Rouse Ball has added *An Essay on Newton's Principia*. Newton's *Principia*, as Mr. Ball justly observes, is the classic of English mathematical writings; and this sound, luminous, and laborious essay ought to be the classical account of the *Principia*. The essay is the outcome of a critical edition of Newton's great work, which Mr. Ball tells us that he once contemplated. It is much to be hoped that he will carry out his intention, for no English mathematician is likely to do the work better or in a more reverent spirit.... It is unnecessary to say that Mr. Ball has a complete knowledge of his subject. He writes with an ease and clearness that are rare.—*The Scottish Leader*.

Le volume de M. Rouse Ball renferme tout ce que l'on peut désirer savoir sur l'histoire des *Principes*; c'est d'ailleurs l'œuvre d'un esprit clair, judicieux, et méthodique.—*Bulletin des Sciences Mathématiques*.

Mr. Ball has put into small space a very great deal of interesting matter, and his book ought to meet with a wide circulation among lovers of Newton and the *Principia*.—*The Academy*.

Admirers of Mr. W.W. Rouse Ball's *Short Account of the History of Mathematics* will be glad to receive a detailed study of the history of the *Principia* from the same hand. This book, like its predecessors, gives a very lucid account of its subject. We find in it an account of Newton's investigations in his earlier years, which are to some extent collected in the tract *de Motu* (the germ of the *Principia*) the text of which Mr. Rouse Ball gives us in full. In a later chapter there is a full analysis of the *Principia* itself, and after that an account of the preparation of the second and third editions. Probably the part of the book which will be found most interesting by the general reader is the account of the correspondence of Newton with Hooke, and with Halley, about the contents or the publication of the *Principia*. This correspondence is given in full, so far as it is recoverable. Hooke does not appear to advantage in it. He accuses Newton of stealing his ideas. His vain and envious disposition made his own merits appear great in his eyes, and be-dwarfed the work of others, so that he seems to have believed that Newton's great performance was a mere expanding and editing of the ideas of Mr. Hooke—ideas which were meritorious, but after all mere guesses at truth. This, at all events, is the most charitable view we can take of his conduct. Halley, on the contrary, appears as a man to whom we ought to feel most grateful. It almost seems as though Newton's physical insight and extraordinary mathematical powers might have been largely wasted, as was Pascal's rare genius, if it had not been for Halley's single-hearted and self-forgetful efforts to get from his friend's genius all he could for the enlightenment of men. It was probably at his suggestion that the writing of the *Principia* was undertaken. When the work was presented to the Royal Society, they undertook its publication, but, being without the necessary funds, the expense fell upon Halley. When Newton, stung by Hooke's accusations, wished to withdraw a part of the work, Halley's tact was required to avert the catastrophe. All the drudgery, worry, and expense fell to his share, and was accepted with the most generous good nature. It will be seen that both the technical student and the general reader may find much to interest him in Mr. Rouse Ball's book.—*The Manchester Guardian*.

Une histoire très bien faite de la genèse du livre immortel de Newton....

Le livre de M. Ball est une monographie précieuse sur un point important de l'histoire des mathématiques. Il contribuera à accroître, si c'est possible, la gloire de Newton, en révélant à beaucoup de lecteurs, avec quelle merveilleuse rapidité l'illustre géomètre anglais a élevé à la science ce monument immortel, les *Principia.—Mathesis*.

NOTES ON THE HISTORY OF

TRINITY COLLEGE, CAMBRIDGE

By W.W. ROUSE BALL.

[*Pp.* xiv + 183.   *Price 2s. 6d. net.*]

MACMILLAN AND CO. Ltd., LONDON AND NEW YORK.

This booklet gives a popular account of the History of Trinity College, Cambridge, and so far as the author knows, it is as yet (1905) the only work published on the subject. It was written mainly for the use of his pupils, and contains such information and gossip about the College and life there in past times as he believed would be interesting to most undergraduates and members of the House.

This modest and unpretending little volume seems to us to do more for its subject than many of the more formal volumes ... treating of the separate colleges of the English universities.... In nine short, extremely readable, and truly informing chapters it gives the reader a very vivid account at once of the origin and development of the University of Cambridge, of the rise and gradual supremacy of the colleges, of King's Hall as founded by Edward II, of the suppression of King's Hall by Henry VIII on December 17, 1546, the foundation of Trinity College by royal charter on December 19, and the subsequent fortunes of the premier college of Cambridge. The subject is in a way treated under the successive heads of the college, but this is quite subordinate to the handling and characterisation of the subject under four great periods—namely, that during the Middle Ages, that during the Renaissance, that under the Elizabethan statutes, and that during the last half-century. The colleges arose from the determination of the University to prevent students who were very young from seeking lodging, whether under the wing of one or other of the religious orders—a circumstance which shows this University to have been an essentially lay corporation. Early in the sixteenth century the college had absorbed all the members of the University, and henceforth the University was little more than the degree-granting body to students who lived and moved and had their educational being under the colleges.... The University finally took the form of an aggregate of separate and independent corporations, with a federal constitution analogous in a rough sort of way to that of the United States of America, and different from similar corporations at Paris by the fact that these latter were always subject to University supervision.... There is a good account of the effort now going on to re-assert the University at the expense of the colleges. No one who begins Mr. Ball's book will lay it down till he has read it from beginning to end.—*The Glasgow Herald.*

It is a sign of the times, and a very satisfactory one, when ... a tutor ... takes the trouble to make the history of his college known to his pupils. Considering the lack of good books about the Universities, we may thank Mr. Ball that he has been good enough to print for a larger circle. Though he modestly calls his book only "Notes," yet it is eminently readable, and there is plenty of information, as well as abundance of good stories, in its pages.—*The Oxford Magazine.*

Mr. Ball has put not only the pupils for whom he compiled these notes, but the large world of Trinity men, under a great obligation by this compendious but lucid and interesting history of the society to whose service he is devoted. The value of his contribution to our knowledge is increased by the extreme simplicity with which he tells his story, and the very suggestive details which, without much comment, he has selected, with admirable discernment, out of the wealth of materials at his disposal. His initial account of the development of the University is brief but extremely clear, presenting us with facts rather than theories, but establishing, with much distinctness, the essential difference between the hostels, out of which the more modern colleges grew, and that monastic life which poorer students were often tempted to join.—*The Guardian.*

An interesting and valuable book.... It is described by its author as "little more than an orderly transcript" of what, as a Fellow and Tutor of the College, he has been accustomed to tell his pupils. But while it does not pretend either to the form or to the exhaustiveness of a set history, it is scholarly enough to rank as an authority, and far more interesting and readable than most academic histories are. It gives an instructive sketch of the development of the University and of the particular history of Trinity, noting its rise and policy in the earlier centuries of its existence, until, under the misrule of Bentley, it came into a state of disorder which nearly resulted in its dissolution. The subsequent rise of the College and its position in what Mr. Ball calls the Victorian renaissance, are drawn in lines no less suggestive; and the book, as a whole, cannot fail to be welcome to every one who is closely interested in the progress of the College.—*The Scotsman.*

Mr. Ball has succeeded very well in giving in this little volume just what an intelligent undergraduate ought and probably often does desire to know about the buildings and the history of his College.... The debt of the "royal and religious foundation" to Henry VIII is explained with fulness, and there is much interesting matter as to the manner of life and the expenses of students in the sixteenth century.—*The Manchester Guardian.*

LICENSING.

End of the Project Gutenberg EBook of Mathematical Recreations and Essays, by
W. W. Rouse Ball

*** END OF THIS PROJECT GUTENBERG EBOOK MATHEMATICAL RECREATIONS ***

***** This file should be named 26839-pdf.pdf or 26839-pdf.zip *****
This and all associated files of various formats will be found in:
        http://www.gutenberg.org/2/6/8/3/26839/

Produced by Joshua Hutchinson, David Starner, David Wilson
and the Online Distributed Proofreading Team at
http://www.pgdp.net

Updated editions will replace the previous one--the old editions
will be renamed.

Creating the works from public domain print editions means that no
one owns a United States copyright in these works, so the Foundation
(and you!) can copy and distribute it in the United States without
permission and without paying copyright royalties. Special rules,
set forth in the General Terms of Use part of this license, apply to
copying and distributing Project Gutenberg-tm electronic works to
protect the PROJECT GUTENBERG-tm concept and trademark. Project
Gutenberg is a registered trademark, and may not be used if you
charge for the eBooks, unless you receive specific permission. If you
do not charge anything for copies of this eBook, complying with the
rules is very easy. You may use this eBook for nearly any purpose
such as creation of derivative works, reports, performances and
research. They may be modified and printed and given away--you may do
practically ANYTHING with public domain eBooks. Redistribution is
subject to the trademark license, especially commercial
redistribution.

*** START: FULL LICENSE ***

THE FULL PROJECT GUTENBERG LICENSE
PLEASE READ THIS BEFORE YOU DISTRIBUTE OR USE THIS WORK

To protect the Project Gutenberg-tm mission of promoting the free
distribution of electronic works, by using or distributing this work
(or any other work associated in any way with the phrase "Project
Gutenberg"), you agree to comply with all the terms of the Full Project
Gutenberg-tm License (available with this file or online at
http://gutenberg.org/license).

Section 1. General Terms of Use and Redistributing Project Gutenberg-tm
electronic works

1.A. By reading or using any part of this Project Gutenberg-tm
electronic work, you indicate that you have read, understand, agree to
and accept all the terms of this license and intellectual property

(trademark/copyright) agreement. If you do not agree to abide by all
the terms of this agreement, you must cease using and return or destroy
all copies of Project Gutenberg-tm electronic works in your possession.
If you paid a fee for obtaining a copy of or access to a Project
Gutenberg-tm electronic work and you do not agree to be bound by the
terms of this agreement, you may obtain a refund from the person or
entity to whom you paid the fee as set forth in paragraph 1.E.8.

1.B. "Project Gutenberg" is a registered trademark. It may only be
used on or associated in any way with an electronic work by people who
agree to be bound by the terms of this agreement. There are a few
things that you can do with most Project Gutenberg-tm electronic works
even without complying with the full terms of this agreement. See
paragraph 1.C below. There are a lot of things you can do with Project
Gutenberg-tm electronic works if you follow the terms of this agreement
and help preserve free future access to Project Gutenberg-tm electronic
works. See paragraph 1.E below.

1.C. The Project Gutenberg Literary Archive Foundation ("the Foundation"
or PGLAF), owns a compilation copyright in the collection of Project
Gutenberg-tm electronic works. Nearly all the individual works in the
collection are in the public domain in the United States. If an
individual work is in the public domain in the United States and you are
located in the United States, we do not claim a right to prevent you from
copying, distributing, performing, displaying or creating derivative
works based on the work as long as all references to Project Gutenberg
are removed. Of course, we hope that you will support the Project
Gutenberg-tm mission of promoting free access to electronic works by
freely sharing Project Gutenberg-tm works in compliance with the terms of
this agreement for keeping the Project Gutenberg-tm name associated with
the work. You can easily comply with the terms of this agreement by
keeping this work in the same format with its attached full Project
Gutenberg-tm License when you share it without charge with others.

1.D. The copyright laws of the place where you are located also govern
what you can do with this work. Copyright laws in most countries are in
a constant state of change. If you are outside the United States, check
the laws of your country in addition to the terms of this agreement
before downloading, copying, displaying, performing, distributing or
creating derivative works based on this work or any other Project
Gutenberg-tm work. The Foundation makes no representations concerning
the copyright status of any work in any country outside the United
States.

1.E. Unless you have removed all references to Project Gutenberg:

1.E.1. The following sentence, with active links to, or other immediate
access to, the full Project Gutenberg-tm License must appear prominently
whenever any copy of a Project Gutenberg-tm work (any work on which the
phrase "Project Gutenberg" appears, or with which the phrase "Project
Gutenberg" is associated) is accessed, displayed, performed, viewed,
copied or distributed:

This eBook is for the use of anyone anywhere at no cost and with
almost no restrictions whatsoever. You may copy it, give it away or
re-use it under the terms of the Project Gutenberg License included
with this eBook or online at www.gutenberg.org

LICENSING. 357

1.E.2. If an individual Project Gutenberg-tm electronic work is derived from the public domain (does not contain a notice indicating that it is posted with permission of the copyright holder), the work can be copied and distributed to anyone in the United States without paying any fees or charges. If you are redistributing or providing access to a work with the phrase "Project Gutenberg" associated with or appearing on the work, you must comply either with the requirements of paragraphs 1.E.1 through 1.E.7 or obtain permission for the use of the work and the Project Gutenberg-tm trademark as set forth in paragraphs 1.E.8 or 1.E.9.

1.E.3. If an individual Project Gutenberg-tm electronic work is posted with the permission of the copyright holder, your use and distribution must comply with both paragraphs 1.E.1 through 1.E.7 and any additional terms imposed by the copyright holder. Additional terms will be linked to the Project Gutenberg-tm License for all works posted with the permission of the copyright holder found at the beginning of this work.

1.E.4. Do not unlink or detach or remove the full Project Gutenberg-tm License terms from this work, or any files containing a part of this work or any other work associated with Project Gutenberg-tm.

1.E.5. Do not copy, display, perform, distribute or redistribute this electronic work, or any part of this electronic work, without prominently displaying the sentence set forth in paragraph 1.E.1 with active links or immediate access to the full terms of the Project Gutenberg-tm License.

1.E.6. You may convert to and distribute this work in any binary, compressed, marked up, nonproprietary or proprietary form, including any word processing or hypertext form. However, if you provide access to or distribute copies of a Project Gutenberg-tm work in a format other than "Plain Vanilla ASCII" or other format used in the official version posted on the official Project Gutenberg-tm web site (www.gutenberg.org), you must, at no additional cost, fee or expense to the user, provide a copy, a means of exporting a copy, or a means of obtaining a copy upon request, of the work in its original "Plain Vanilla ASCII" or other form. Any alternate format must include the full Project Gutenberg-tm License as specified in paragraph 1.E.1.

1.E.7. Do not charge a fee for access to, viewing, displaying, performing, copying or distributing any Project Gutenberg-tm works unless you comply with paragraph 1.E.8 or 1.E.9.

1.E.8. You may charge a reasonable fee for copies of or providing access to or distributing Project Gutenberg-tm electronic works provided that

- You pay a royalty fee of 20% of the gross profits you derive from
     the use of Project Gutenberg-tm works calculated using the method
     you already use to calculate your applicable taxes. The fee is
     owed to the owner of the Project Gutenberg-tm trademark, but he
     has agreed to donate royalties under this paragraph to the
     Project Gutenberg Literary Archive Foundation. Royalty payments
     must be paid within 60 days following each date on which you
     prepare (or are legally required to prepare) your periodic tax
     returns. Royalty payments should be clearly marked as such and
     sent to the Project Gutenberg Literary Archive Foundation at the

address specified in Section 4, "Information about donations to the Project Gutenberg Literary Archive Foundation."

- You provide a full refund of any money paid by a user who notifies you in writing (or by e-mail) within 30 days of receipt that s/he does not agree to the terms of the full Project Gutenberg-tm License. You must require such a user to return or destroy all copies of the works possessed in a physical medium and discontinue all use of and all access to other copies of Project Gutenberg-tm works.

- You provide, in accordance with paragraph 1.F.3, a full refund of any money paid for a work or a replacement copy, if a defect in the electronic work is discovered and reported to you within 90 days of receipt of the work.

- You comply with all other terms of this agreement for free distribution of Project Gutenberg-tm works.

1.E.9. If you wish to charge a fee or distribute a Project Gutenberg-tm electronic work or group of works on different terms than are set forth in this agreement, you must obtain permission in writing from both the Project Gutenberg Literary Archive Foundation and Michael Hart, the owner of the Project Gutenberg-tm trademark. Contact the Foundation as set forth in Section 3 below.

1.F.

1.F.1. Project Gutenberg volunteers and employees expend considerable effort to identify, do copyright research on, transcribe and proofread public domain works in creating the Project Gutenberg-tm collection. Despite these efforts, Project Gutenberg-tm electronic works, and the medium on which they may be stored, may contain "Defects," such as, but not limited to, incomplete, inaccurate or corrupt data, transcription errors, a copyright or other intellectual property infringement, a defective or damaged disk or other medium, a computer virus, or computer codes that damage or cannot be read by your equipment.

1.F.2. LIMITED WARRANTY, DISCLAIMER OF DAMAGES - Except for the "Right of Replacement or Refund" described in paragraph 1.F.3, the Project Gutenberg Literary Archive Foundation, the owner of the Project Gutenberg-tm trademark, and any other party distributing a Project Gutenberg-tm electronic work under this agreement, disclaim all liability to you for damages, costs and expenses, including legal fees. YOU AGREE THAT YOU HAVE NO REMEDIES FOR NEGLIGENCE, STRICT LIABILITY, BREACH OF WARRANTY OR BREACH OF CONTRACT EXCEPT THOSE PROVIDED IN PARAGRAPH F3. YOU AGREE THAT THE FOUNDATION, THE TRADEMARK OWNER, AND ANY DISTRIBUTOR UNDER THIS AGREEMENT WILL NOT BE LIABLE TO YOU FOR ACTUAL, DIRECT, INDIRECT, CONSEQUENTIAL, PUNITIVE OR INCIDENTAL DAMAGES EVEN IF YOU GIVE NOTICE OF THE POSSIBILITY OF SUCH DAMAGE.

1.F.3. LIMITED RIGHT OF REPLACEMENT OR REFUND - If you discover a defect in this electronic work within 90 days of receiving it, you can receive a refund of the money (if any) you paid for it by sending a written explanation to the person you received the work from. If you received the work on a physical medium, you must return the medium with

your written explanation. The person or entity that provided you with
the defective work may elect to provide a replacement copy in lieu of a
refund. If you received the work electronically, the person or entity
providing it to you may choose to give you a second opportunity to
receive the work electronically in lieu of a refund. If the second copy
is also defective, you may demand a refund in writing without further
opportunities to fix the problem.

1.F.4. Except for the limited right of replacement or refund set forth
in paragraph 1.F.3, this work is provided to you 'AS-IS' WITH NO OTHER
WARRANTIES OF ANY KIND, EXPRESS OR IMPLIED, INCLUDING BUT NOT LIMITED TO
WARRANTIES OF MERCHANTIBILITY OR FITNESS FOR ANY PURPOSE.

1.F.5. Some states do not allow disclaimers of certain implied
warranties or the exclusion or limitation of certain types of damages.
If any disclaimer or limitation set forth in this agreement violates the
law of the state applicable to this agreement, the agreement shall be
interpreted to make the maximum disclaimer or limitation permitted by
the applicable state law. The invalidity or unenforceability of any
provision of this agreement shall not void the remaining provisions.

1.F.6. INDEMNITY - You agree to indemnify and hold the Foundation, the
trademark owner, any agent or employee of the Foundation, anyone
providing copies of Project Gutenberg-tm electronic works in accordance
with this agreement, and any volunteers associated with the production,
promotion and distribution of Project Gutenberg-tm electronic works,
harmless from all liability, costs and expenses, including legal fees,
that arise directly or indirectly from any of the following which you do
or cause to occur: (a) distribution of this or any Project Gutenberg-tm
work, (b) alteration, modification, or additions or deletions to any
Project Gutenberg-tm work, and (c) any Defect you cause.

Section 2. Information about the Mission of Project Gutenberg-tm

Project Gutenberg-tm is synonymous with the free distribution of
electronic works in formats readable by the widest variety of computers
including obsolete, old, middle-aged and new computers. It exists
because of the efforts of hundreds of volunteers and donations from
people in all walks of life.

Volunteers and financial support to provide volunteers with the
assistance they need, is critical to reaching Project Gutenberg-tm's
goals and ensuring that the Project Gutenberg-tm collection will
remain freely available for generations to come. In 2001, the Project
Gutenberg Literary Archive Foundation was created to provide a secure
and permanent future for Project Gutenberg-tm and future generations.
To learn more about the Project Gutenberg Literary Archive Foundation
and how your efforts and donations can help, see Sections 3 and 4
and the Foundation web page at http://www.pglaf.org.

Section 3. Information about the Project Gutenberg Literary Archive
Foundation

The Project Gutenberg Literary Archive Foundation is a non profit
501(c)(3) educational corporation organized under the laws of the
state of Mississippi and granted tax exempt status by the Internal

Revenue Service. The Foundation's EIN or federal tax identification
number is 64-6221541. Its 501(c)(3) letter is posted at
http://pglaf.org/fundraising. Contributions to the Project Gutenberg
Literary Archive Foundation are tax deductible to the full extent
permitted by U.S. federal laws and your state's laws.

The Foundation's principal office is located at 4557 Melan Dr. S.
Fairbanks, AK, 99712., but its volunteers and employees are scattered
throughout numerous locations. Its business office is located at
809 North 1500 West, Salt Lake City, UT 84116, (801) 596-1887, email
business@pglaf.org. Email contact links and up to date contact
information can be found at the Foundation's web site and official
page at http://pglaf.org

For additional contact information:
    Dr. Gregory B. Newby
    Chief Executive and Director
    gbnewby@pglaf.org

Section 4. Information about Donations to the Project Gutenberg
Literary Archive Foundation

Project Gutenberg-tm depends upon and cannot survive without wide
spread public support and donations to carry out its mission of
increasing the number of public domain and licensed works that can be
freely distributed in machine readable form accessible by the widest
array of equipment including outdated equipment. Many small donations
($1 to $5,000) are particularly important to maintaining tax exempt
status with the IRS.

The Foundation is committed to complying with the laws regulating
charities and charitable donations in all 50 states of the United
States. Compliance requirements are not uniform and it takes a
considerable effort, much paperwork and many fees to meet and keep up
with these requirements. We do not solicit donations in locations
where we have not received written confirmation of compliance. To
SEND DONATIONS or determine the status of compliance for any
particular state visit http://pglaf.org

While we cannot and do not solicit contributions from states where we
have not met the solicitation requirements, we know of no prohibition
against accepting unsolicited donations from donors in such states who
approach us with offers to donate.

International donations are gratefully accepted, but we cannot make
any statements concerning tax treatment of donations received from
outside the United States. U.S. laws alone swamp our small staff.

Please check the Project Gutenberg Web pages for current donation
methods and addresses. Donations are accepted in a number of other
ways including checks, online payments and credit card donations.
To donate, please visit: http://pglaf.org/donate

Section 5. General Information About Project Gutenberg-tm electronic
works.

LICENSING.

Professor Michael S. Hart is the originator of the Project Gutenberg-tm
concept of a library of electronic works that could be freely shared
with anyone. For thirty years, he produced and distributed Project
Gutenberg-tm eBooks with only a loose network of volunteer support.

Project Gutenberg-tm eBooks are often created from several printed
editions, all of which are confirmed as Public Domain in the U.S.
unless a copyright notice is included. Thus, we do not necessarily
keep eBooks in compliance with any particular paper edition.

Most people start at our Web site which has the main PG search facility:

     http://www.gutenberg.org

This Web site includes information about Project Gutenberg-tm,
including how to make donations to the Project Gutenberg Literary
Archive Foundation, how to help produce our new eBooks, and how to
subscribe to our email newsletter to hear about new eBooks.

Printed in Great Britain
by Amazon